ACS SYMPOSIUM SERIES **435**

Liquid-Crystalline Polymers

R. A. Weiss, EDITOR
University of Connecticut

C. K. Ober, EDITOR
Cornell University

Developed from a symposium sponsored
by the Division of Polymer Chemistry, Inc.,
and the Division of Polymeric Materials: Science and Engineering
at the 198th National Meeting
of the American Chemical Society,
Miami Beach, Florida,
September 10–15, 1989

American Chemical Society, Washington, DC 1990

Library of Congress Cataloging-in-Publication Data

Liquid-crystalline polymers
R. A. Weiss, editor; C. K. Ober, editor.

p.   cm.—(ACS Symposium Series, 0097–6156; 435)

"Developed from a symposium sponsored by the Division of Polymer
Chemistry, Inc., and the Division of Polymeric Materials: Science and
Engineering at the 198th Meeting of the American Chemical Society,
Miami Beach, Florida, September 10–15, 1989."

Includes bibliographical references and indexes.

ISBN 0–8412–1849–8
1. Polymer liquid crystals—Congresses.

I. Weiss, R. A., 1950–   . II. Ober, C. K., 1954–   .
III. American Chemical Society. Division of Polymer Chemistry.
IV. American Chemical Society. Division of Polymer Materials: Science
and Engineering. V. American Chemical Society. Meeting (198th: 1989:
Miami Beach, Fla.). VI. Series

QD923.L533   1990
668.9—dc20   90–41743
                                    CIP

The paper used in this publication meets the minimum requirements of American National
Standard for Information Sciences—Permanence of Paper for Printed Library Materials, ANSI
Z39.48–1984.   ∞

ACS Symposium Series

M. Joan Comstock, *Series Editor*

1990 ACS Books Advisory Board

Foreword

THE ACS SYMPOSIUM SERIES was founded in 1974 to provide a medium for publishing symposia quickly in book form. The format of the Series parallels that of the continuing ADVANCES IN CHEMISTRY SERIES except that, in order to save time, the papers are not typeset, but are reproduced as they are submitted by the authors in camera-ready form. Papers are reviewed under the supervision of the editors with the assistance of the Advisory Board and are selected to maintain the integrity of the symposia. Both reviews and reports of research are acceptable, because symposia may embrace both types of presentation. However, verbatim reproductions of previously published papers are not accepted.

Contents

SYNTHESIS OF SIDE-CHAIN LIQUID-CRYSTALLINE POLYMERS

PHYSICS OF LIQUID-CRYSTALLINE POLYMERS: NETWORKS

PHYSICS OF LIQUID-CRYSTALLINE POLYMERS:
TEXTURE AND STRUCTURE

PHYSICS OF LIQUID-CRYSTALLINE POLYMERS: TRANSITIONS AND PROPERTIES

APPLICATIONS OF LIQUID-CRYSTALLINE POLYMERS: RHEOLOGY AND PROCESSING BEHAVIOR

Preface

DURING THE 1980s, liquid-crystalline polymers (LCPs) captured the imagination of scientists and engineers with their technological potential and scientific challenges. Although fibers made from lyotropic LCPs, notably poly(phenylene terephthalamide), i.e., KEVLAR, were already successful commercial products by the beginning of the decade, the commercialization of thermotropic, melt-processable LCPs, first by Dart Industries and later by other U.S. and European companies, provided the stimulus for the intense activity in research and development that followed. Research continued to be directed toward attaining high stiffness and strength from these materials, but new high-tech thrusts also developed for applications such as optical memory storage, holographic imaging, and nonlinear optics.

Major symposia focusing on LCPs have been held often in the past 20 years although the subject, LCPs, is sufficiently broad that a single symposium cannot cover all of its aspects. The objective of the latest LCP symposium, held at the 198th National Meeting of the American Chemical Society, was to bring together experts from around the world covering the fields of synthesis, physics, processing, and applications of LCPs to discuss the state of the art and recent advances in these subjects. That symposium was an unqualified success, judging by the number of papers contributed and the high attendance.

This book evolved from that ACS symposium, and the chapters herein provide a broad view of the direction LCP research is taking in both industry and academe as we enter the 1990s. A wide range of topics has been covered, including synthesis of main-chain and side-chain LCPs, structural characterization of LCPs, rheology and processing, and applications such as electro-optics and "self-reinforcing" blends. We expect that this material will be of interest to academic and industrial scientists alike, whether their primary interests are in fundamental science or the development of the next generation of LCPs.

We thank the authors for their diligence and cooperation in ensuring timely publication of this volume. We are especially grateful to Nancy Borman for her invaluable assistance in coordinating the editing and preparation of this book.

Acknowledgment is made to the Donors of the Petroleum Research Fund, administered by the American Chemical Society, for partial support of the symposium. Finally, we would like to acknowledge the support of the following organizations, which were instrumental in putting together the symposium: Amoco, Granmont, Inc., Hoechst Celanese, Inc., ICI Advanced Materials, Institute of Materials Science (University of Connecticut), Materials Science and Engineering (Cornell University), and Tennessee Eastman.

R. A. WEISS
Polymer Science Program
and Department of Chemical Engineering
University of Connecticut
Storrs, CT 06269–3136

C. K. OBER
Department of Materials Science
and Engineering
Cornell University
Ithaca, NY 14853–1501

June 15, 1990

Chapter 1

Current Topics in Liquid-Crystalline Polymers

C. K. Ober[1] and R. A. Weiss[2]

[1]Department of Materials Science and Engineering, Cornell University, Ithaca, NY 14853
[2]Polymer Science Program and Department of Chemical Engineering, University of Connecticut, Storrs, CT 06269–3136

This chapter provides an overview of current researches on liquid crystalline polymers (LCP's). Topics include syntheses of main-chain and side-chain LCP's, structured characterization of LCP's and LCP networks and rheology and processing. Applications of LCP/polymer blends as self-reinforced polymers and electro-optical meterials are also discussed.

Liquid crystal is a term that is now commonly used to describe materials that exhibit partially ordered fluid phases that are intermediate between the three dimensionally ordered crystalline state and the disordered or isotropic fluid state. Phases with positional and/or orientational long-range order in one or two dimensions are termed mesophases. As a consequence of the molecular order, liquid crystal phases are anisotropic, i.e., their properties are a function of direction.

Although the technical applications of low molar mass liquid crystals (LC) and liquid crystalline polymers (LCP) are relatively recent developments, liquid crystalline behavior has been known since 1888 when Reinitzer (1) observed that cholesteryl benzoate melted to form a turbid melt that eventually cleared at a higher temperature. The term liquid crystal was coined by Lehmann (2) to describe these materials. The first reference to a polymeric mesophase was in 1937 when Bawden and Pirie (3) observed that above a critical concentration, a solution of tobacco mosaic virus formed two phases, one of which was birefringent. A liquid crystalline phase for a solution of a synthetic polymer, poly(γ-benzyl-L-glutamate), was reported by Elliot and Ambrose (4) in 1950.

Modern-day interest in LCPs had its origin with the molecular theories of Onsager (5) and Flory (6). They predicted that rod-like molecules would spontaneously order above a critical concentration that depended on the aspect ratio of the molecule. These theories were later expanded to include other effects such as polydispersity (7) and partial rigidity (8).

0097–6156/90/0435–0001$06.00/0
© 1990 American Chemical Society

In the past 35 years the volume of published literature and patents on LCs and LCPs has grown substantially, and the subject has been the topic of a number of books (9-14). Undoubtedly, the most important event contributing to the growth of the field was the development and subsequent commerciallization of high strength fibers from poly(p-phenylene terephalamide), PPTA, by DuPont de Nemours Co. in the 1970's (15). This spawned a tremendous growth in the field that has continued unabated for the past 10-15 years and has led to numerous developments in both new materials and applications, as well as in the underlying science.

Liquid crystalline order is a consequence solely of molecular shape anisotropy, such as found in rigid rod-shaped molecules or relatively stiff chain segments with an axial ratio greater than three (6). For example, a succession of para-oriented ring structures is widely used to prepare LCPs. Liquid crystalline phases do not depend on intermolecular associations, but occur as a result of intermolecular repulsions. That is, units of two molecules cannot occupy the same space. For rod-like molecules, or chains with rigid segments, there is a limit to the number of molecules that can arrange randomly in solution or the melt. When this *critical concentration* is exceeded, either a crystalline or an ordered, liquid crystalline phase forms. The rigid unit responsible for the liquid crystalline behavior is referred to as the mesogen.

For many rigid polymers, such as PPTA, the critical concentration is achieved with solutions, and these materials are classified as being lyotropic. Liquid crystalline phases are not observed in the bulk for lyotropic LCPs, primarily because the melting points of these materials are generally so high that they degrade before they melt. The melting point can be depressed by introducing a degree of flexibility into the polymer chain, such as with structures that provide a kink or swivel to the chain or with flexible spacer groups that separate the rigid chain segments. In that case, the critical concentration for forming a liquid crystalline phase is high, usually requiring bulk polymer, and anisotropic melts are obtained. These materials are thermotropic.

The first thermotropic LCPs were reported in the mid-1970's by Roviello and Sirigu (15) and Jackson and Kuhfuss (16). Since then, a large number of LCPs have been reported; an excellent, though now dated, review of main-chain (i.e., the mesogen is in the polymer backbone) thermotropic LCPs was published by Ober et al. (17). In the 1980's, several thermotropic aromatic copolyester LCPs were commercialized.

Side-chain LCPs can be prepared by attaching a mesogen pendant to a flexible polymer backbone (18). These materials often have optical properties similar to low molar mass LCs and have generated interest for such applications as non-linear optics, filters and optical storage devices.

Research on LCPs is diversified, covering topics in chemistry, physics and engineering. This chapter is not intended to be a comprehensive review of the state-of-the-art for LCPs. Instead, we seek to provide a broad overview of the current trends in LCP research. Many of these subjects are treated in more detail in the following chapters of this book.

Synthesis of LCP's

The synthesis of liquid crystalline polymers has evolved over the last decade to include mesophase forming polymers with a wide variety of structural features and linking groups.   Work has continued on developing novel structures which exhibit either thermotropic or lyotropic behavior. Once it was discovered that it was possible to place the components of low molar mass mesogens in polymer chains and retain the mesophase forming character of these structures, early synthetic studies concentrated on incorporating these mesogenic groups into either a side chain or the main chain.

Recently, new thermotropic polymers have been developed with mesogenic cores arranged not only in the main chain or the side chain, but with rigid mesogenic groups parallel to the chain (19). Even discotic, or disk-shaped mesogenic cores have been incorporated into polymers (20). Typical LCP structures are shown schematically in Figure 1 and include those forms mentioned above as well as network polymers - elastomers and thermosets. While much of the current synthetic activity has concentrated on improving the understanding of structure-property relationships, other new LCPs have been prepared to test specific aspects of the physical character of LC polymers, e.g. polymers with precisely ordered chain sequences (21).

The commonly accepted ideas of what constitutes a mesogenic group have been challenged by researchers who have incorporated phenylene rings linked by non-rigid methylene-ether groups. Mesophase behavior has been reported for polymers of this type without spacers (22), and new developments show that even the presence of a spacer does not prevent mesophase formation by these non-rigid mesogenic groups. An example is provided in the chapter by Jonsson et al. in this book. Also described in this book are newly developed hydrocarbon LCP's which have no polar functional groups, see the chapter by Sung et al. Polymers made with linking groups other than the esters and amides have been prepared and the use of ester-sulfides are discussed in the chapter by Chiellini et al. Inorganic components such as silicon and phosphorus have also been used to prepare main-chain LCP's as shown in the contribution by Paleos et al.

In the following two sections, a number of recent developments in the syntheses of main-chain and side-chain LCP's are discussed. Other synthetic aspects are also found in many of the chapters concerned with physical characteristics of LCP's.

Main Chain LC Polymers.   New thermotropic copolyesters with either random or ordered mesogenic sequences have been reported with a wide range of mesophase behaviors.   Recent developments in this field have included the use of naphthalene, stilbene and related structures in addition to the traditional phenylene groups to produce the required rigid main chain, and these are described in chapters by Jin, Jackson and Morris, and Skovby et al. Efforts have been undertaken to control transition temperatures and solubility through the use of either substituents or changes in the monomer sequence distribution. Successful application of these efforts have led to the commercialization of several thermotropic aromatic copolyesters (23,24).

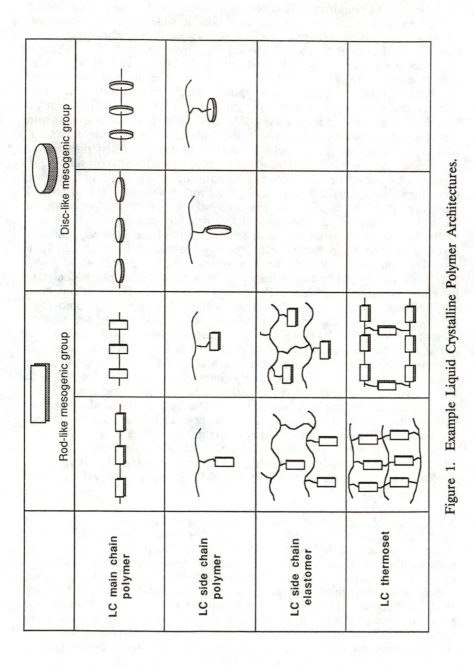

Figure 1. Example Liquid Crystalline Polymer Architectures.

Random monomer sequences lead to very complicated liquid crystalline behavior and produce complex morphologies in the polymer solid (25).

Several fundamental studies have shown the importance of monomer sequence distribution on mesophase behavior (26). Simply changing the direction of ester linkages in a chain affects the transition temperatures, the range of the mesophase stability and, in some cases, even the mesophase texture (27). Polyester chains are susceptible to transesterification, which raises the question of which sequence structure is actually responsible for the properties observed for a given polymer. A recent study of aromatic LC polymers by neutron scattering indicates that transesterification occurs in the mesophase at rates twice that in poly(ethylene terephthalate) (28). Such behavior has also been observed to occur in other aromatic polyesters where rapid sequence redistribution was detected by nmr, see for example, the chapters by Jin and Economy et al. The temperature dependence of this effect has not been fully explored, and it may not be as pronounced in those polymers which exhibit mesophase behavior at lower temperatures, for example, those with aliphatic spacers.

Recent developments in the substitution of completely aromatic LC polyesters have produced polymers which show improved solubilities and reduced transition temperatures (29). The presence of these side groups provides a method for producing polymers that are compatible with other similarly modified polymers. In this way, blends of rigid and flexible polymers can be prepared. Substituents have included alkyl, alkoxy (30) and phenyl alkyl groups (31), some of which lead to mesophases that have been reported as being "sanidic" or board-like. This approach has been used with both polyesters and polyamides and has lead to lyotropic and thermotropic polymers depending on the particular composition used. Some compositions even show the ability to form both lyotropic and thermotropic mesophases (32).

In addition to preparing completely new LC structures, new strategies are being applied to existing mesogenic structures to enhance their performance. Liquid crystalline segments are being combined with other high performance polymers such as polysulfone or poly(etherketone) to form polymers with unusual properties. Linking groups, such as carbonate, are being evaluated in LC polymers in conjunction with ester linkages in the development of more readily processable LCPs that show no reduction in thermal stability, see chapters by Kawabe et al. and Lai et al.

It is well known that below a critical molar mass, the properties of polymers are very sensitive to changes in molar mass. Very little, however, is known about the molar mass dependence of the properties of LCPs. This is partly due to the difficulties in controlling molar mass in the reactions usually used to synthesize LCPs and partly due to the lack of good molar mass data on most LCPs, which are often soluble only in fairly harsh solvents. The chapter by Kim and Blumstein describes the preparation of series of main-chain LCPs varying in molar mass and the effect of molar mass on the thermal properties.

Side Chain LC Polymers. Some of the earliest LC polymers were those that contained mesogenic groups pendant to a flexible polymer chain. Many of the property differences observed between side chain LCP's and main chain LCP's are due to the fact that the mesogenic group has much greater mobility in the side chain, resulting in a polymer chain that is also much less restricted in its motions. Recently, very novel LCP's of this type were synthesized with unusual pendant structures. Examples include the introduction of crown ethers or the use of half disk and half rod-like mesogenic cores as side chain mesogens, described in the chapter by Keller and coworkers. Discotic mesogenic groups have also been attached to polymer chains and novel studies with electron acceptor molecules have shown the ability of charge transfer complexes to form compatible blends (33). Even ionic linkages have been employed to connect mesogenic cores to a polymer backbone (34).

Polymers with long, chiral mesogenic structures have been used to study the effect of molecular structure on the cooperativity of motion between mesogenic side groups (35). Studies using coordinated orientation of mesogenic groups in side chain LC polymers to photochemically crosslink the polymers, described in the chapter by Noonan and Caccamo, showed that the LC environment enhanced crosslinking.

One of the important features of side chain LCP's is that the mesogenic group, not the main chain, determines the mesophase order. Therefore, a wide variety of backbone polymers can be used with this strategy for LCP formation. The mobility of the side-chains also allows them to respond to applied fields more quickly than main-chain LCPs. Mesogenic groups have been attached to methacrylate, acrylate, siloxane and poly(phosphazene) chains (36,37). Synthetic approaches involve either the formation of polymer from monomers containing the mesogenic groups already present or, alternatively, the attachment of mesogenic groups to a preformed polymer chain. Using the latter approach, side chain LC polymer segments have been introduced into diblock copolymers. Examples of several of these types of polymers and approaches are described in chapters by Keller et al., Singler et al., and Adams and Gronski.

Combined Side Chain-Main Chain LC Polymers. Among the recently developed LCP's with intriguing possibilities for novel behavior are those which contain mesogenic cores in both the side chain and main chain (38). These polymers have been prepared in both linear thermoplastic and crosslinked elastomeric forms (39). The presence of mesogenic cores in both side and main chain positions can have a significant effect on the properties and organization of the mesophase. Nematic phases only occur when there is a mismatch between the length of the spacers in the side chain and the main chain. Otherwise, smectic mesophases predominate. In these structures, the lateral mesogenic group folds up to lie parallel to the mesogenic group in the polymer chain. Recent results indicate that the mesogenic group itself is somewhat tilted with respect to the chain direction (40).

LC Polysurfactants. Micelles and vesicles are forms of mesomorphic structures that have until recently been restricted to the world of small molecules. New studies have resulted in the synthesis of micellar polymers linked together after formation of an organized structure. Polymerizable amphiphiles have been synthesized where the amphiphilic moiety is either within the polymer main chain or attached as a side chain via the hydrophilic or hydrophobic end to the polymer backbone (41). Detailed investigations of binary phase diagrams of monomeric and polymeric surfactants in aqueous solution have proven that the LC side chain polysurfactants and the amphiphilic copolymers form liquid crystalline phases in water and that the phase structure of the polymers is similar to the phase structure of conventional low molar mass lyotropic liquid crystals (42). These polymers show phase transitions associated with small molecule micelles, but form micelles that are much more stable in structure. Potential applications of these materials exist in the biomedical field and include the ability to deliver drugs in a controlled fashion.

Physics of LCP's

Rigid LC Networks. The study of entirely rigid networks composed of LC materials is an area of research that has recently been revived. In de Gennes' early work on the theory of LC polymers he suggested that an elastomeric network could trap the mesophase order in a polymeric structure (43). It has since been found that crosslinking is not necessary, but some early research on LC networks showed interesting effects of the crosslinking on mesophase order in rigid networks (44). More recent work used this approach to solve the problem of poor compressive modulus in LC polymers by crosslinking aligned fibers using UV-radiation (45).

Recent theory suggests that if polymers were prepared with high functionality crosslinks connected by mesogenic structures, extremely rigid, high strength networks with low thermal expansion coefficient would result (46). Questions such as the effect of the mesophase structure and its orientation on network behavior remain unanswered (47). Such materials could lead, however, to extremely strong thermosets useful as the matrix materials in advanced composites. Efforts to prepare rigid networks based on LC monomers are described in the chapter by Hoyt et al. and monomer syntheses for LC thermoset coatings were reported by Kangas et al. (48).

LC Elastomers. Elastomeric LC networks have potential for translating mechanical deformation into optical or electrical changes, see the chapter by Zentel et al. Elastomeric network polymers are unique because of the presence of both the network and the mesogenic groups. The behavior of these components must be accounted for to fully understand the behavior of such materials. The coupling of the orientation of the network and the mesogenic structures will differ markedly depending on the type of LC elastomer, see Figure 1. For example, in side chain elastomers, the mesogenic groups are fairly mobile (19), whereas in main chain elastomers the arrangement of the mesogen is coupled to the conformation of the main chain. In combined, main-chain/side-chain elastomers, both types of mesogenic groups must interact in a coordinated fashion (39).

The ability to align LC elastomers with mechanical forces means that applications where LC alignment is necessary may be accessible through orientation of such materials. Applications such as waveguides are possible. In addition, the ability to transform a turbid LC into a transparent monodomain means that mechanically switched-optical devices may be possible. Finally, elastomers with smectic C* mesophases may be useful as piezoelectric materials (39).

Biphasic Behavior. An understanding of the extended biphasic structure of LCP's during the clearing transition is critical to their applications. Biphasic behavior implies the presence of two distinct phases between the melting and clearing transitions, these two phases being typically the mesophase and the isotropic melt. In most cases the biphase can be explained on the basis of a distribution of molar masses in an LC polymer. This distribution will lead to a range of clearing temperatures for a given polydisperse polymer. Experiments which have elegantly demonstrated this point were made using LC polymers in which the isotropic and mesomorphic melts were allowed to separate due to density differences (49). Upon analysis, the nematic component was shown to be of a higher molar mass than the isotropic phase.

Recently, theoretical and experimental work has demonstrated that the effect of chemical heterogeneity as a cause of the biphase must be considered (50). Polymers with similar molar masses will have broader clearing ranges if the chemical composition of the polymer is disordered. Such chemical heterogeneity has been shown to cause a low degree of crystallinity at low temperatures (51) and an extended clearing transition (26) at higher temperatures (the nematic/isotropic biphase) in random copolyesters. Statistical treatment of random copolymerization suggests that the probability of sequences matching in two different chains, and hence the degree of crystallinity, increases as the chain length is lowered.

Low molar mass LC polymers would be expected to be biphasic over a wide temperature range, as both the chemical heterogeneity and the molar mass effects on the clearing temperature are large for short chains. A further complication in understanding this behavior arises in cases where the components form mesophases of different types as discussed in a chapter by Ober et al.

Structure. The traditional methods for identifying and characterizing the texture of LC phases are light microscopy and x-ray diffraction (52). Electron microscopy has also been used when it is possible to freeze the mesophase structure in the solid state (53). The chapter by Viney reviews the optical microscopy technique as applied to LCPs.

Applications of LCPs usually depend on their solid state structure and properties. Very often the design of LCPs involves a compromise between chain rigidity and processability---i.e., either a sufficiently low melting point for thermotropic LCPs or solubility for lyotropic LCPs. Numerous semi-flexible LCPs have been synthesized by introducing flexible spacer groups or lateral substituents into otherwise rigid chains. One important question is how these groups affect the molecular packing in the solid state, since that undoubtedly affects the mechanical and physical properties of the LCP. This subject is addressed in the chapters by Biswas et al. and Azaroff et al.

Polymorphism in PPTA fibers is discussed in the chapter by Rutledge and Suter. The effect of the chemical structure on the theoretical properties of LCPs is discussed in the chapter by Dowell.

Engineering with LCPs

The largest market for LCPs is structural fibers, principally PPTA. World-wide production of LCP fibers in 1989 was estimated to be about 94 million pounds with an average market value of nearly $10/lb. (54). The desireable characteristics of PPTA fibers are high strength, low density, nonabrasiveness and dimensional and thermal stability. Applications of LCP fibers include protective fabrics (e.g., bullet-proof vests and gloves), high strength fabrics (conveyer belts, sails and inflatable boats), industrial fibers (rope and thread), rubber reinforcement (radial tires), plastics reinforcement (structural composites), and asbestos replacement (brake linings, clutch facings, gaskets and packing) (55).

In contrast to the relatively mature markets of LCP fibers, there are few established applications of thermotropic LCPs. In 1989, the world-wide consumption of melt processable LCPs was only about 10 million pounds (valued at about $10/lb.) of which about 5 million pounds was used in cookware (54). Nearly all of the thermotropic LCP consumption was for injection molded parts. Although thermotropic LCPs exhibit exceptional mechanical properties when oriented, anisotropy is actually a weakness in injection molded parts, resulting in poor properties transverse to the machine direction and weak weld lines. These facts coupled with their high cost have limited the growth of markets for molded-LCPs. Usually LCPs are filled with glass, which reduces the anisotropy of the molded part, though it also reduces the modulus and strength associated with a highly oriented LCP. Where LCPs have been successful are applications where low viscosities are needed to fill complex molds and thin walled-parts and where dimensional stability at elevated temperatures and low thermal expansion are required. For example, molded LCPs have been used in electrical applications, primarily as surface-mounted and fiber-optic connectors. Other uses for thermotropic LCPs include applications where chemical resistance is needed (e.g., tower packings, pumps and valves) and mechanical components requiring good wear resistance (pulleys, bushings and seals) (55).

Rheology. The rheological behavior of LCPs is poorly understood (56), though in recent years, significant advances have been made in theory (57). Below a critical concentration, i.e., at concentrations sufficiently low that the solution forms a single, isotropic phase, the solution viscosity of lyotropic LCPs increases with increasing concentration. Above the critical concentration, an anisotropic, liquid crystalline phase forms and a dramatic drop in viscosity occurs. A similar phenomenon was observed as composition was varied for PET/HBA thermotropic LCPs (16). In that case, the melt viscosity increased with increasing mole fraction of the more rigid HBA monomer, but the viscosity decreased once a critical composition was reached. This composition corresponded to the onset of liquid crystalline behavior. Onogi and Asada (58) proposed a three-zone flow curve for LCPs consisting of (1) a shear-thinning region at low shear rates, (2) a Newtonian viscosity region at intermediate shear rates and (3) a second shear-thinning region at higher shear rates. These have been observed in the rheological behavior of both lyotropic and thermotropic LCPs. Atlthough, it is generally believed that the first region of shear-thinning, which is not usually observed in other non-LC polymers, is a consequence of some the domain structure of the melt, understanding of this phenomenon is incomplete.

The melt viscosity of LCPs is sensitive to thermal and mechanical histories. Quite often, instrumental influences are important in the value of viscosity measured. For example, the viscosity of HBA/HNA copolyesters are dependent on the die diameter in capillary flow (59). LCP melts or solutions are very efficiently oriented in extensional flows, and as a result, the influence of the extensional stresses at the entrance to a capillary influence the shear flow in the capillary to a much greater extent than is usually found with non-LC polymers.

Thermotropic LCPs have high melt elasticity, but exhibit little extrudate swell. The latter has been attributed to a yield stress and to long relaxation times (60). The relaxation times for LCPs are normally much longer than for conventional polymers. Anomalous behavior such as negative first normal stress differences, shear-thickening behavior and time-dependent effects have also been observed in the rheology of LCPs (56). Several of these phenomena are discussed for poly(benzylglutamate) solutions in the chapter by Moldenaers et al.

Processing. Relatively little has been published on the processing of thermotropic LCPs. The morphology of melt-processed articles is dependent on the deformation and thermal histories. Extensional flows produce fibrillar structures with high orientation in the machine direction. Flows with complicated stress distributions and temperature gradients, such as encountered in injection molding, yield complicated morphologies.

Melt extrusion of LCP rods and films was discussed by Chung (61) and by Ide and Chung (62), respectively. Considerably more work has been done on injection molding of LCPs, but the reults were usually less conclusive due to the complex thermomechanical histories and the resultant complex morphologies (24,63-66). The anisotropic shrinkage of molded LCP parts is discussed in the chapter by Frayer and Huspeni.

Blends. There has been considerable research in recent years on polymer blends that contain an LCP. This subject was recently reviewed by Dutta et al. (67). The addition of an LCP to another thermoplastic melt effectively lowers the melt viscosity and improves processability. In addition, if the flow field contains an extensional stress component, the LCP dispersed phase is extended into a fibrous morphology and oriented in the flow direction. This microstructure can be retained in the solidified blend to provide self-reinforcement.

The preparation and properties of blends of an HBA/HNA LCP with poly(phenylene sulfide) and poly(etherether-ketone) is described in the chapter by Baird et al. LCP/LCP blends is the subject of the chapter by

DeMeuse and Jaffe. In another chapter, Brostow et al. discuss the phase behavior of binary and ternary mixtures containing an LCP component. The kinetics of thermally induced phase separation of an HBA/PET LCP blended with poly(ether imide) are described by Zheng and Kyu.

Dispersions of low molar mass LCs in amorphous polymers (PDLC) represent a new class of electro-optical materials. PDLC-based devices operate on the principle of electrically modulating the difference between the refractive indices of the LC and the polymer to control the scattering of light. This subject is reviewed in the chapter by West et al. A new development in this field is to use blends of an LC with an LCP to increase the angle of view of the device.

Conclusions

Liquid crystalline polymers have captured the excitment and imagination of contemporary polymer scientists and engineers. These materials exhibit many unique properties that present not only challanges for basic research, but also numerous technological opportunities. Many questions concerning the physics of LCPs need to be answered before these materials achieve widespread use. In addition, much work is needed to fully understand their rheology and processing behavior in order that the outstanding mechanical properties of LCPs may be exploited in molded articles. This chapter briefly described many of the areas of research currently being pursued. Considerably more detail is given in the following chapters.

Literature Cited

1. Reinitzer, F. Monatsh. Chem. 1888, 9, 421.
2. Lehmann, O. Z. Kristallogr. Kristallgem. Kristallphys. Kristallchem. 1890, 18, 464.
3. Bawden, F. C.; Pirie, N. W. Proc. R. Soc., Ser. B 1937, 123, 1274.
4. Elliot, A.; Ambrose, E. J. Discuss. Faraday Soc. 1950, 9, 246.
5. Onsager, L. Ann. N.Y. Acad. Sci. 1949, 51, 627.
6. Flory, P. J. Proc. Royal Soc. London 1956, 234A, 73.
7. Flory, P. J.; Abe, A. Macromolecules 1978, 11, 119.
8. Flory, P. J. Macromolecules 1978, 11, 1138.
9. Thermotropic Liquid Crystals; Gray, G. W., Ed.; John Wiley: New York, 1987.
10. Recent Advances in Liquid Crystalline Polymers; Chapoy, L. L., Ed.; Elsevier Applied Sciences Publ.: London, 1984.
11. Polymeric Liquid Crystals; Blumstein, A., Ed.; Plenum Press: New York, 1985.
12. Ciferri, A.; Krigbaum, W. R.; Meyer, R. B. Polymer Liquid Crystals; Academic Press: New York, 1982.
13. Chandrasekhar, S. Liquid Crystals; Cambridge Univ. Press: Cambridge, 1977.
14. de Gennes, P. G. The Physics of Liquid Crystals; Clarendon Press: Oxford, 1975.
15. Roviello, A.; Sirigu, A. J. Polym. Sci.: Polym. Lett. 1975, 13, 455.

16. Jackson, W. J.; Kuhfuss, H. F. J. Polym. Sci.: Polym Chem. 1976, 14, 2043.
17. Ober, C. K.; Jin, J.-I.; Lenz, R. W. In Liquid Crystal Polymers-I, Adv. in Polym. Sci., Vol. 59, Springer-Verlag, New York, 1984, p 103.
18. Finkelmann, H. In ref. 9, p 145.
19. Finkelmann, H. Agnew. Chem. Int. Ed. Engl. 1987, 26, 816.
20. Kreuder, W.; Ringsdorf, H.; Tschirner, P. Makromol. Chem. Rapid Commun. 1985, 6, 367.
21. Moore, J. S.; Stupp, S. I. Macromolecules 1988, 21, 1217.
22. Percec, V.; Yourd, R. Macromolecules 1988, 21, 3379.
23. Ober, C. K.; Bluhm, T. L. In Current Topics in Polymer Science, Vol. 1; Inoue, S.; Utracki, L. A.; Ottenbrite, R. M., Ed.; Hanser: New York, 1987.
24. Calundann, G. W.; Jaffe, M. Proc. Robert A. Welch Conf. Chem. Res., XXVI. Synth. Polym., Houston, TX, 1982.
25. Donald, A. M.; Viney, C.; Windle, A. H. Polymer 1983, 10, 434.
26. Stupp, S. I.; Moore, J. S.; Martin, P. G. Macromolecules 1988, 21, 1228.
27. Ober, C.; Lenz, R. W.; Galli, G.; Chiellini, E. Macromolecules 1983, 16, 1034.
28. MacDonald, W. A.; McClean, G.; McLenaghan, A. D. W.; Richards, R. W. MRS Meeting Abstracts, Boston, 1989.
29. Ballauff, M.; Schmidt, G. F. Makromol. Chem. Rapid Commun. 1987, 8, 93.
30. Ringsdorf, H.; Tschirner, P.; Hermann-Schönherr, O.; Wendorff, J. H. Makromol. Chem. 1987, 188, 131.
31. Freund, L.; Jung, H.; Niebner, N.; Sonan, F.; Schmidt, H. W.; Wicker, M. Makromol. Chem. in press
32. Hose, H.; Ringsdorf, H.; Tschirner, P.; Wüstefeld, R. Polym. Preprints (ACS Div., Polym. Chem.) 1989, 30(2), 478.
33. Ebert, M.; Ringsdorf, H.; Wendorff, H. J.; Wüstefeld, R. Polym. Preprints (ACS Div., Polym. Chem.) 1989, 30(2), 479.
34. Kato, T.; Fréchet, J. M. J. Macromolecules 1989, 22, 3819.
35. Hong, X. J.; Stupp, S. I. Polym. Preprints (ACS Div., Polym. Chem.) 1989, 30(2), 469.
36. Percec, V.; Tomazos, D. J. Polym. Sci.: Polym. Chem. 1989, 27, 999.
37. Singler, R. E.; Willingham, R. A.; Lenz, R. W.; Furukawa, A.; Finkelmann, H. Macromolecules 1987, 20, 1726.
38. Reck, B.; Ringsdorf, H. Makromol. Chem. Rapid Commun. 1985, 6, 291.
39. Zentel, R. Agnew. Chem. Int. Ed. Engl. Adv. Mater. 1989, 101, 1437.
40. Voight-Martin, I. G.; Durst, H.; Reck, B.; Ringsdorf, H. Macromolecules 1988, 21, 1620.
41. Ringsdorf, H.; Schlarb, B.; Tyminski, P. N.; O'Brien, D. F. Macromolecules 1988, 21, 671.
42. Ringsdorf, H.; Schmidt, G.: Schneider, J. Thin Solid Films 1987, 152, 207.
43. de Gennes, P. S. Phys. Letters 1969, A28, 725.

44. Blumstein, A.; Blumstein, R.; Clough, S.; Hsu, E. Macromolecules 1975, 8, 73.
45. Lin, C. H.; Maeda, M.; Tayebi, A.; Blumstein, A. Polym. Preprints. (ACS Div., Polym. Chem.) 1989, 30(2), 459.
46. Boué, F.; Edwards, S. F.; Vilgis, T. A. J. Phys. 1988, 49, 1635.
47. Ballauff, M. Agnew. Chem. Int. Ed. Engl. Adv. Mater. 1989, 28, 1130.
48. Kangas, S. L.; Menzies, R. H.; Wang, D.; Jones, F. N. Polym. Preprints (ACS Div., Polym. Chem.), 1989, 30(2), 462.
49. D'Allest, J. F.; Sixou, P.; Blumstein, A.; Blumstein, R. B. Mol. Cryst. Liq. Cryst. 1988, 157, 229.
50. Stupp, S. I. Polym. Preprints (ACS Div., Polym. Chem.) 1989, 30(2), 509.
51. Hanna, S.; Windle, A. H. Polymer 1988, 29, 207.
52. Noel, C. In Polymeric Liquid Crystals, Blumstein, A., Ed., Plenum Press 1985. p21
53. Thomas, E. L.; Wood, B. A. Faraday Disc. Chem. Soc., 1985, 79, 229.
54. Kaplan, S., YGB-119 Liquid Crystals; Bus. Comm. Co., Norwalk, CT., 1990
55. Chung, T.-S.; Calundann, G. W.; East, A. J. In Handbook of Polymer Science and Technology, Vol. 2, Chermisinoff, N. P., Ed.,; Marcel Dekker, Inc., NY, p. 625.
56. Wissbrun, K. F. J. Rheol., 1981, 25, 619.
57. Larson, R. G. Constitutive Equations for Polymer Melts and Solutions, Butterworths, Boston, 1988.
58. Onogi, S.; Asada, T. In Rheology, Vol. 1, Astarita, G.; Marucci, G.; Nicolais, L., Eds., Plenum Press, New York, 1980, p. 127.
59. Wissbrun, K. F.; Kiss, G; Cogswell, F. N. Chem. Eng. Commun. 1987.
60. Jerman, R. E.; Baird, D. G. J. Rheol., 1981, 25, 275.
61. Chung, T. S., J. Polym. Sci., Polym. Lett., 1986, 24, 299.
62. Ide, Y.; Chung, T. S. J. Macromol. Sci. Phys. 1984, B23, 497.
63. Duska, J. J. Plast. Eng. 1986, 12, 39.
64. Joseph, E. J.,; Wilkes, G. L.; Baird, D. G. Polym. Eng. Sci. 1985, 25, 377.
65. Garg, S. K.; Kenig, S. In High Modulus Polymers; Zachariades, A. E.; Porter, R. S., Eds. Marcel Dekker, Inc., New York, 1988, p.71.
66. Chung, N. M.S. Thesis, University of Connecticut, 1989.
67. Dutta, D.; Fruitwala, H.; Kohli, A.; Weiss, R. A. Polym. Eng. Sci. in press.

RECEIVED April 24, 1990

SYNTHESIS OF MAIN-CHAIN
LIQUID-CRYSTALLINE POLYMERS

Chapter 2

Polyesters of 4,4′-Biphenyldicarboxylic Acid and Aliphatic Glycols for High-Performance Plastics

W. J. Jackson, Jr., and J. C. Morris

Research Laboratories, Eastman Chemical Company, Eastman Kodak Company, Kingsport, TN 37662

Thermotropic liquid crystalline polyesters (LCP's) were prepared from the dimethyl ester of 4,4'-biphenyldicarboxylic acid (BDA) and aliphatic glycols containing 2 to 10 methylene units and compared with similar trans-4,4'-stilbenedicarboxylic acid (SDA) LCP's. The BDA homopolyesters prepared with 1,4-butanediol or 1,6-hexanediol and their copolyesters with ethylene glycol, 1,4-butanediol, or 1,6-hexanediol had suitable melting points and melt processing characteristics for injection-molding applications. It was necessary, however, to injection mold most of the BDA compositions in their isotropic state (in contrast to the SDA LCP's, which had higher Ti's). Also, because of the apparent smectic nature of the polymer melts, melt viscosities were appreciably higher than those of similar SDA LCP's, which were nematic. The BDA LCP's which gave the highest properties, copolyesters of ethylene glycol and 1,4-butanediol, were injection molded in their anisotropic state.

The lowest cost process for preparing all-aromatic liquid crystalline polyesters involves the reaction of aromatic carboxylic acids with acetates of aromatic hydroxy compounds; a recent history (1) describes the development of these LCP's. Because acetic acid is evolved in the process and reaction temperatures are above 300°C, expensive corrosion-resistant reactors must be installed for commercial production. In our latest paper (2) of this LCP series, we described a number of aliphatic-aromatic LCP's which can be produced in conventional polyester reactors and injection molded to give plastics with very high mechanical properties, heat-deflection temperatures (HDT's), and solvent resistance. These LCP's (Ia) were prepared by the reaction of the dimethyl ester of

NOTE: This chapter is part 13 in the series "Liquid Crystal Polymers."

0097–6156/90/0435–0016$06.00/0
© 1990 American Chemical Society

trans-4,4'-stilbenedicarboxylic acid (SDA) with certain aliphatic glycols. Also it was known that homopolyesters of 4,4'-biphenyldicarboxylic acid (BDA) and aliphatic glycols containing from 2 to 10 methylene units (Ib, n = 2 to 10) exhibit thermotropic liquid crystallinity (3-6). The objective of this paper is to describe LCP's (Ib) prepared from the potentially low cost dimethyl ester of BDA and the aliphatic glycols which were most effective in giving SDA copolyesters suitable for high performance plastics.

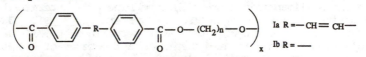

Of the injection-molded SDA homopolyesters, the 1,4-butanediol (BD) and 1,6-hexanediol (HD) compositions had the highest tensile and flexural properties and HDT's (2). The ethylene glycol (EG) homopolyester melted too high (Tm 418°C) to be injection molded without excessive thermal decomposition. When the SDA/BD homopolyester was modified with either EG or HD, tensile strengths were about twice as high (up to 41,000 psi) as those of the homopolyester, and HDT's ranged up to over 260°C at 264 psi stress. When 1,3-propanediol or 1,5-pentanediol was the modifier, kinks were introduced in the polymer chains because of the odd number of methylene units in the glycol, as was discussed by Watanabe and Hayashi for BDA LCP's (6). Consequently, the tensile and flexural properties were greatly decreased because the amount of extended chain orientation that could be attained was limited. Also the HDT's were greatly decreased because of the decrease in crystallinity. When the SDA/HD homopolyester was modified with EG or BD, tensile strengths about 50 to 100% higher (up to 35,000 psi) than those of the homopolyester were obtained, and HDT's ranged up to over 260°C at 264 psi. Thus, the glycols of greatest interest for use in preparing the BDA LCP's were EG, BD, and HD.

Experimental

Dimethyl 4,4'-biphenyldicarboxylate was obtained from Ihara Chemical Industry Co., Ltd. The polyesters were prepared from aliphatic glycols and the dimethyl ester of BDA by procedures similar to those described earlier for our SDA polyesters (7) but using titanium tetraisopropoxide (100 ppm titanium) as the sole catalyst. Depending upon the polymer melting point (Tm), the final reaction temperatures were 260 to 300°C. If the polymer solidified because of a high Tm or if a higher molecular weight was desired, the polymer was ground to pass a 3-mm screen, dried at 100-120°C, and solid-state polymerized by heating for 3-4 hr. at 220-240°C/0.5 mm. The molar amounts of the glycol components in the final copolyesters were determined in 70/30 hexafluoroisopropanol/deuterated methylene chloride by proton NMR. Inherent viscosities (I.V.'s) were measured at 25°C in a 25/40/35 wt % mixture of phenol/p-chlorophenol/tetrachloroethane at a polymer concentration of 0.1 g/100 mL.

Absolute molecular weight determinations were made using a
Brookhaven Photon Correlation Spectrometer. The polymer was
dissolved in a solution of o-chlorophenol at 110°C and clarified by
filtration through a 0.2 micron pore size Teflon membrane filter.
The molecular weights were determined at 25°C by Zimm plot analysis
of measurements of the concentration and angular dependence of
632.8 nm wavelength scattered light intensity.

The thermal properties (DSC second cycle), melt viscosities,
and properties of test bars injected into unheated molds in a 1-oz
Watson-Stillman injection-molding machine were determined as
described earlier for the SDA copolyesters (2, 7, 8). The glass
transition temperatures (Tg's) were determined on 1/16-in. thick
injection-molded bars at 4°C/min and a frequency of 0.3 Hz with a
Mark IV Dynamic Mechanical Thermal Analyzer from Polymer
Laboratories, Inc.

Results and Discussion

Figure 1 compares the Tm's and isotropic transition temperatures
(Ti) from DSC scans of homopolyesters we prepared from the dimethyl
esters of SDA and BDA and glycols containing 2 to 10 methylene
groups. Except for the BDA/1,5-pentanediol Tm (122°C), all of the
Tm's and Ti's of the BDA polyesters are somewhat higher than those
of similarly prepared polyesters reported by earlier investigators
(3-6), probably because our I.V.'s were higher (0.9 to 1.3). Also
some of our compositions were monotropic because of the increased
I.V.'s and Tm's whereas the I.V.'s reported for these compositions
by the earlier investigators were 0.2 to 0.4. The decreased length
of the rigid mesogenic biphenyl unit, compared to the stilbene
unit, decreased the thermal stabilities of the mesophases (lower
Ti's). All of these BDA homopolyesters are reported to form
thermotropic smectic phases (3-6). In addition, Krigbaum and
co-workers (4) modified the BDA/EG homopolyester with 30 and 50 mol
% terephthalic acid (T) and also BDA/HD with 20 mol % T and "found
no evidence for a nematic phase in any of the copolymers."

Like the SDA/EG homopolyester, the BDA/EG homopolyester (Tm
352°C) also melted too high to be injection molded without
excessive thermal decomposition. The other BDA homopolyesters had
suitable Tm's and, as has been discussed, those of particular
interest were the BD and HD homopolyesters and copolyesters.

BDA/1,6-Hexanediol/1,4-Butanediol Copolyesters.

Figure 2 shows the
effect of BD content on the Tm and Ti values of BDA/HD/BD
copolyesters. Modification of the BDA/HD homopolyester with BD
introduced disorder in the crystalline polymer and, consequently,
the Tm's decreased until sufficient BD was present to start
increasing the order as the BDA/BD homopolyester composition was
approached. The Ti values, on the other hand, increased
continuously because the liquid crystalline mesophase became more
stable as the six-carbon HD component of the BDA/HD homopolyester
was replaced by the less flexible four-carbon BD component.

Table I shows the effect of BD content on the properties of
injection-molded BDA/HD/BD polyesters. The actual polymer melt
temperatures during molding were between the temperatures listed

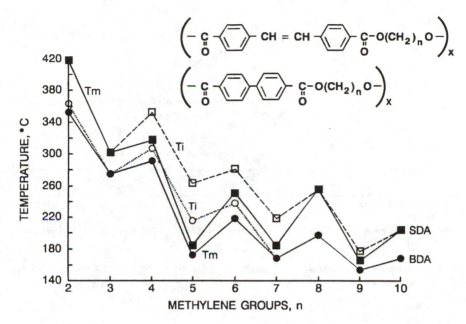

Figure 1.   DSC thermal transitions of homopolyesters of SDA (squares) and BDA (circles); Tm = solid symbols, Ti = open symbols.

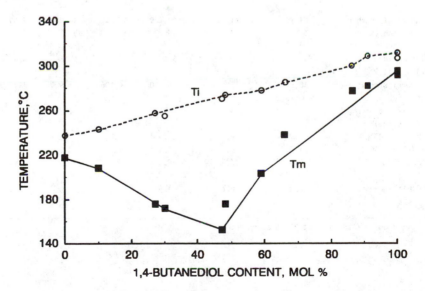

Figure 2.   Effect of BD content on Tm and Ti of BDA/HD/BD copolyesters.

Table I. Effect of Composition on Properties of Injection-Molded
BDA/1,6-Hexanediol/1,4-Butanediol Polyesters

	1,4-Butanediol Content, Mol %					
	0	27	47	66	91	100
Polymer DSC Endotherms (Tm/Ti,°C)	210,219/240	176/258	152/270	238/285	282/309	291/307[a]
Polymer DSC Exotherms on Cooling (°C)	223/159	237/159,111	252/128	269/164	295/218	283/241
Molding Temperature (Barrel/Nozzle, °C)	255/240	260/250	285/270	290/275	330/315	345/330
Polymer I.V.						
Before Molding	--	1.18	1.41	1.09	1.40	1.27
After Molding	1.27	1.20	1.31	1.01	1.34	1.22
Heat-deflection Temp (°C)						
At 66 psi	204	143	114	205	255	273
At 264 psi	190	109	73	119	158	222
Tensile Strength (psi)	21,900	21,300	17,400	17,600	14,000	9,800
Elongation (%)	6	12	19	3	5	4
Flexural Modulus (105 psi)	9.5	5.9	5.6	6.8	5.1	4.4
Flexural Strength (psi)	18,500	12,300	10,700	14,100	13,900	11,100
Izod Impact Strength (ft-lb/in.)						
Notched, 23°C	1.9	9.4	12.9	2.5	1.2	1.2
Notched, -40°C	1.8	9.4	9.1	1.7	0.7	--
Unnotched, 23°C	9.1	33.3	49.8	18.4	8.9	7.9
Mold Shrinkage (%)	0.0	0.7	1.4	0.1	0.6	1.0

[a] Obtained 301/318 on first DSC heating scan.

for the injection molding machine barrel, which was heated, and the
nozzle, which was not externally heated. (The nozzle was heated
only by the polymer which was forced through it by a hydraulic
ram.) It was possible to mold the highest melting composition in
the table (and in the other two tables) because of the small
molding machine and very short molding cycle. The HD
homopolyester, also included in Table I, has the best overall
properties. If a higher HDT is required, particularly if a stress
of 66 psi is sufficient, the higher melting compositions containing
the highest levels of BD have an advantage, but they do have lower
tensile and flexural (stiffness) properties. At lower I.V.'s (1.0
to 1.2), the BDA/HD homopolyester gave appreciably lower mechanical
properties. For comparison, the SDA/HD/BD copolyesters (2) had
HDT's of 190°C to over 260°C (limit of oil bath) at 66 psi and
160°C to over 260°C at 264 psi, tensile strengths of 29,000-35,000
psi, elongations of 10-28%, flexural moduli of 6.2-8.8 x 10^5 psi,
flexural strengths of 15,000-22,000 psi, notched Izod impact
strengths at 23°C of 4.7-7.8 ft-lb/in., and unnotched Izod impact
strengths of 16-44 ft-lb/in.

BDA/1,6-Hexanediol/Ethylene Glycol Copolyesters. By modifying
BDA/HD with EG instead of the more flexible BD used in Figure 2,
higher Tm's and Ti's were obtained (Figure 3).

    Since many of the DSC scans did not show Tg's of the
compositions, Tg's were determined by dynamic mechanical thermal
analysis (DMTA) with injection-molded bars. The tan δ
and storage modulus (log E') curves obtained for the BDA/HD
homopolyester are illustrated in Figure 4. Figure 5 is a plot of
the Tg's (tan δ peaks) obtained with BDA/HD/EG copolyesters.
Because the order due to both liquid crystallinity and
three-dimensional crystallinity reduces the mobility of the
amorphous regions of a polymer, as in the BDA/HD homopolyester, the
"effective Tg" is higher than it would be in an amorphous,
unordered polymer. Modification with EG decreased the order due to
three-dimensional crystallinity, and the Tg initially decreased.
As the six-carbon HD component was further replaced by the less
flexible two-carbon EG component, the Tg increased, as expected.
It is of interest that extrapolation of the upper three points in
the figure to zero EG gives a value of 24°C for the BDA/HD
homopolyester, and extrapolation to 100 mol % EG gives a value of
92°C for the BDA/EG homopolyester. A similar plot was also
obtained with BDA/HD/BD LCP's: minimum Tg of 50°C at 24 mol % BD
content and maximum Tg of 79°C at 100 mol % BD content.

    Table II shows the effect of EG content on the properties of
injection-molded BDA/HD/EG polyesters. As in the case of the
BD-modified compositions in Table 1, the BDA/HD homopolyester had a
higher HDT at 264 psi than any of the copolyesters, but the two
higher melting copolyesters containing the highest EG levels had
higher HDT's at 66 psi. Compared to the BD-modified copolyesters,
these EG-modified copolyesters had the higher tensile strengths and
flexural moduli. The SDA/HD/EG copolyesters studied earlier had
HDT's of 162-232°C at 66 psi stress and 60-207°C at 264 psi stress,
tensile strengths of 25,000-34,000 psi, elongations of 15-30%,
flexural moduli of 6.4-8.9 x 10^5 psi, notched Izod impact strengths

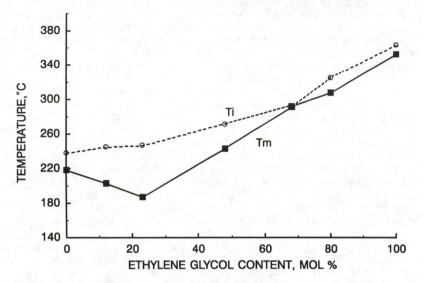

Figure 3.   Effect of EG content on Tm and Ti of BDA/HD/EG copolyesters.

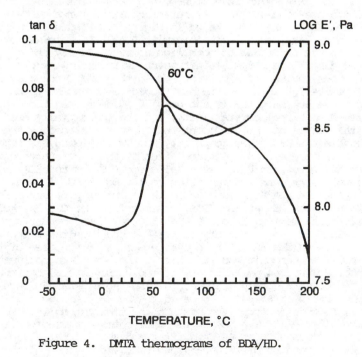

Figure 4.   DMTA thermograms of BDA/HD.

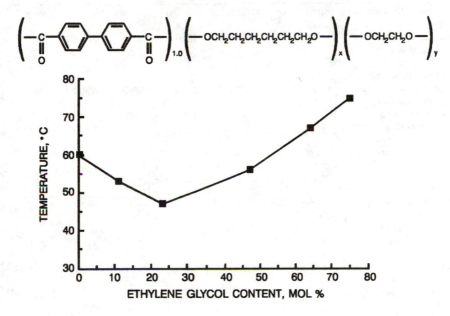

Figure 5.   Tg's of BDA/HD/EG copolyesters from DMTA thermograms.

Table II. Effect of Composition on Properties of Injection-Molded
BDA/1,6-Hexanediol/Ethylene Glycol Polyesters

	Ethylene Glycol Content, Mol %					
	0	12	23	48	68	80
Polymer DSC Endotherms (Tm/Ti,°C)	210,219/240	189,203/245	187/247	243/272[a]	292[b]	308c/325[a]
Polymer DSC Exotherms on Cooling (°C)	223/159	226/157,106	229/-	252/-	272/252	289/284[c]
Molding Temperature (Barrel/Nozzle, °C)	255/240	255/245	275/255	280/265	310/290	345/330
Polymer I.V.						
Before Molding	--	1.16	1.14	1.05	1.05	1.10
After Molding	1.27	1.15	1.14	1.05	0.93	0.73
Heat-deflection Temp (°C)						
At 66 psi	204	177	136	112	231	251
At 264 psi	190	124	84	86	71	137
Tensile Strength (psi)	21,900	22,100	18,500	29,600	37,800	27,300
Elongation (%)	6	9	9	22	16	11
Flexural Modulus (105 psi)	9.5	10.0	8.9	12.4	11.6	11.5
Flexural Strength (psi)	18,500	16,000	12,200	15,800	20,700	20,800
Izod Impact Strength (ft-lb/in.)						
Notched, 23%	1.9	2.6	3.2	7.9	3.5	1.9
Notched, -40°C	1.8	4.4	2.6	6.2	3.9	1.9
Unnotched, 23°C	9.1	18.5	25.6	43.2	22.1	4.0
Mold Shrinkage (%)	0.0	0.0	0.0	0.0	0.0	0.3

[a] First DSC heating scan (no Tm on second scan).
[b] Monotropic.
[c] Shoulder.

at 23°C of 5.5-15.5 ft-lb/in., and unnotched Izod impact strengths of 24-43 ft-lb/in.

The melt viscosity of the BDA/HD homopolyester was too high to measure in the anisotropic state because of the relatively low isotropic transition temperature (Ti 240°C) and the smectic nature of the melt, but modification of the polyester with 50 mol % EG increased the Ti to 272°C. Figure 6 compares the apparent melt viscosities of the homopolyester with those of two BDA/50 EG/50 HD copolyesters at 260°C and also polyethylene terephthalate (PET, grade for blowing bottles) at 300°C. At the higher shear rates which are used for melt spinning and injection-molding, the LCP copolyesters had lower melt viscosities at 260°C than the PET at 300°C. The BDA/HD homopolyester, which has a lower Tg (Figure 5) than PET (80°C), also had a somewhat lower melt viscosity at 300°C than the PET.

BDA/HD homopolyesters and BDA/50 EG/50 HD copolyesters having the same I.V.'s had essentially the same weight average molecular weights by gel permeation chromatography in 70/30 methylene chloride/hexafluoroisopropanol. As for absolute molecular weights, at an I.V. of 1.2 the homopolyester in Figure 6 had an absolute weight average molecular weight of 68,000 by angular dependent light scattering in o-chlorophenol. At an I.V. of 1.8, which corresponds to the higher I.V. copolyester in Figure 6, the molecular weight was 143,000. The PET in Figures 6 and 7 had an absolute molecular weight of 51,000.

The lower I.V. BDA/50 EG/50 HD copolyester in Figure 6 (I.V. 1.09, molecular weight 60,000 determined as above) was used to compare the apparent melt viscosities at 260°C, 280°C, and 300°C with the viscosities at 300°C of three SDA/20 EG/80 BD copolyesters (Tm 276°C) and PET (Figure 7). These SDA LCP's were nematic (threaded texture under crossed polarizers), and the 1.43 I.V. copolyester had an absolute molecular weight of 140,000 by angular dependent light scattering in o-chlorophenol. Even though the SDA LCP's have a Tg similar to that of PET (80°C) and higher than that of the BDA LCP (56°C by DMTA and 46°C by DTA) and have higher molecular weights, they have appreciably lower melt viscosities because they are anisotropic (and nematic) at 300°C (Ti 357°C) whereas the BDA LCP is anisotropic (and apparently smectic) at 260°C (Ti 272°C) but isotropic at 280°C and 300°C. At the highest shear rate (10,670 sec^{-1}) BDA/50 EG/50 HD had a slightly lower melt viscosity in its anisotropic state at 260°C than its isotropic state at 300°C, and at 280°C the viscosity was even lower, perhaps because the high shear rate induced liquid crystallinity, which reduced the viscosity. Thus the melt processability of these BDA LCP's in their anisotropic state is excellent at high shear rates but inferior to the processability of the nematic SDA LCP's in Figure 7.

By polarizing light microscopy we saw the typical focal conic smectic texture of the BDA/HD homopolyester mesophase but observed no distinctive smectic or nematic structure of the BDA/50 EG/50 HD copolyester cooled from the isotropic melt and held in the mesophase many hours. Also on an x-ray diffractometer trace of the copolyester we did not see the characteristic peak due to the spacing between the layers of the smectic mesophase but did see

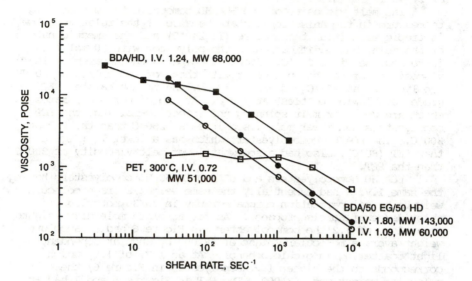

Figure 6. Apparent melt viscosity of BDA/HD and BDA/50 EG/50 HD at 260°C and PET at 300°C.

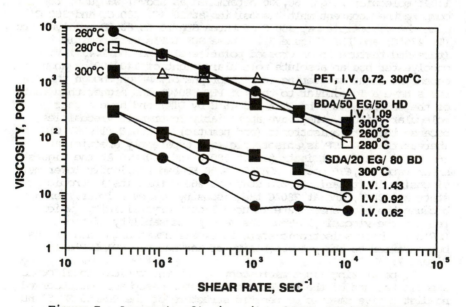

Figure 7. Apparent melt viscosity of PET, BDA/50 EG/50 HD, and SDA/20 EG/80 BD.

this peak (very small) at 18A with the homopolyester, as was also observed by Krigbaum and Watanabe (5). Thus we have not been able to demonstrate that this BDA copolyester is actually smectic.

BDA/Ethylene Glycol/1,4-Butanediol Copolyesters. The BDA/EG/BD copolyesters appeared to solidify when prepared by conventional procedures with a metal bath temperature of 300°C, the maximum temperature which was used in order to avoid any thermal decomposition. If a large excess (250% over theoretical) of glycols or a long ester interchange time was used, it was possible to prepare in the melt a narrow range of BDA/EG/BD copolyesters containing about 60-75 mol % EG and 25-40 mol % BD components. Apparently because of their smectic nature, the copolyesters containing glycol levels outside of these ranges appeared to become solid or semisolid, even though DSC Tm's of some of the compositions were appreciably below 300°C.

     In Table III, which lists the DSC transitions of the polyesters, footnote (d) notes that BDA/BD modified with 50 mol % EG gave DSC endotherm peaks at 251°C and 331°C on the first heating scan, but on a hot stage the copolyester softened at 245°C and did not become sufficiently molten to flow until 300°C. (During preparation in a 300°C metal bath, the polymer melt temperature was about 290°C and the polymer appeared to solidify.) Since no indication of the melting behavior at 300°C was given on the DSC scan, the endotherm at 251°C apparently was due to the formation of a very viscous smectic mesophase, and it was necessary for the polymer to be heated to 300°C for the viscosity to be reduced sufficiently for the polymer to flow when the cover glass on the hot stage was pressed. The second DSC endotherm at 331°C was obviously the isotropic transition, because the polymer gave a clear isotropic melt on the hot stage at 332°C (and was no longer birefringent under crossed polarizers). The endotherms obtained on the second DSC scan appeared at somewhat lower temperatures, perhaps because of some thermal decomposition which occurred during the first DSC scan to 350°C.

     Footnote (f) shows that solid-state polymerized BDA/65 EG/35 BD gave similar results, but an additional endotherm was present in the region where the polymer was sufficiently molten to flow. When this polymer was prepared in the melt, however, these softening and melting endotherms were not present because of a broad exotherm in this area [footnote (e)]. We have been unable to demonstrate by polarizing light microscopy or by x-ray diffraction studies that these copolyesters actually are smectic. (We had the same problems discussed earlier for the BDA/EG/HD copolyesters.)

     Table III also shows the effect of EG content on the properties of injection-molded BDA/EG/BD polyesters. The two copolyesters which contained 65 and 73 mol % EG and did not solidify during preparation were molded in their anisotropic state (Ti's above molding temperatures), and it is significant that these two LCP's have the highest tensile, flexural, and impact properties (and the highest HDT's at 264 psi stress). The other four compositions in the table became solid or semisolid during preparation at 300°C, and additional molecular weight buildup was achieved by solid-state polymerization to give I.V.'s comparable to

Table III. Effect of Composition on Properties of Injection-Molded
BDA/Ethylene Glycol/1,4-Butanediol Polyesters

Ethylene Glycol Content, Mol %

	0[a]	27[a]	50[a]	65	73	79[a]
Polymer DSC Endotherms (Tm/Ti,°C)	291/307[b]	239/299[c]	246/311[d]	300[e,f]/321	314[e]/326	336[e]/338
Polymer DSC Exotherms on Cooling (°C)	283/241	269/166	288/203	301/254	307/276	312/303
Molding Temperature (Barrel/Nozzle, °C)	345/330	335/320	325/310	305/290	325/310	345/330
Polymer I.V.						
Before Molding	1.27	1.11	1.17	0.96	1.05	1.02
After Molding	1.22	1.08	0.97	0.87	0.92	0.94
Heat-deflection Temp (°C)						
At 66 psi	273	237	234	247	266	306
At 264 psi	222	191	215	235	234	210
Tensile Strength (psi)	9,800	18,400	24,100	31,200	39,500	13,500
Elongation (%)	4	5	7	7	10	3
Flexural Modulus (10^5 psi)	4.4	5.8	10.4	14.7	12.8	6.7
Flexural Strength (psi)	11,100	14,900	18,600	24,800	23,600	14,200
Izod Impact Strength (ft-lb/in.)						
Notched, 23°C	1.2	0.9	3.0	11.4	6.2	0.5
Unnotched, 23°C	7.9	8.7	10.1	22.9	24.5	3.7
Mold Shrinkage (%)	1.0	0.7	0.6	0.4	0.3	1.0

[a] Solidified during preparation with 300°C bath temperature; solid-state polymerized in a second step.
[b] Obtained 301/318 on first DSC heating scan.
[c] Obtained 254/322 on first DSC heating scan; softened on hot stage at 235°C, molten at 310°C, and clear isotropic melt at 325°C.
[d] Obtained 251/331 on first DSC heating scan; softened on hot stage at 245°C, molten at 300°C, and clear isotropic melt at 332°C.
[e] Hot stage value where completely molten; broad exotherm instead of Tm endotherm on DSC thermograms of first and second heating scans.
[f] Obtained 263,299/341 on first DSC heating scan of a different sample which was solid-state polymerized (I.V.1.22); endotherms corresponded, respectively, to softening (on hot stage), melting, and clearing; broad exotherm instead of first two endotherms on second DSC heating scan.

those of the melt-prepared copolyesters. Except for BDA/BD modified with 50 mol % EG (Ti 331°C on first DSC scan), it was necessary to mold three of these compositions in their isotropic state, and they gave the lowest tensile, flexural, and impact properties.

As with SDA/BD modified with terephthalic acid, modification of BDA/BD with 10 mol % T instead of a second glycol (EG or HD) did significantly increase the tensile properties but greatly decreased the HDT (only 137°C at 264 psi).

Conclusions

Because of their relatively low Ti's, most of the BDA compositions were in their isotropic state when injection-molded whereas all of the SDA compositions (2) were anisotropic when molded (Ti's 281-352°C). Figure 8 compares the effects on tensile strength when BDA/HD and SDA/HD were modified with EG and BD. Each of the plotted values represents an average from five 1/16-in. thick tensile bar specimens injection molded on the same machine (Tables I and II for BDA LCP's). The BDA/HD/BD compositions had the lowest tensile strengths, but the strengths ranged up to twice as high as those of unreinforced performance plastics which are not liquid crystalline. The highest average value, however, was for a BDA/HD/EG copolyester (37,800 psi). The highest tensile strengths, particularly for the SDA LCP's, were generally associated with the higher elongations, which permitted the high strengths to be attained. Because of their high crystallinity, the homopolyesters gave the lowest elongations and, consequently, generally lower tensile strengths than their copolyesters.

Figure 9 compares the HDT's of BDA/HD modified with EG and BD (Tables I and II). Because modification initially introduced disorder in the crystalline polymer, the Tm's and degree of crystallinity decreased and, consequently, the HDT's decreased; high levels of modification reduced the disorder and, thereby, increased the Tm's, crystallinity, and HDT's as the homopolyester compositions were approached. Lower HDT's were obtained at the 264 psi stress level than at 66 psi, of course, and it is interesting that the two 66 psi curves are near each other, as are the two 264 psi curves.

Figure 10 compares the HDT's at 264 psi of the LCP's in Figure 8 (SDA/HD and BDA/HD modified with EG and BD). EG significantly reduced the crystallinity and, therefore, the HDT's of both the BDA and the SDA copolyesters, whereas BD had less adverse effect on the more crystalline SDA copolyesters. Another reason for the higher HDT's of the SDA/HD/BD LCP's, compared to the corresponding BDA/HD/BD LCP's, is 20 to 30°C higher Tm's in the SDA compositions. The SDA copolyesters also have higher HDT's than the BDA copolyester when the Tm's are similar. All of these polyesters were injected into room temperature molds, and the less crystalline, lower melting BDA copolyesters probably would give higher HDT's if molded into hot molds (so that higher levels of crystallinity could develop as the polymers cooled more slowly). However, if a stress of 66 psi is sufficient for the desired application, HDT's above 200°C can be attained with several of the

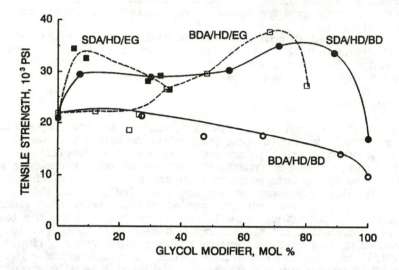

Figure 8. Effect on tensile strength of SDA/HD (solid symbols) and BDA/HD (open symbols) modified with EG (squares) and BD (circles).

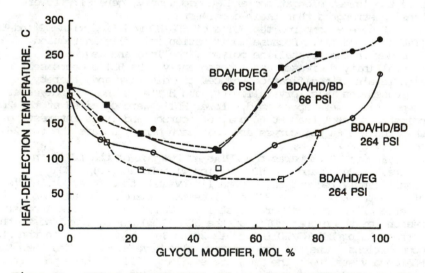

Figure 9. Effect on HDT at 66 psi (solid symbols) and 264 psi (open symbols) of BDA/HD modified with EG (squares) and BD (circles).

BDA polyesters molded even in unheated molds (Figure 9). Of these BDA compositions, the higher tensile and flexural properties were obtained with the BDA/HD/EG copolyesters (Table II).

Figure 11 compares the HDT's at 264 psi of SDA/BD and BDA/BD modified with EG and HD and injected into unheated molds. Of the LCP's which can be prepared by melt polymerization at 300°C and injection molded at about 300°C, the SDA copolyesters have the highest HDT's (up to about 265°C), but BDA/65 EG/35 BD does have an HDT of 235°C at 264 psi. This composition also has an appreciably higher flexural modulus (14.7 x 10^5 psi vs about 9 x 10^5 psi for the SDA compositions) and tensile, flexural, and notched Izod impact strengths comparable to those of these SDA copolyesters and the BDA LCP's in Tables I and II. Its melt viscosity at 300°C is higher than that of the highest I.V. SDA/20 EG/80 BD in Figure 7 but lower than that of the PET. Further study of this LCP is planned.

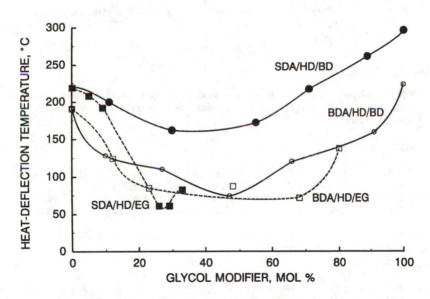

Figure 10. Effect on HDT at 264 psi of SDA/HD (solid symbols) and BDA/HD (open symbols) modified with EG (squares) and BD (circles).

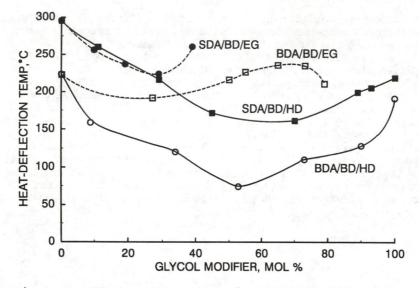

Figure 11. Effect on HDT at 264 psi of SDA/BD (solid symbols) and BDA/BD (open symbols) modified with EG (dashed lines) and BD (solid lines).

Acknowledgments

We are indebted to J. R. Bradley, who prepared many of the polyesters, and to G. B. Caflisch, who determined the absolute molecular weights.

Literature Cited

1. Jackson, W. J., Jr. Mol. Cryst. Liq. Cryst 1989, 169, 23.
2. Jackson, W. J., Jr.; Morris, J. C. in Frontiers of
    Macromolecular Science; Saegusa, T.; Higashimura, T.; Abe, A.,
    Eds.; Blackwell Scientific Publications: London, 1989; pp
    405-410.
3. Meurisse, P.; Noel, C.; Monnerie, L.; Fayolle, B. Brit.
    Polymer J. 1981, 13, 55.
4. Krigbaum, W. R.; Asrar, J.; Toriumi, H.; Ciferri, A.; Preston,
    J. J. Polym. Sci. Polym. Lett. Ed. 1982, 20, 109.
5. Krigbaum, W. R.; Watanabe, J. Polymer 1983, 24, 1299.
6. Watanabe, J.; Hayashi, M. Macromolecules 1988, 21, 278.
7. Jackson, W. J., Jr.; Morris, J. C. J. Appl. Polym. Sci. Appl.
    Polym. Symp. 1985, 41, 307.
8. Jackson, W. J., Jr.; Morris, J. C. J. Polym. Sci., Polym Chem.
    Ed. 1988, 26, 835.

RECEIVED February 28, 1990

Chapter 3

Thermotropic Aromatic Copolyesters Having Ordered Comonomer Sequences

Syntheses and Properties

Jung-Il Jin

Chemistry Department, College of Sciences, Korea University, Seoul 136–701, Korea

Multi-step synthetic methods for the preparation of various liquid crystalline, aromatic copolyesters having ordered, regular comonomer sequences have been developed. These copolyesters exhibit significantly different thermal transition behavior compared with the corresponding random copolyesters having the same overall compositions. Crystallinity as well as liquid crystallinity also were found to depend greatly on the comonomer sequence. The copolyesters of regular sequences undergo rapid sequence randomization at the temperatures higher than their melting points. This article reviews synthetic strategy, properties, and sequence randomization of aromatic copolyesters having ordered comonomer sequences.

There has been a great deal of interest in thermotropic, liquid crystalline polymers in the past twenty years or so since the discovery of useful materials based on them. Many critical factors such as structure of mesogenic units, presence and structure of flexible spacers or rigid kinks, molecular weight and its distribution, and thermal history influence thermal, physical and thermotropic properties of liquid crystalline polymers(1-13).

Among the many thermotropic polymer compositions reported, only aromatic copolyesters are presently of commercial importance. We and others(14-20) recently observed a strong dependence of thermal and crystallyzing properties of aromatic copolyesters on the comonomer sequence. We also observed that the copolyesters having ordered sequences undergo rapid sequence randomization above their melting points(T_m)(16), although it is rather slow below T_m. In this paper we would like to review synthetic methods for the preparation of copolyesters having regular costructural units and compare their properties with those of the random counterparts. Qualitative description of thermal randomization also will be described for a couple of selected systems.

0097–6156/90/0435–0033$06.00/0

Synthetic Strategy

Preparation of copolyesters having ordered comonomer sequences requires a multi-step synthetic route where a preprepared dimeric or
trimeric compound with a tailored sequential structure is polymerized
with a monomeric or multimeric compound(14-23). The latter can simply be a linking compound. The reactions that one can utilize in
such polymerization processes are limited only to those which gua-
rantee the maintenance of the sequence order set in the dimeric or
trimeric monomers. In other words, only the two terminal functional
groups should participate in the polymerization leaving the internal
links undisturbed. Reactions between a diacid dichloride and a diol
is a representative example that satisfies this requirement(21,24).
Direct polycondensation under a mild condition between a dicarboxylic
acid and a diol in the presence of novel condensing agent(s) is another excellent example that has many advantages over the acid chloride method(25-35). A wide range of condensing agents has been reported. The combinations of $SOCl_2$/pyridine, $C_6H_5POCl_2$/pyridine, and
$P(C_6H_5)_3/C_2Cl_6$/base are some of the examples.
        Conventional, high temperature (trans)esterification reactions,
as one can expect, are not suitable for the synthesis of the copolyesters we are concerned with(36). But polycondensation reactions
between activated monomers may be adopted if the reaction proceeds
reasonably fast even at a relatively low temperature(37).

Types of Ordered Sequence Copolyesters

Depending on the sequential structure of the monomer(s) used in the
polymerization reactions the resulting copolyesters can have several
variations in comonomer sequence:

$$A-B-A + C \longrightarrow \cdots-A-B-A-C-A-B-A-C-\cdots \tag{1}$$

1:1 alternating copolyester

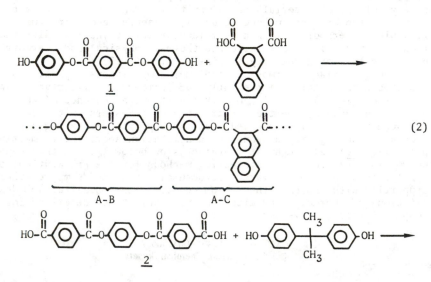

$$\tag{2}$$

A-B   A-C

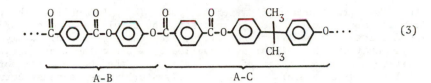

$$\text{(3)}$$

A–B A–C

$$A\text{-}B \longrightarrow \ldots\text{-}A\text{-}B\text{-}A\text{-}B\text{-}A\text{-}B\text{-}\ldots \tag{4}$$

1:1 alternating copolyester

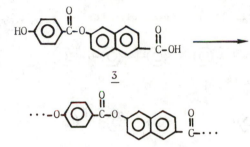

$$\text{(5)}$$

The examples given in Equations 2(38) and 3(21) are 1:1 alternating copolyesters consisting of two different repeating units linked to each other alternatively. In order to prepare the desired polyesters the triad monomer has to be synthesized first via a multi-step route that will be described in the next section.

The example given in Equation 5(39) is a special kind copolyester of alternating sequence. Since the two monomers, 4-hydroxybenzoic acid and 6-hydroxy-2-naphthoic acid, can undergo homopolymerization, one first has to synthesize the starting dyad monomer and then polymerize it to obtain the final, alternating copolyester. Otherwise, the copolyester chains will consist of random chain segments consisting of the two monomer units(40,41).

There is another interesting type of 1:1 alternating copolyester that can be prepared by the reaction given by Equation 1:

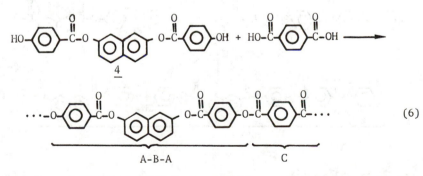

$$\text{(6)}$$

A–B–A C

The polymer of Equation 6(15) has a peculiar comonomer sequence compared with polymers in Equations 2 and 3. In the latter the two repeating units, A-B and A-C, are of the same type esters, whereas in

the former the two are of different type. The A-B-A unit in the former is a diol moiety and, in contrast, the C unit is a dicarboxylic
acid moiety. This difference arises from the presence of 4-oxyben-
zoyl unit in the polymer of Equation 6.   Since the two   repeating
units, A-B-A and C, are of different types the polymer is better
called just a regular sequence copolyester where each structural
unit appears regularly along the chain.

Polymerization between two triad type monomers can lead to a
2:1 alternating copolymer or to an 1:1:1 alternating terpolyester(39).

$$A-B-A \ + \ B-C-B \longrightarrow \cdots-A-B-A-B-C-B-A-B-A-B-C-B-\cdots \qquad (7)$$

2:1 alternating copolyester

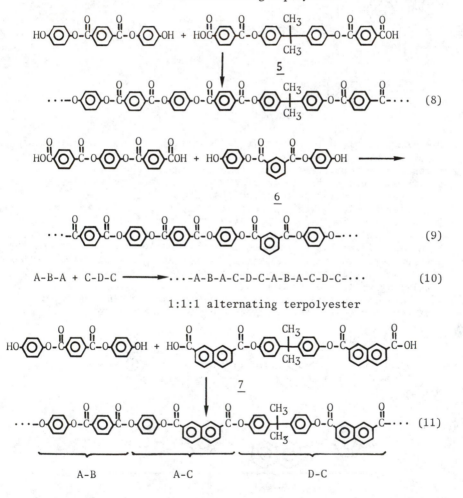

$$A-B-A \ + \ C-D-C \longrightarrow \cdots-A-B-A-C-D-C-A-B-A-C-D-C-\cdots \qquad (10)$$

1:1:1 alternating terpolyester

Another type of copolyesters that can be obtained by polymerization of two triad monomers and have the structural feature simi-
liar to that of the polymer of Equation 6 is shown below(39).

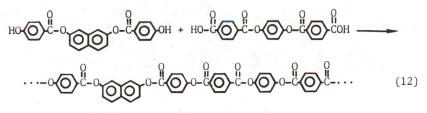

$$\cdots (12)$$

a regular sequence copolyester

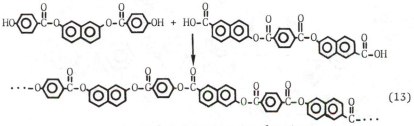

$$\cdots (13)$$

a regular sequence copolyester

These two copolyesters also are of regular sequence in which each structural unit appears in a periodic interval along the chain.

When a unsymmetrical dyad type monomer is polymerized with another reactant, the comonomer sequence is not perfectly ordered (14):

$$\text{A-B} + \text{C} \longrightarrow \cdots\text{-A-B-C-A-B-C-B-A-C-}\cdots \qquad (14)$$

semiregular sequence copolyester

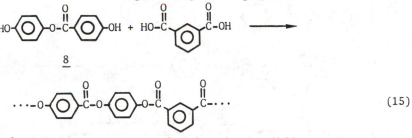

$$\cdots (15)$$

Since the A-B monomer can attack C in the two different directions, A-B $\longrightarrow$ C and B-A $\longrightarrow$ C, the polymer chain to be formed are not completely regular in monomer sequence although the C unit appears at every third positions. Similar situation arises when the dyad monomer A-B is polymerized with a triad comonomer C-D-C. Polymerization of A-B with another unsymmetrical dyad C-D lead to an irregular sequence copolyester:

$$\text{A-B} + \text{C-D} \longrightarrow \cdots\text{-A-B-C-D-A-B-D-C-B-A-C-D-B-A-D-C-}\cdots \qquad (16)$$

Synthesis of Dyad and Triad Monomers

The dyad or triad compounds utilized in the illustrative polymerizations given above are not commercially available and, therefore,

they have to be synthesized via multi-step routes. Three representative examples are shown below in chemical equations:

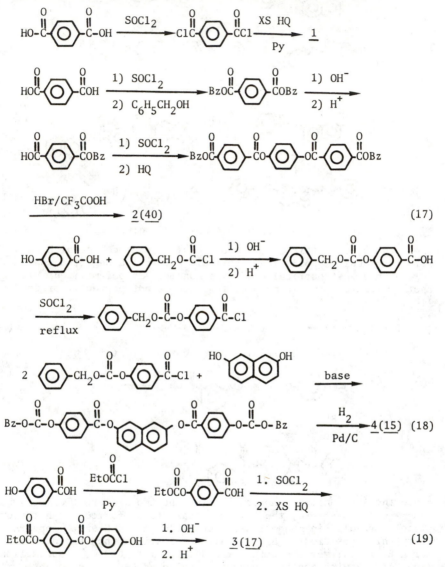

Hydroquinone(HQ) or a naphthalenediol in the above reactions
can be replaced with other diols such as biphenols and bisphenols.
In the same manner, other dicarboxylic acid such as cyclohexane
dicarboxylic acid, isophthalic acid, naphthalene dicarboxylic acids
can be used in place of terephthalic acid. Of course, these sub-
stitutes should not destroy the linearity of the polymer chains if
one is to obtain thermotropic compositions. 6-Hydroxy-2-naphthoic
acid is a substitute for 4-hydroxybenzoic acid in Reactions 18 and
19.

If copolyesters having flexible spacers are to be prepared, monomers containing spacers should be included in the polymerizations. Some of the examples are(6,17,18,42).

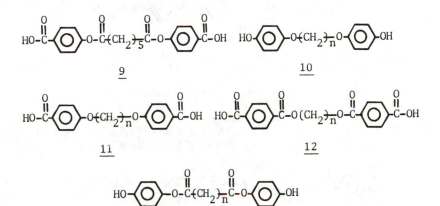

9

10

11

12

13

Properties of Ordered Sequence Copolyesters

There are not yet many reports describing the properties of sequentially ordered copolyesters. The reasons seem to be two-fold: 1) Although relatively straightforward reactions are involved in each synthetic steps, preparation of the dyad and triad monomers is rather tedious and time-consuming and 2) as will be explained later, thermal stability of ester bonds formed is rather limited, which makes the study of the properties of the copolyesters difficult. The latter reason may cause significant degree of artefacts in the experimental data or observations. Therefore, comparison of the properties of sequentially ordered copolyesters with those of random counterparts can be made only based on limited experimental information gathered up to now by us(14-16) and others(17-20).

As far as thermal transitions are concerned, there is not any distinct difference in the glass transition temperature(T_g) of ordered and random copolyesters. The melting temperature, T_m, however, depends very strongly on the sequence order(14-21). Very often the T_m value of an ordered sequence copolymer is higher even by about 100°C than that of the random counterpart(15,21,43). This must be due to more effective chain packing in the former than in the later. For example, the T_m value of the sequentially regular copolyester shown in Equation 6 is 290°C, while that of the random copolymer is only 191°C. The mesophase-to-isotropic phase transition (T_i), i.e., isotropization or clearing temperature, of sequentially ordered, liquid crystalline copolyesters may be higher or lower than that of random ones. When the copolymer's chain consists of linear and bent structural units and if the monomer for the linear repeating unit is capable of homopolymerization as 4-hydroxybenzoic acid or 6-hydroxy-2-naphthoic acid, the T_i value for the ordered sequence copolyester is lower as exhibited by the following copolyester(43,44):

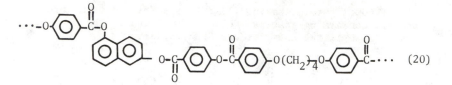

$$T_m; \text{ ordered } 229°C \qquad T_i; \text{ ordered } 280°C$$
$$\text{random amorphous} \qquad \text{random } 312°C$$

The T_i value of the above copolyester is 280°C, whereas the value for the random copolymer is about 30°C higher, 312°C.

However, for the copolyesters consisting of more or less lienar costructural units, the T_i values for the ordered copolymers appear to be slightly higher. Examples are given below(21,44):

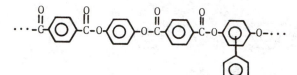

(21)

$$T_m; \text{ ordered } 323°C \qquad T_i; \text{ ordered } 365°C$$
$$\text{random } 296°C \qquad \text{random } 349°C$$

(22)

$$T_m; \text{ ordered } 267°C \qquad T_i; \text{ ordered } 351°C$$
$$\text{random } 175°C \qquad \text{random } 285°C$$

Such a difference can be understood by considering the possible shape of copolymers' chains. In the copolyesters containing both linear and nonlinear repeating units, the chain of an ordered sequence copolymers are bent regularly by the kinked units present at a fixed intervals along the chain. This would result in a relatively short segments of linear portion interrupted by the bent units. In contrast, there should be a broad distribution in the length of linear segments in random copolymers. The presence of longer linear portion in the chain of the random copolyesters is expected to improve the thermal stability of the mesophase and, thus, increase T_i as actually observed.

In addition to the difference in transition temperatures, DSC thermograms of ordered sequence copolyesters show sharper melting as well as isotropization transition(17,20,21). Although part of such a difference may come from the difference in the molecular weight and its distribution, it is our conjecture that the difference comes mainly from the difference in structural regularity of the chains. The existence of a broad distribution in the length of the linear segments along the chains as in random copolyesters will broaden the temperature range for isotropization, which in turn, will broaden the DSC peak for isotropization.

Another point to be noted is the fact that the mesophase tempe-rature ranges, i.e., $\Delta T_{LC} = T_i - T_m$, are broader for the random copoly-esters than for the ordered sequence ones. This phenomenon is par-ticularly evident when a copolyester is composed of linear and non-linear units, as shown for the copolymers of Equations 20 and 22.

The sequentially ordered copolyesters not only reveal much higher degree of crystallinity than the random ones, but also possess different crystal structures when compared with those of random co-polymers(14,15). When an ordered sequence copolymer sample is heated on a DSC instrument, very often a crystallization exotherm appears before melting due to higher crystallizing tendency. As pointed out above, regular chain structure promotes easier chain packing. For example, the copolyester of Equation 22 exhibits about 25% of degree of crystallinity, whereas the random one reveals less than 10%(21,43 44). An extreme example can be found in the polymer of Equation 20. The random copolymer is amorphous in contrast to the semicrystalline nature of the ordered sequence copolymer. The degree of crystalli-nity of the latter is about 15%(43,44).

Liquid Crystalline Properties

Mesophase-forming ability of a copolyester strongly depends on its comonomer sequence. For example, the ordered sequence copolyester shown below is not liquid crystalline, whereas its random counter-part is nematic(4,15):

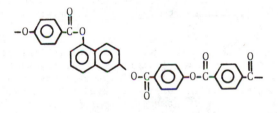

(23)

Such a difference can be ascribed to the fact that the random co-polymer's chain has long enough linear segments consisting of dime-ric or longer repeated 4-oxybenzoyl structure and terephthaloyl unit. Therefore, the random copolyester chains are able to have long enough rigid, linear segments that are essential to be thermotropic.

On the other hand, in the ordered sequence copolymer of Equa-tion 23, all of the 4-oxybenzoyl moieties exist in a monomeric unit and, thus, every linear triad segments consisting of 4-oxybenzoyl-terephthaloyl-4-oxybenzoyl links are interrupted by the bent 1,6-naphthalene moieties that destroy the linear geometry of the chain. Although we have discussed only one representative example, similar phenomenon and explanations should apply to other systems of the same structural characters.

Sequence Changes in Melts

It is well known that esters can undergo various interchange reac-tions such as intermolecular acidolysis and alcoholysis, and ester-

ester exchange. Since the terminal groups and internal ester bonds in polyesters can undergo similar reactions, the sequentially ordered copolyesters are expected to experience sequence randomization or changes when subjected to a thermal treatment at a high temperature, especially above $T_m(45)$.

The interchange reactions between two homopolyesters have been studied by many different groups(45-49). For example, Yamadera and Murano(46) observed that a random copolyester was obtained after 3 hours when poly(ethylene terephthalate) and poly(ethylene sebacate) were allowed to react at 276 °C. A similar observation was made by Zachmann et al.(47) for a mixture of deuteriated and nondeuteriated PET.

However, interchange reactions appear to be rather slow at temperatures below $T_m(15)$. We(50) observed that thermal treatment of some of liquid crystalline, aromatic copolyesters at the temperatures substantially lower than T_m did not lead to any changes in the comonomer sequence even after a prolonged period of time. A copolyester especially of 4-hydroxybenzoic acid, however, can undergo a special type of sequence changes below T_m, which is called the crystallization induced reaction(51).

Recently we subjected the ordered sequence copolyester of Equation 6 to thermal interchange reactions at 300 °C without adding any external catalyst and observed how its C-13 NMR spectrum changed with time(16). The T_m of the original copolyester was 290 °C. The C-13 NMR spectral analysis aided by DSC and X-ray analyses lead to the conclusion that the sample treated for 30 minutes at 300 °C became amorphous due to high degree of sequence randomization. In fact, the C-13 NMR spectrum of the sample treated only for 4 minutes at 300 °C was already quite different from that of the original sample.

Figure 1 compares how the C-13 NMR spectrum of the sequentially ordered copolyester of Equation 22 changes when it is thermally treated at 280 °C under a N_2 atmosphere that is slightly higher than its T_m of 267 °C. Up to 2 minutes there is not observed any change in the spectrum. The spectrum of the sample treated for 4 minutes, however, starts to reveal substantial changes compared with that of the original sample. The changed spectra are not exactly the same, but similar to that of the random copolyester separately prepared. These observations clearly demonstrate how fast interchange reactions can occur for the ordered sequence aromatic copolyesters above their mp's resulting in sequence randomization and also how much one should be cautious when their properties in melts such as melt viscosity, liquid crystallinity, thermal transition temperatures, processing characteristics, etc. are to be discussed. Due to the sequence changes at high temperatures thermal history of a sample has to be carefully controlled in order to make a reliable characterization. Last but not least, presently there are a great deal of interests in the polyblends consisting of a liquid crystalline polyester and a non-liquid crystalline polyester. In studying such systems, on has to be extremely careful on the probable intermolecular ester exchanges that will lead to compositional changes(49,52).

Acknowledgments

It is gratefully acknowledged that this work was supported by the Ministry of Education of the Republic of Korea through a 1989's Re-

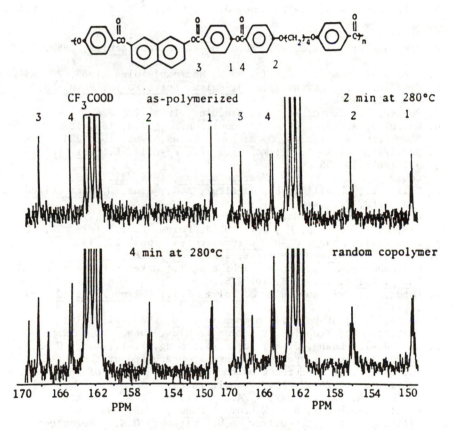

Figure 1. Changes in C-13 NMR spectra of the copolyester of Equation 22 after heat-treatment at 280°C (solvent; CF_3COOD: $CDCl_3 = 1:1 (v/v)$).

search Fund for Advanced Materials. The author thanks Prof. Hong-Ku
Shim of the Korea Institute of Technology for his help in C-13 NMR
studies and coworkers, Byung-Wook Jo, Jin-Hae Chang, Ki-Youn Sung and
Chung-Seock Kang for their contribution.

Literature Cited

1. Jin, J.-I.; Antoun, S.; Ober, C.; Lenz, R.W. Br. Polym. J., 1980,
   12, 132.
2. Ober, C.K.; Jin, J.-I.; Zhou, Q.-F.; Lenz, R.W. Adv. Polym. Sci.,
   1984, 59, 102.
3. Lenz, R.W.; Jin, J.-I. Macromolecules, 1981, 14, 1405.
4. Jin, J.-I.; Chang, J.-H.; Shim, H.-K. Macromolecules, 1989, 22,
   93.
5. Jin, J.-I.; Choi, E-J.; Jo, B.-W. Macromolecules, 1987, 20, 934.
6. Griffin, A.C.; Havens, S.J. J. Polym. Sci., Polym. Phys. Ed.,
   1981, 19, 951.
7. Percec, V.; Yourd, R. Macromolecules, 1988, 22, 524.
8. Roviello, A.; Sirigu, A. Macromol. Chem., 1982, 185, 183.
9. Blumstein, A.; Thomas, O. Macromolecules, 1982, 15, 1264.
10. Meurisse, P.; Nöel, C.; Monnerie, L.; Fayolle, B. Brit. Polym.
    J., 1981, 13, 55.
11. Cowie, J.M.G.; Wu, H.H. Macromolecules, 1988, 21, 2865.
12. Galli, V.; Chiellini, E.; Angeloni, A.S.; Laus, M. Macromolecu-
    les, 1989, 22, 1120.
13. Uryu, T.; Kato, T. Macromolecules, 1988, 21, 378.
14. Jin, J.-I.; Lee, S.-H.; Park, H.-J. Polym. Bull., 1988, 20, 19.
15. Jin, J.-I.; Chang, J.-H. Macromolecules, 1989, 22, 4402.
16. Jin, J.-I. Polymer Preprints, 1989, 30(2), 481.
17. Krigbaum, W.R.; Kotek, R.; Ishihara, T.; Hakemi, H.; Preston, J.
    Europ. Polym. J., 1984, 20, 225.
18. Kotek, R.; Krigbaum, W.R. J. Polym. Sci., Polym. Phys. Ed., 1988,
    26, 173.
19. Moore, J.S.; Stupp, S.I. Macromolecules, 1988, 21, 1219.
20. Martin, D.G.; Stupp, S.I. Macromolecules, 1988, 21, 1222.
21. Jin, J.-I.; Chang, J.-H.; Jo, B.-W.; Sung, K.-Y.; Kang, C.-S.
    Makromol. Chem. Symposium, in press; presented at the IUPAC
    Macromolecular Conference on the Molecular Design of Functional
    Polymers, Jun 26-28, 1989, Seoul, Korea.
22. Bilibin, A. Yu; Tenkovtsev, A.V.; Piraner, O.N.; Pashkovsky,
    E.E.; Skorokhodov, S.S. Makromol. Chem., 1985, 186, 1525.
23. Bilibin, A. Yu.; Tenkovtsev, A.V.; Piraner, O.N.; Skorokhodov,
    S.S. Vysokomol. Soyed., 1984, A26(12), 2570.
24. Morgan, P.W. 'Condensation Polymers: By Interfacial and Solution
    Methods', Interscience Publishers, N.Y., 1965, pp.115-161 and
    pp.325-391.
25. Yasuda, S.; Wu, G.-C.; Tanaka, H.; Sanui, K.; Ogata, N. J.
    Polym. Sci., Polym. Chem. Ed., 1983, 21, 2609.
26. Tanaka, H.; Wu, G.-C.; Iwanaga, Y.; Sanui, K.; Ogata, N. Polym.
    J., 1982, 14, 331 and 635.
27. Higashi, F.; Mashimo, T.; Takahashi, I. J. Polym. Sci., Polym.
    Chem. Ed., 1986, 24, 97 and 1697.
28. Ueda, M.; Kano, T. Makromol. Chem., Rapid Commun., 1984, 5, 833;
    6, 847.

29. Yamazaki, N.; Higashi, F. Tetrahedron, 1974, 30, 1323.
30. Higashi, F.; Akiyama, N.; Takahashi, I.; Koyama, T. J. Polym. Sci., Polym. Chem. Ed., 1984, 22, 1653 and 3607.
31. Ueda, M. J. Synth. Org. Chem.(Japan), 1990, 48(2), 144.
32. Higashi, F.; Kobayashi, A. J. Polym. Sci., Polym. Chem. Ed., 1988, 27, 507.
33. Ueda, M.; Honnma, T. Makromol. Chem. 1989, 190, 1507.
34. Higashi, F.; Ozawa, M.; Mochizuki, A. J. Polym. Sci., Polym. Chem. Ed., 1988, 26, 3071.
35. Higashi, F. J. Synth. Org. Chem.(Japan), 1989, 47, 994.
36. Moore, J.S.; Stupp, S.I. Macromolecules, 1987, 20, 273.
37. Beiersdorf, P. Brit. Pat., 846044(1960).
38. Kang, C.-S.; Jin, J.-I. Unpublished results.
39. This polymer has not been yet reported.
40. Gutierrez, G.A.; Blackwell, J. Macromolecules, 1984, 17, 2744.
41. Blackwell, J.; CKeng, H.M.; Biswas, A. Macromolecules, 1988, 21, 39.
42. Hasslin, H.W.; Dröscher, M.; Wegner, G. Makromol. Chem., 1980, 181, 301.
43. Jo, B.-W.; Sung, K.-Y.; Choi, J.-K.; Chang, J.-H.; Jin, J.-I. Polymer(Korea), 1989, 13, 675.
44. Jin, J.-I.; Jo, B.-W.; Sung, K.-Y.; Chang, J.-H. Unpublished results.
45. Kotliar, A.W. J. Polym. Sci., Macromol. Revs., 1981, 16, 367.
46. Yamadera, R.; Murano, M. J. Polym. Sci., 1976, A-15, 2259.
47. Kugler, J.; Gilmer, J.W.; Wiswe, D.; Zachmann, H.-G.; Hahn, K.; Fischer, E.W. Macromolecules, 1987, 20, 1116.
48. Aden, M.A.; Gilmer, J.W. Polymer Preprints, 1987, 28(2), 304.
49. Suzuki, T.; Tanaka, H.; Nishi, T. Polymer, 1989, 30, 1287.
50. When the ordered sequence copolyester of Equation 22 was subjected to a thermal treatment at 200°C, its C-13 NMR spectrum did not change even after one hour.
51. Lenz, R.W.; Jin, J.-I.; Feichtinger, K.A. Polymer, 1983, 24, 327.
52. Laivins, G.V. Macromolecules, 1989, 22, 3974.

RECEIVED May 2, 1990

Chapter 4

All-Aromatic Liquid-Crystalline Polyesters of Phenylhydroquinone with Ether and Ketone Linkages

Michael H. B. Skovby, Claus A. Heilmann[1], and Jørgen Kops

Institut for Kemiteknik, Technical University of Denmark,
DK–2800 Lyngby, Denmark

The modification of polyterephthalates of phenylhydroquinone by substituting the terephthalic acid for
kinked 3,4'- and 4,4'-dicarboxydiphenylether and
–ketone is a suitable way of obtaining melt processable thermotropic polyesters with melting transitions
in the range 200-300°C. The all-aromatic polyesters
with ether linkages show excellent thermal stability
surpassing those with ketone linkages. Fibers were
spun from the liquid crystalline melts. E' moduli
ranging from 30-50 GPa and break tenacities in the
range 400-600 MPa were found for these fibers. The
melt flow properties were independent of the type of
linkage, 3,4' or 4,4'. A large increase in the viscosities were found in case of high degrees of substitution when isotropic melt state were approached.

Recent work in the field of thermotropic liquid crystal (LC)
polymers, and the introduction into the commercial market of various
polymers of this type has shown that in order to achieve high performance characteristics with regards to mechanical and thermal
properties the polymers should most often have an all–aromatic
structure. In our studies, we have been interested in investigating
the effect of compositional variations in some closer defined fully
aromatic polyesters on the properties. The main objective has been
to achieve thermotropic behavior at processing temperatures normal
for engineering plastics and at the same time thermal stability
under processing conditions and in connection with use at elevated
temperatures. Linear 1,4-linked aromatic polyesters are known to
have too high melting points for melt processing. Various changes
in the structure may lower the melt transition and the basic structural modifications in this respect have been summarized by Griffin

[1]Current address: Röhm Gmbh, Kirschenallee, Postfach 4242, D–6100 Darmstadt 1,
Federal Republic of Germany

0097–6156/90/0435–0046$06.00/0
© 1990 American Chemical Society

and Cox (1). The commonly used method to build-in flexible chain
segments, most often consisting of methylene groups, into an aromatic polyester (2) was not considered to be suitable in our case.
Thus, if the melt processing range for the aromatic polyester was to
be lowered to a desirable range of less than 300°C, while maintaining high performance characteristics, including good thermal stability, the flexible chain segments should be avoided, since they
would represent weak links in the polymers. Substitution of the
aromatic rings was chosen as a way of depressing the melting ranges.
In this connection the work by W.J. Jackson, Jr. (3) is essential
concerning substituted hydroquinones. It was obvious from this work
that particularly phenylhydroquinone (PHQ) is able to lower the melt
transition while good thermal stability is maintained. This monomer
has been used throughout the present work as the diol component in
the polyester syntheses. Concerning the diacid component,
terephthalic acid (TA) was chosen due to good properties and general
availability, however, in order to lower the melting to below 300°C
for the polymers kinked dicarboxylic acids have been substituted for
the terephthalic acid to various extents.

We already have reported on the replacement of the
terephthalic acid with kinked diphenylether dicarboxylic acids (4).
3,4'- and 4,4'-Dicarboxydiphenylether (3,4'-O and 4,4'-O) were synthesized and all-aromatic polyesters were prepared represented by
structure 1. These polyesters were thermotropic with melt transi-

tions decreasing to about 200°C with increasing replacement of the
terephthalic acid with the kinked monomers. The polymers generally
were thermally stable without measurable weight loss until well over
400°C. We wish here to supplement our previous studies with rheological measurements and fiber spinning of the polymers, including
some measurements of fiber properties.

In addition 3,4'- and 4,4'-dicarboxydiphenylketone (3,4'-K and
4,4'-K) have also been synthesized as reported in preliminary
fashion (5). The thermal properties of the polyesters prepared with
these monomers, represented by structure 2, will be reported and
compared with those of polymers 1.

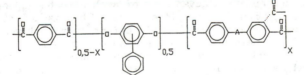

1: A = -O- , 2: A = -CO- and x = molfraction

Experimental
Monomers

3,4' and 4,4'-Dicarboxydiphenylether (3,4'-O and 4,4'-O) have been
prepared by oxidation of the corresponding dimethyldiphenylether as
described (4).
3,4'- and 4,4'-Dicarboxydiphenylketone (3,4'-K and 4,4'-K)
have been synthesized by dichromate oxidation of the corresponding
dimethyldiphenylketones which in turn were obtained as described
(5).

Terephthalic acid (purity >99%) was always used without
further purification.

Phenylhydroquinone (Aldrich) was distilled and acetylated and
recrystallized from a 80/20 pentane/chloroform mixture (4).

Polymers

All polymerizations were carried out by melt acidolysis based on a
procedure described in a U.S. Patent (6). A general procedure was
used for all the experiments (4).

Three larger 100g batches of polymers with the composition
PHQ, TA, 4,4'-O = 0.5/0.35/0.15 were prepared for fiber and melt
flow studies. The polymerization procedure for these was as
follows: Temp(^{o}C)/time(min)/pressure (mmHg) = 290/60/760;
320/45/760; 320/70/10; 340/60/0.1. The three batches were ground
and melt blended in a Brabender at 300°C for 15 min under nitrogen
purge.

Thermal Analysis

The thermal transitions of liquid crystal polymers are normally
studied by a combination of differential scanning calorimetry (DSC)
and visual observations on a hot-stage polarizing microscope. While
DSC is run in a recording manner and the actual transi-tions are
determined from the recorded curve, analysis by microscopy is
generally based on subjective evaluations and descriptions of the
observations. It appeared to us that certain advantages could be
derived from running the hot-stage polarizing microscopy in a recording manner in order to obtain a light transmission curve with
relation to the birefringence and determine the transition temperatures on the basis of this curve. Although subjective evaluations
still is required if the texture of the thermotropic phase is to be
established, a recording of the degree of light transmission could
be much more rational and lead to better and perhaps standardized
ways of determining the transitions temperatures.

Despite the apparent advantages of such a technique only very
few references are found in the literature. This is surprising in
view of the very large number of publications in recent years on
liquid crystal polymers and of course also on low molecular weight
liquid crystals. We wish here to illustrate this thermo-optical
analysis (TOA) as a tool for characterization of liquid crystal
polymers.

Originally, TOA was devised and used for analyses of chain
mobility in polymers (7) and polymer blends (8,9) by monitoring
birefringence disappearance during programmed heating in scratches
scribed on film surfaces. Lenz (10,11) has reported the use of TOA
in the study of liquid crystal polymers, and recently we have used
the technique in determining melt- and isotropization temperatures
for thermotropic fully aromatic liquid crystal polymers (4,5).

DSC thermograms were recorded on either a DuPont 900 or a
Stanton Redcroft STA 785 + CPC 706 instrument with a heating rate of
20°C/min. TGA thermograms were determined with the Stanton Redcroft
instrument. Thermal transitions were also studied using a Linkam

PR600 hot stage (HS) in connection with a Reichert-Jung Microstar 110 microscope equipped with a Photodyne 22xL photometer and a recorder. The heating rate was set to 20°C/min. The crystalline to nematic transition is reported as the temperature corresponding to a 20% increase in light transmission and similarly the nematic to isotropic transition temperature as that corresponding to an 80% decrease in light transmission from the maximum value. This was based on obtaining the best correspondance with values determined by DSC.

Rheological Measurements

Viscosity measurements at low shear rates were made with a Rheometrics System 4 or Rheometrics RMS-800/RDS-II spectrometer in either cone and plate (cone angle = 0.1, radius = 25 mm) or plate and plate (radius = 25 mm, distance = 1 mm). For higher shear rates viscosity measurements were made with an Instron capillary viscometer. A capillary with L/D = 51.4 and D = 0.889 mm was used. Sample discs were prepared by compression molding of the predried (at least 3 days at 125°C in vacuum) copolyesters at 250°C. Prepared discs were dried at 125°C in vacuum for 5 hr before use. The correction for non-Newtonian behavior was applied, but entrance pressure corrections were neglected.

Melt spinning of fibers

Melt spinning of two polyesters: a) 15 mol% 4,4'-dicarboxy diphenyl ether modified poly(phenyl-1,4-phenylene therephthalate) and b) 20 mol% p-hydroxybenzoic acid modified poly(phenyl-1,4-phenylene terephthalate) was carried out using an Instron capillary viscometer. The die had a diameter of 1.24 mm and L/D = 40.99. The draw ratios (DR) were calculated as surface area ratios of the fibers to the Instron plunger area (0.7125 cm^2). The resulting fiber was cooled by a continuous stream of dry nitrogen approx. 2.5 cm below the capillary die. In order to avoid moisture, the poly-ester was added to the barrel through a closed box purged with a continuous stream of nitrogen. The DR values were varied by varying the speed of a system of specially constructed take-up rolls, and fiber diameters were determined by optical microscopy.

Mechanical properties

Dynamic mechanical properties were determined with a Polymer Laboratories Dynamic Mechanical Thermal Analyzer (DMTA) using the tensile mode. The fiber length was in all cases 20 mm at an initial 0.5% elongation. The heating rate was set to 5°C/min. Break tenacities were measured on an Instron tensile tester model 1130 using a sample length of 25.4 mm and a strain rate of 0.508 mm/min (0.02 in./min). All reported break-tenacities and moduli are the mean values of four measurements.

Results and Discussion
Thermal Properties

The thermal transitions for polymers of structure $\underset{\sim}{1}$ and $\underset{\sim}{2}$ are summarized in Tables I and II and the transition temperatures for for-
mation of nematic melts (T_{KN}) are plotted in Figure 1.

The kinked dicarboxy compounds obviously has considerable
capability for decreasing the melt transitions with increasing substitution for the terephthalic acid. A decrease to below $300^{\circ}C$ is
already achieved with the diphenyl ether derivatives-(3,4'-0 and
4,4'-0) at a degree of substitution corresponding to 30%. For the
corresponding diphenyl ketone derivatives (3,4'-K and 4,4'-K) a
somewhat higher degree of substitution is required, 35-40%, to get a
comparable melt depression. The more stiff structure of the ketones
compared to the ethers may be connected with this kind of behaviour.
However, further increase in the amounts has a very drastic effect
on the melt transition.

The actual type of kinked monomer with regards to the connect-
ing linkage between the phenyl rings or the positions of substitution has little influence on the quantitative effect on the transitions at high levels of addition. Previously, we have found (4)
that the melt transition of the homopolymer of hydroquinone and
3,4'-0 was much lower than the homopolymer made with 4,4'-0. This
may, however, be due to an effect from the unsymmetrical monomer
3,4'-0 which may be incorporated in a head-to-tail or a random
fashion in contrast to the symmetrical monomer 4,4'-0 in combination
with the unsubstituted hydroquinone. The effect of the diphenyl
ether linkage has been investigated before in case of preparation of
polyesters of chlorohydroquinone and 4,4'-0/TA in different ratios
(1, 12-13). For this polymer system it was also possible to obtain
nematogenic compositions that were melt processable in the range
250-300°C.

The thermal stability have been investigated for the polymers
and this property is very important in relation to the study of the
rheological properties which are reported in this paper. Generally,
the diphenyl ether modified polymers were more thermally stable than
the corresponding polymers with ketone structure, since a significant weight loss occurred at around a 50° C higher temperature.
Typical TGA curves are illustrated in Figure 2.

An example on a TOA thermogram is shown in Figure 3. The
polymer was prepared with TA/4,4'-K = 0,285/0,215. By DSC a melt
transition of $272^{\circ}C$ was recorded. The TOA curve indicates a phase
change in the thermotropic interval, which could not be detected by
DSC. The initial melting to the nematic state corresponds to the
first increase in the light intensity. However, a second transition
is clearly indicated at a higher temperature. The various curves
were recorded for different starting light intensities corresponding
to different sample thicknesses.

Polyesters for fibers and flow studies

We report here studies on polyesters A, A1, A2, B1, B2 (see Table I)
and the big batch called BB1. Polyester BB1 had the same composition as B1, but an inherent viscosity could not be determined

Table I. Thermal Transitions of Polyesters 1

Nr.	Composition Mol Fraction		I.V a)	DSC °C			Hot Stage, TOA °C	
	TA	3,4'-O		T_g	T_{KN}	T_{NI}	T_{KN}	T_{NI}
A	0,5	0	–	146	342	423	350	400
	0,45	0,05	–	131	326	425	333	>490
A1	0,35	0,15	0.92	124	281	325	278	350
A2	0 25	0,25	0.45	133	210	–	205	405
	0,15	0,35	0.38	122	–	–	198	–
	TA	4,4'-O						
	0,45	0,05	1.00	149	318	415	320	480
B1	0,35	0,15	1.00	135	280	350	275	331
B2	0,25	0,25	0.63	128	225	288	255	<300
	0,15	0,35	0.55	138	195	–	210	–

a) Measured in o-chlorophenol/chloroform = 1/3 at 30°C with 0.5g polymer/100 ml solvent.

Table II. Thermal Transition of Polyesters 2

Composition Mol Fraction		DSC °C			Hot Stage, TOA °C	
TA	3,4'-K	T_g	T_{KN}	T_{NI}	T_{KN}	T_{NI}
0,45	0,05	132	318	455	–	480
0,35	0,15	138	372	–	372	382
0,30	0,20	138	284	–	278	332
0,25	0,25	139	–	–	260	295
0,15	0,35	145	–	–	185	219
0,05	0,45	142	–	–	–	
TA	4,4'-K					
0,45	0,05	140	330	450	337	447
0,35	0,15	135	372	–	379	425
0,33	0,17	140	290	–	305	432
0,30	0,20	118	<200	385	223	380
0,25	0,25	115	185	–	193	375
0,15	0,35	105	173	–	172	209
0,05	0,45	110	–	–	171	171

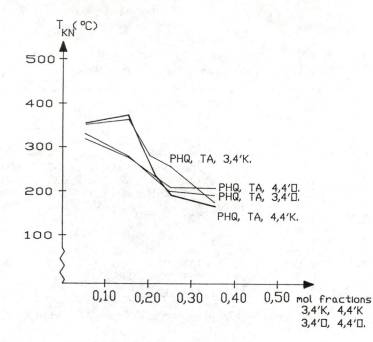

Figure 1. Melting points of 3,4'- and 4,4'-dicarboxy diphenyl
ether and 3,4'- and 4,4'-dicarboxy diphenyl ketone modified
polyesters.

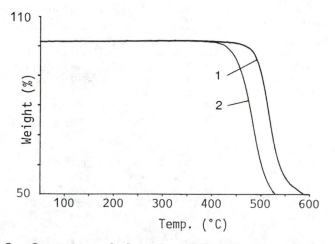

Figure 2. Comparison of thermal stability by thermogravimetric
analysis (TGA) between a 4,4'-dicarboxy diphenyl ether and ketone
containing polyester.
1: PHQ/TA/4,4'-O = 0.5/0.25/0.25
2: PHQ/TA/4,4'-K = 0.5/0.25/0.25.

because of only partial solubility in the solvent, which indicates higher molecular weight than polyester B1.

Fibers

The spinning of fibers from the nematic liquid crystalline state may at least in principle result in fibrous structures exhibiting nearly perfect molecular orientation. Imperfections such as chain ends should then be randomly distributed. A large amount of work has been performed in recent years on semicommercialized LCP's and also on more research based LCP's (13-16).

    Fibers made from thermotropic LCP's by melt spinning are highly oriented, but their mechanical properties are unfortunately still not competitive to the corresponding rigid main-chain aromatic polyamides. The disadvantage of the main chain aromatic polyamides is the lack of melting and therefore non-applicability for injection-molding, and selfreinforced composites (17).

    For fiber studies polyester BB1 was chosen. A method to measure birefringence developed by Yang and co-workers (18) was applied on polymer BB1 fibers and showed no dependence on DR, but remained constant 0.32 within experimental error. Similar observations were made by McIntyre and co-workers with a somewhat similar LC co-polyester with chloro instead of phenyl-hydroquinone and 25 mol% 4,4'-dicarboxydiphenylether (13).

    A polyester first prepared by Jackson (3) (20 mol% modified p-hydroxybenzoic acid poly(phenyl-1,4-phenylene terephthalate)) was spun at 320°C and fibers were characterized "as-made". The dynamic tensile modulus as a function of DR is seen in Figure 4 and shows that very high values are obtained at even low DR-values.

    Polyester BB1 which was spun at 300°C gave much lower values (Figure 5), which is not that surprising, due to the highly disrupted chemical nature of the polyester chains. Although the onset melting point was as low as 253°C, it was not possible to produce regular and reproducable fibers before the temperature reached 300°C.

    The break tenacity for "as-made" fibers was 420 MPa (curve 1, Figure 6), but could be improved to 580 MPa (curve 2, Figure 6) by heat treating at 235°C for 3 hr. Heat treatment was performed under nitrogen by winding fibers around a U-shaped metal device which applied a small tension to the fibers. Heat treatment increased the elongation at break, but it was in all cases below 3.5%. A typical stress-strain curve for DR = 500 fibers is seen in Figure 6, and it shows a slight decrease of modulus with increasing strain.

    The DMTA behaviour of BB1 (Figure 7) showed a large drop in modulus starting at approx. 130°C and a peak of tanδ at 140°C, which gives a quite broad range of the glass transition. However, no significant decrease of modulus is observed before 140°C, which corresponds to the glass transition determined by DSC. A small drop of modulus and a small tanδ peak was observed at 90°C, which perhaps may be attributed to some extent of blockiness in the polyester backbone.

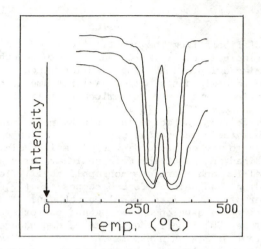

Figure 3.   Thermo optical analysis (TAO) of polyester:
PHQ/TA/4,4'-K = 0.5/0.285/0.215 with different sample
thicknesses.

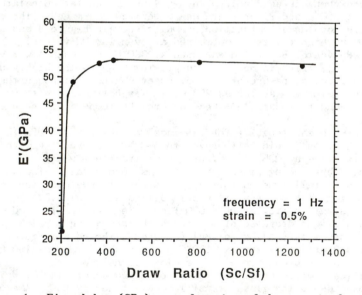

Figure 4.   E'-modulus (GPa) as a function of draw ratio for 20
mol% p-hydroxy benzoic acid modified poly(phenyl-1,4-phenylene
terephthalate) (not listed in tables) fibers.   Spinning
temp. = 320°C.

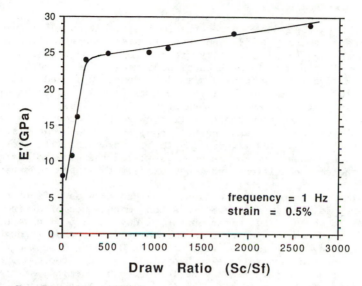

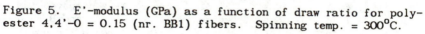

Figure 5. E'-modulus (GPa) as a function of draw ratio for poly-ester 4,4'-O = 0.15 (nr. BB1) fibers. Spinning temp. = 300°C.

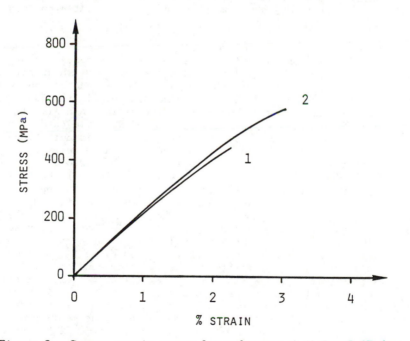

Figure 6. Stress strain curve for polyester 4,4'-O = 0.15 (nr. BB1) fiber. Draw ratio = 500. 1) 'as-made'. 2) Heat treated at 235°C for 3 hr.

Melt Flow Behaviour

Much work has been done on the rheological behaviour of liquid crystalline polymers (19–25), and recently also theoretical approaches
have been made (26). Generally, thermotropic (and also lyotropic)
LCP's exhibit a variety of unusual phenomena. These include shear
thinning viscosity, viscosity time dependence (27), and little or no
extrudate swelling. Furthermore these materials tend to be highly
sensitive to the thermal and mechanical history (28). As a result
many confusing and even contradictionary results occur. As an
example Wissbrun, Kiss and Cogswell (29) report both capillary and
dynamic torsional flow viscosities to be increasing with decreasing
gap spacing. However, Kalika et al. studied the same polyester
under the same conditions and found exactly the opposite behaviour
(30).
     The viscoelastic properties of polyester A as a function of
frequency is seen in Figure 8. A strong dependence of drying conditions is obvious, but after approx. 2 days at $125^{\circ}C$ in vacuum no
significant changes occurred. Shear thinning response throughout
the frequency range is observed, and with G' being almost constant
at low frequencies a solid like behaviour is indicated.
     In order to investigate the influence of chemical structure on
the melt viscosity, polyesters A1,A2,B1 and B2 were compared in the
steady mode in plate and plate-geometry at $300^{\circ}C$. A2 and B2 should
be in the isotropic region, whereas A1, B1 would be in the nematic
state. Prior to steady shear measurements a transient step from
$\dot{\gamma} = 0.01$ sec^{-1} (60 sec); $\dot{\gamma} = 0.1$ sec^{-1} (60 sec); $\dot{\gamma} = 1$ sec^{-1} (60
sec) to $\dot{\gamma} = 10$ sec^{-1} (30 sec) was performed, and steady state was
achieved after approx. 30 sec. Overshoot (yield stress) was predominant in A1, B1. The viscosities as a function of shear-rate is
seen in Figure 9 and shows similar behaviour of A1 and B1, but with
differences between A2 and B2. A2 and B2 are both in the isotropic
state and not surprisingly show higher viscosities than A1, B1. The
difference between the viscosities for A2 and B2 could be due to
different I.V's (0.45 and 0.55, respectively), and is not interpreted as a difference between the 3,4'-O and 4,4'-O moiety.
     Polyester BB1 was run twice in steady mode at $290^{\circ}C$
(Figure 10), and shows that the orientational effect of the first
run has a drastic effect on steady shear viscosity. In the first
run the log viscosity vs. log shear rate had a slope of −0.92 (solid
like behaviour, yield stress), but in the second run a pseudo–Newtonian plateau was reached from approx. 1 sec^{-1}. Capillary viscosity values corresponded reasonably well with the second run steady
shear data. The slope at high shear rates was close to −0.91 which
corresponds nicely to the first-run steady shear run. All this
could suggest, that this system is not completely melted, but still
has some solid like regions incorporated. At $300^{\circ}C$ capillary viscosity data showed an almost pseudo-Newtonian plateau. This corresponds quite well to the fact that fiber spinning as mentioned
earlier was difficult and almost impossible below $290^{\circ}C$, but easy at
$300^{\circ}C$. At an apparent shear rate of 100 sec^{-1}, a die-swell was
found to be approximately 0.95.

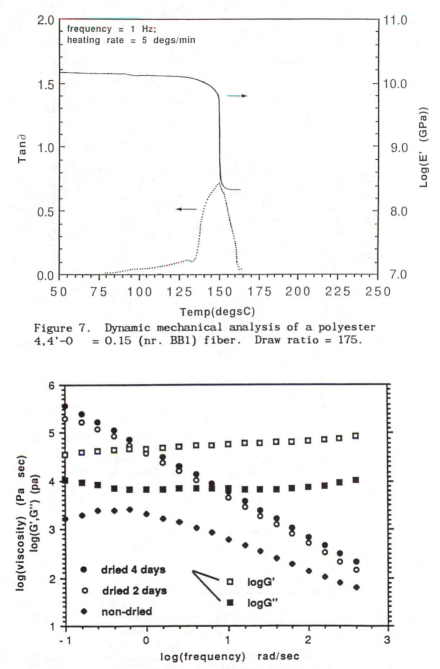

Figure 7. Dynamic mechanical analysis of a polyester
4,4'-O   = 0.15 (nr. BB1) fiber. Draw ratio = 175.

Figure 8. Complex viscosity, storage modulus (G') and loss
modulus (G") vs. frequency for poly(phenyl-1,4-phenylene
terephthalate) (polyester nr. A). G' and G" correspond to the 4
days dried sample. Parallel plates. Strain = 3% and
Temp. = 355°C.

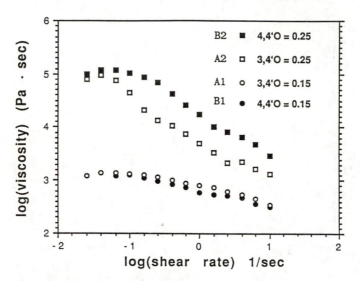

Figure 9.   Steady shear viscosities of 3,4'-O and 4,4'-O containing polyesters (see Table I) as a function of shear rate in parallel plates.   Temp. = 300°C.

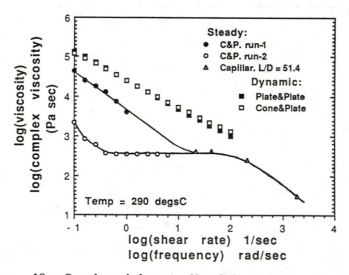

Figure 10.   Steady and dynamic flow behaviour of polyester 4,4'-O = 0.15 (nr. BB1) at 290°C.   Strain = 3%

The complex viscosity of BB1 was measured at 290°C in both cone and plate and parallel plates geometry and showed no significant differences. The general trend that systems containing some kind of microstructures in the melt (suspensions) show complex viscosities higher than the steady shear viscosities, was also true in this case. A similar observation was made with an Eastman copolyester (31).

No influence of strain on the viscoelastic properties was found. A strain sweep to 14% oscillatory strain was performed with frequencies ranging from 1 rad/sec to 10 rad/sec, and gave an essentially linear viscoelastic respons.

Figure 11 shows the viscoelastic properties of BB1 at 5% strain, and shear thinning is again observed over the entire frequency range. The shapes and magnitudes of G' and G" could suggest solidlike behaviour, which is a result of high relaxation times.

Conclusion

The thermal characterization of 3,4'- and 4,4'-dicarboxy diphenyl ether containing polyesters has shown better thermal-stability than the corresponding 3,4'- and 4,4'-dicarboxy diphenyl ketone polyesters. Furthermore, higher degrees of substitution with the kinked monomer is necessary in case of the ketone containing polyesters to achieve the same depression of the melt transitions, compared to the ether polyesters. Crystalline-nematic transitions may be as low as approx. 200°C.

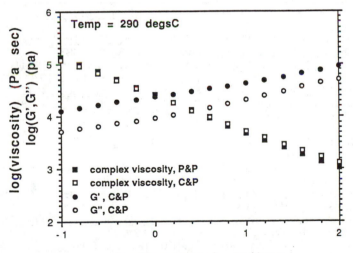

Figure 11. Complex viscosity, storage modulus and loss modulus vs. frequency at 290°C. Polyester 4,4'-O = 0.15 (nr. BB1). Strain = 5%.

No differences have been found between the 3,4'- and 4,4'-dicarboxy diphenyl ether substituted poly(phenyl-1,4-phenylene terephthalate) with regards to the steady shear viscosity behaviour, but a dramatic increase in viscosity was observed as the degree of substitution was high enough to result in a isotropic melt. Generally large yield stress effects were dominant in the nematic melts, but they were strongly pre-history dependent. A three region flow curve for 15 mol % modified poly(phenyl-1,4-phenylene terephthalate) was probably due to a not completely molten system. Dynamic viscosity measurements showed strong pseudoplastic behaviour. Strain and time dependence phenomena were not observed.

Fibers with E' modulus ranging from 30 to 50 GPa could very easily be produced. Break tenacities very generally around 400 MPa, but could be increased to approx. 600 MPa by heat treatment. The draw ratio had no significant effect on the fiber-birefringency but remained constant at 0.32, within experimental error.

Acknowledgment

The financial support by The Danish Technical Research Council and The Danish Research Academy is gratefully acknowledged.

Literature Cited

1.   Griffin, B.P.; Cox, M.K. Brit. Polym. J. 1980, 147.
2.   Ober, C.K.; Jin, J.-I.; Lenz, R.W. Adv. Polym. Sci. 1984, 59, 103.
3.   Jackson, Jr., W.J. Contemporary Topics in Polymer Science 1984, 7, 177.
4.   Skovby, M.H.B.; Lessél, R.; Kops, J. J. Polym. Sci. Part A Polym. Chem. 1989, 27. In press.
5.   Heilmann, C.A.; Skovby, M.H.B.; Kops, J. Polym. Prepr. Am. Chem. Soc., Div. Polym. Chem. 1989, 30(2), 483.
6.   Irwin, R.S. (to DuPont) U.S.Patent 4, 487, 916 (1984).
7.   Kovacs, A.J.; Hobbs, S.Y. J. Appl. Polym. Sci. 1972, 16, 301.
8.   Shulz, A.R.; Gendron, B.M. J. Appl. Polym. Sci. 1972, 16, 461.
9.   Shulz, A.R.; Gendron, B.M. J. Polym. Sci. Symp. 1973, 43, 89.
10.   Ober, C.; Lenz, R.W.; Galli, G.; Chiellini, E. Macromolecules 1983, 16, 1034.
11.   Bhattacharya, S.K.; Lenz, R.W. J. Polym. Sci., Polym. Phys. Ed., In press.
12.   McIntyre, J.E; Maj, P.E.P.; Sills, S.A.; Tomka, J.G., Polymer 1987, 28, 1971-76.
13.   McIntyre, J.E; Maj, P.E.P.; Sills, S.A.; Tomka, J.G., Polymer 1988, 29, 1095-1100.
14.   Muramatsu, H.; Krigbaum, W.R., J. Polym. Sci. Polym. Phys. 1986, 24, 1695-1711.
15.   Nobile, M.R.; Amendola, E.; Nicolais, L., Polym. Eng. Sci. 1989, 29(4), 244-257.
16.   Dibenedetto, A.T.; Nicolais, L.; Amendola, E.; Carfagna, C.; Nobile, M.R., Polym. Eng. Sci. 1989, 29(3), 153-162.
17.   Kiss, G.; Polym. Eng. Sci., 1987, 27(6).
18.   Yang, H.H.; Chouinard, M.P.; Lingg, W.J., J. Polym. Sci. Polym. Phys. Ed. 1982, 20, 981-87.

19.   Berry, G.C.; Cotts, P.M.; Chu, S.G., <u>British Polym. J.</u> 1981, <u>13</u>, 47–54.
20.   Wissbrun, K.F., <u>British Polym. J.</u> dec. 1980, 163–69.
21.   Prasadarao, M.; Pearce, E.M.; Han, C.D., <u>J. Appl. Polym. Sci.</u>, 1982, <u>27</u>, 1343–1354.
22.   Wissbrun, K.F.; Griffin, A.C., <u>J. Polym. Sci. Phys. Ed.</u>; 1982, <u>20</u>, 1835–45.
23.   Tuttle, J.R.; Bartony, H.E.; Lenz, R.W., <u>Polym. Eng. Sci.</u> 1987, <u>27</u>(15), 1156–63.
24.   Zhou, Z.; Wu, X.; Wang, M., <u>Polym. Eng. Sci.</u> 1988, <u>28</u>(3), 136–42.
25.   Mantia, F.P.; Valenza, A., <u>Polym. Eng. Sci.</u> 1989, <u>29</u>, 625–31.
26.   Larson, R.G.; Mead, D.W.; <u>J. Rheology</u> 1989, <u>33</u>(2), 185–206.
27.   Lin, Y.G.; Winter, H.H., <u>Macromolecules</u> 1988, <u>21</u>, 2439–43.
28.   Done, D.; Baird, D.G.; <u>Polym. Eng. Sci.</u> 1987, <u>27</u>(11), 816–22.
29.   Wissbrun, K.F.; Kiss, G.; Cogswell, F.N., <u>Chem. Eng. Comm.</u> 1987, <u>53</u>, 149–73.
30.   Kalika, D.S.; Nuel, L.; Denn, M.M., <u>J. Rheol.</u> 1989, <u>33</u>(7), 1059–1070.
31.   Lee, B.L.; <u>Polym. Eng. Sci.</u> 1988, <u>28</u>(17), 1107–1114.

RECEIVED February 28, 1990

Chapter 5

Thermotropic Main-Chain Polyethers Based on Bis(4-hydroxyphenoxy)-*p*-xylene

Håkan Jonsson, Ulf W. Gedde, and Anders Hult

Department of Polymer Technology, Royal Institute of Technology, S—100 44 Stockholm, Sweden

Liquid crystalline main-chain polyethers based on bis(4-hydroxy-phenoxy)-*p*-xylene, i.e. a non-rigid rodlike mesogen, and dibromo-alkanes have been synthesized using a phase-transfer-catalyzed Williamson ether synthesis. The molecular weight of the polyethers follows a distinct odd-even dependence due to differences in solubility of the polymer during the polymerization. A difference in order between polymers having an odd or an even numbered spacer group was revealed by DSC and X-ray diffraction. The isotropic-smectic mesophase transition was studied at isothermal conditions, and it was shown that the transition which exhibits all the characteristics of a nucleation-controlled process, follows the Avrami equation with an exponent near 2.8. Dielectric relaxation measurements revealed a glass and a subglass process.

The traditional way to make mesogenic units used in liquid crystalline main-chain polymers, is to connect two or more aromatic or cycloaliphatic rings in the para position by a short link. Groups used for this purpose include ester, imino, vinylene, stilbene, azo, and azoxy (1). They are all stiff with restricted rotation and give the mesogen rigidity and molecular parallelism (linear macroconformation), together with a large aspect ratio and anisotropic molecular polarizability.

Recently it has been shown that it is possible to replace the rigid link in the mesogen in a liquid crystalline main-chain polymer by a more flexible group. The units used this far are methylol (2-3) and ethane (4). Since neither the methylene nor ethane group contain any multiple bonds the mesogens linked by them can adopt different conformations and can therefore not be classified as rigid rod mesogens. Instead Percec and Yourd have introduced the terms *flexible rod-like mesogens* or *mesogenic units based on conformational isomerism* for this group of mesogens (2). Of the two conformations of lowest energy, the extended anti and the nonlinear (bent) gauche, it is only the former that can give rise to liquid crystallinity. The fact that it has been shown possible to make liquid crystalline polymers having flexible rod-like mesogens instead of the traditional rigid rodlike ones, is very important since it increases the available number of links and mesogens and therefore also the number of possible new polymers.

0097–6156/90/0435–0062$06.00/0

Most of the liquid crystalline main-chain polymers synthesized are polyesters. For the present study where isothermal phase transition kinetics studies should be performed, this was a less suitable alternative since they undergo thermally induced reactions, i. e. transesterifications, and for this reason instead polyethers were chosen. Polyethers also have other advantages like lower transition temperatures and higher solubilities.

This paper presents some of our results on the synthesis and structure of thermotropic main-chain liquid crystalline polyethers based on bis(4-hydroxy-phenoxy)-*p*-xylene. It also deals with two areas in the field of liquid crystalline polymers that have received only little attention, namely the dielectric relaxation (5-10 and Gedde, U.W.; Liu, F.; Hult, A.; Gustafsson, A.; Jonsson, H.; Boyd, R.H. Polymer submitted) and the kinetics of isotropic-mesomorphic state transitions (11-14, 32). They are both very important for the understanding of the nature of the mesomorphic state in polymers and for the understanding of similarities and differences of physical phenomena between liquid crystalline and semi-crystalline polymers.

Synthesis and Structure

Mesogen Studies. The conformational characteristics of the anticipated mesogenic group (diphenoxyxylene unit) was investigated via the synthesis and conformational studies of a model compound, α,α'-diphenoxy-*p*-xylene.

Attempts to determine the dimensions of the unit cell by use of X-ray diffraction were not successful. It was, however, concluded that the cell is large and triclinic. Computer-based conformational analysis showed that the energy difference between the extended lowest-energy isomer and the non-linear conformer (secondary energy minimum) was 16.2 kJ/mol.

Polymer Synthesis. The polymers were made by a phase-transfer-catalyzed Williamson ether synthesis as described in the literature (15-20), and one reason for this choice of polyetherification method was that only electrophilic chain ends are produced (15, 20-22), which gives the polymer a well-defined structure. This was important for the transition kinetics measurements presented later in this paper.

Bis(4-hydroxyphenoxy)-*p*-xylene was synthesized and used as the nucleo-philic monomer in the polyetherification with a dibromoalkane using tetrabutylammo-nium bromide as the phase-transfer catalyst (Figure 1). The length of the dibromo-alkane was varied between 7 and 12 methylene units. The notation HPX-C7 refers to a polymer synthesized from bis(4-hydroxyphenoxy)-*p*-xylene and 1,7-dibromoheptane. One polymer, HPXB-C10, was synthesized in a manner different from the others. Instead of using bis(4-hydroxyphenoxy)-*p*-xylene as the nucleophilic monomer and polymerizing it with an alkylic bromide, 1,10-bis(4-hydroxyphenoxy)decane was polymerized with a benzylic bromide, i. e. α,α'-dibromo-*p*-xylene (Figure 2). The structure of polymer HPXB-C10 is identical with that of polymer HPX-C10, except that HPXB-C10 has benzylic bromides as chain ends. This synthetic route was chosen in an attempt to increase the rate of reaction and the molecular mass of the resulting polymer through the use of a more reactive bromide in the polymerization step.

The polymers are soluble in hot solvents such as 1,1,2,2-tetrachloroethane, *o*-dichlorobenzene, trichlorobenzene, and nitrobenzene. Polymers having an odd number of methylene units in the spacer have a higher solubility than those having even-numbered spacers. The solubility also tends to increase with increasing length of the spacer group. This is probably due to differences in order of the different samples in the semicrystalline phase. As a result of this variation in solubility, the molecular mass of the polymers display a clear odd-even dependence (Figure 3). This demonstrates the importance of having a good solvent as the organic phase, in order to avoid precipitation and early termination of the growing chain, in a phase-transfer-catalyzed polycondensation. This molecular mass determining effect may also be the reason for the absence of any significant difference in molecular mass between samples HPX-C10 and HPXB-C10 (M_n=7 300 and 7 600 respectively), despite the difference in reactivity between the monomers used.

Polymer Structure. A typical DSC thermogram is shown in Figure 4. Polymer HPXB-C10, having benzylic bromine chain ends, was degraded in the first heating run and no clear transitions could be recorded. For polymer HPX-C10, having the same repeating unit, no such degradation was observed. The degradation of polymer HPXB-C10 must therefore be due to the lower stability of the chain ends, and indeed, there is a large difference in dissociation energy between a benzylic and a linear aliphatic C-Br bond, 230 kJ/mole ($C_6H_5CH_2$-Br) (23) and 289 kJ/mole (C_3H_7-Br) (24) respectively.

The thermal transitions exhibit a pronounced odd-even effect (Figure 5). Samples having an even numbered spacer exhibit both higher melting and higher isotropization temperatures. The transition temperatures also decrease with increasing length of the spacer group. The isotropization process was further verified by hot-stage polarized microscopy and the melting transition by TMA. The TMA experiments were performed on about 1-mm-thick specimens held in small aluminum containers. At the temperature indicated the probe fast and fully penetrated the specimen and thus proved the melting of the polymer.

In Table I the changes in enthalpies and entropies of melting and isotropization are summarized, and two main effects can be observed: (1) the values of Δh and Δs associated with isotropization are larger than those for melting; (2) the enthalpy and entropy changes associated both with melting and isotropization are larger for the even-numbered samples than for the odd ones.

Table I. Enthalpy and Entropy Changes of the Melting and Isotropization Transitions, together with the Crystallinity Index[a]

Polymer	Δh_m	Δh_i	Δs_m	Δs_i	Cryst. Index
HPX-C7	9.0	24.1	19.7	50.3	0.33
HPX-C8[a]	-	-	-	-	0.40
HPX-C9	8.4	23.9	18.7	50.7	0.34
HPX-C10	15.7	26.8	34.3	55.3	0.40
HPX-C11	8.6	24.0	19.3	51.5	0.36
HPX-C12	19.6	37.7	42.8	78.6	0.40

a) Δh in kJ/mru·K and Δs in J/mru·K (mru= moles of repeating unit). b) No registration in the DSC was possible for polymer HPX-C8 after the first heating scan due to thermal degradation.

The first effect implies that the difference in order between the semicrystalline state and the mesophase is smaller than the difference between the mesophase and the isotropic melt. This indicates that the mesophases involved are highly ordered and the high values of the enthalpies and entropies may also be taken as another indication supporting this assumption. The photomicrograph in Figure 6 clearly shows that the mesophase is highly ordered. The mosaic domain is characterized by a uniform

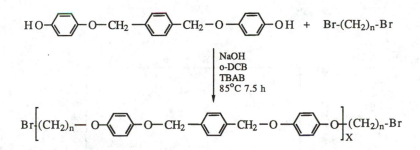

Figure 1. Synthesis of α,ω-bis(bromoalkoxy) polyethers of bis-(4-hydroxy-phenoxy)-*p*-xylene.

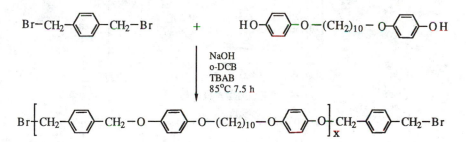

Figure 2. Synthesis of the polyether of bis(4-hydroxyphenoxy)decane and α,α'-dibromo-*p*-xylene.

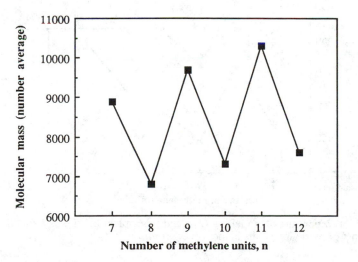

Figure 3. Molecular weight as a function of the number of methylene units in the spacer group.

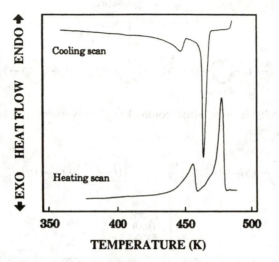

Figure 4. DSC thermogram of HPX-C12. Second cooling and third heating scan.

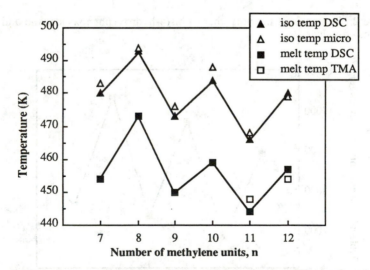

Figure 5. Temperatures for melting (T_m) and isotropization (T_i) as a function of the number of methylene units in the spacer group.

orientation of the refractive index ellipsoid and disclinations of strength ±1/2 have not been observed. The mosaic structure displays all the characteristics of an ordered smectic structure (25). The polarized light microscopy characterization of the polymer will be discussed in more detail later in this paper.

The second effect, the well-known odd-even dependence, may be explained in terms of different degrees of order for polymers having odd or even numbers of methylene units in the spacer. The overall order in the semicrystalline state and in the mesophase is larger in the even-numbered samples than in the odd-numbered.

The assumption of different degrees of order between polymers having odd or even numbers of methylene units in the spacer is verified by the X-ray diffraction patterns shown in Figure 7. Crystallinity index data were determined on the basis of these diffraction patterns and the data are presented in Table I. The crystallinity index was calculated according to Equation 1, in which A_c is the area of the sharp Bragg reflections corresponding to the crystalline part of the polymer, and A_a is the area of the broad amorphous peak.

$$W_c = \frac{A_c}{A_c + A_a}$$ (1)

It is evident that samples with a high Δh_m (Δs_m) also have a high crystallinity value. The Bragg spacings also show a clear and perfectly alternating odd-even dependence with odd-numbered polymers having spacings at 0.47, 0.45, 0.37, 0.32, and 0.23 nm and even-numbered polymers at 0.45, 0.43, 0.40, 0.37, 0.32, and 0.22 nm. As can be seen, the 0.40 nm spacing is not observed in the odd-numbered samples, indicating a different molecular arrangement of the polymer chains.

Kinetics of Phase Transitions

The Effect of Polymer Heterogeneity on the Enthalpy. The kinetics of the isotropic-smectic phase transition were studied for two of the polymers; HPX-C9 and HPX-C11, and in Figure 8 a summary of the calorimetric data for the former is presented. The behaviour of the HPX-C11 polymer was similar. Two processes are in fact revealed by these data: *(a)* at high temperatures ($T \geq 471.9$ K) the two processes have approximately the same induction time and rate and are therefore not resolved. *(b)* at intermediate temperatures ($468.9 \leq T \leq 470.9$ K) the two processes are clearly resolved and the heat evolved in the slower process is 30-50% of that evolved in the more rapid process. There is a tendency for an increase in Δh° for the slow process with increasing temperature; for HPX-C9, Δh° was equal to 13 kJ/kg at 468.9 K and 20 kJ/kg at 470.9 K. The same trend was observed for HPX-C11. *(c)* at low temperatures ($T \leq 467.9$ K), the slower process was much retarded and not observed within the experimental window.

The solid line in Figure 9 shows the total heat (Δh°) involved in the two processes. It is clear that Δh° decreases with increasing temperature over the whole temperature range, although there is a narrow temperature region, for HPX-C9 between 467.9 K and 469.9 K, in which Δh° increases with increasing temperature. The reason for this increase in Δh° is related to the definition of this quantity and to the limited experimental potential for detecting very slow processes. At temperatures less than 467.9 K, the slower process is not recorded and the Δh° value measured is only that associated with the rapid process, whereas at higher temperatures the Δh° values partially or fully include the contribution from the second slower process. Hence, if the total contribution from the second slower process were included in Δh° at the lower temperatures, the broken curve in Figure 9 would be obtained, on the assumption that Δh° for the slower process is 20 kJ/kg throughout the experimental temperature range.

Figure 6. Polarized photomicrograph showing the formation of the mosaic structure in polymer HPX-C11 during a 0.5 K/min cooling scan from isotropic melt at 455 K.

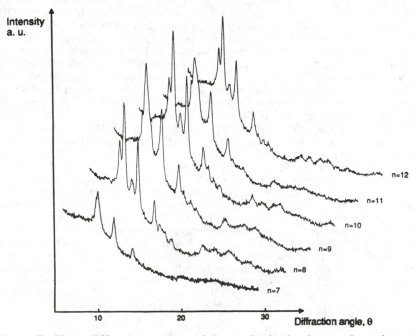

Figure 7. X-ray diffraction patterns of the synthesized polymers. Intensity as a function of the diffraction angle θ.

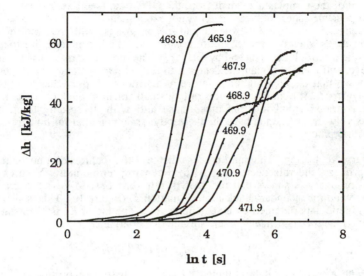

Figure 8. Exothermal heat (Δh) developed in HPX-C9 as a function of time (t) at different temperatures (in K).

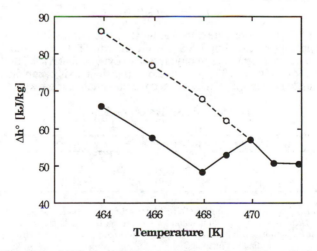

Figure 9. Total heat ($\Delta h°$) recorded under isothermal conditions in the DSC for HPX-C9 plotted versus temperature. Measured values ($\bullet$); corrected values ($\circ$).

The same monotonic decrease in the total $\Delta h°$ with increasing temperature was obtained for HPX-C11.

There are at least two plausible reasons for this decrease in $\Delta h°$, both related to the multicomponent character of the polymer. In the first place, the chain length poly-dispersity of the samples is evident from the GPC data (see Experimental Section). Another possible source of heterogeneity, not experimentally verified, is a variation in the sequence distribution of mesogenic and spacer groups in different molecules. This variation arises when two spacer groups couples via an ether linkage, due to displace-ment of the electrophilic chain end by OH- (22). The molecular mass distribution itself is however sufficient to explain the decrease in $\Delta h°$ with temperature. Percec et al. (20) have shown that the clearing temperature of a similar LC polyether increases with molecular mass. At high temperatures only a small fraction, the longer chains, of the sample is transformed into the mesophase and with decreasing temperature a progressively larger fraction have the thermodynamic potential for transformation to the mesomorphic state.

Secondary Nucleation. The rapid process displays the features of a nucleation cont-rolled process, the rate decreasing with increasing temperature (Figures 10-11). Regression analysis shows that the growth rate data best follows the growth rate equation valid for secondary nucleation (Equation 2) (Figure 10). This is also the case for the kinetic data obtained by DSC (Figure 11). The factor K_g is a constant which depends on the mechanism and on the surface free energies.

$$G = G_o \exp\{-K_g/[T(T°_i-T)]\} \tag{2}$$

The growth rate data obtained by DSC is treated according to Equation 3 (Figure 11), which is a DSC equivalent of Equation 2 ($t_{0.4}$ is the time for 40% conversion of the rapid process).

$$(t_{0.4})^{-1} = C \exp\{-K_g/[T(T°_i-T)]\} \tag{3}$$

The K_g values presented in Table II demonstrate the similarity of the exothermal process recorded by DSC and the growth of the mesomorphic domains recorded by polarized light microscopy. It is also evident that K_g is 20-30% lower for HPX-C9 than for HPX-C11. The growth along the chain axis occurs by a mechanism with a K_g value which is 2/3 of the value for the perpendicular growth.

Table II. Phase Transition Kinetics

Polymer	K_g (DSC)[a]	$<K_g>$ (micr)[b]	K_{gc} (micr)[c]	K_{gp} (micr)[d]
HPX-C9	21 600	23 600	-	-
HPX-C11	26 000	28 400	23 000	33 800

a) From DSC according to Equation 3. b) Mean value, from polarized light microscopy according to Equation 2. c) Growth in chain axis direction, from polarized light microscopy according to Equation 2. d) Growth perpendicular to chain axis direction, from polarized light microscopy according to Equation 2.

Rearrangement of the Mesophase. The slower process displays a temperature dependence opposite to that of the rapid process, the rate increasing with increasing temperature (Figure 12). This process can not thus be controlled by nucleation, but is

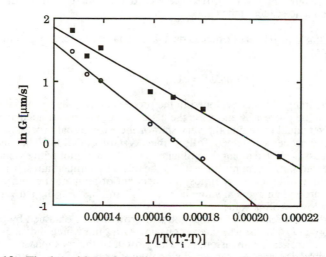

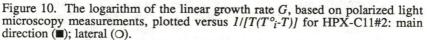

Figure 10. The logarithm of the linear growth rate G, based on polarized light microscopy measurements, plotted versus $1/[T(T^{\circ}_i-T)]$ for HPX-C11#2: main direction (■); lateral (O).

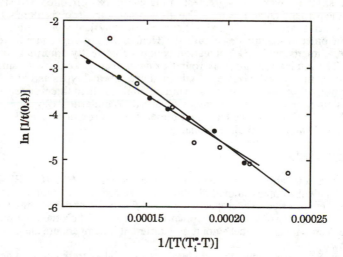

Figure 11. The logarithm of the reciprocal time for 40% conversion of the "rapid" process plotted versus $1/[T(T^{\circ}_i-T)]$ for HPX-C9 (●) and HPX-C11#1 (O).

possibly a rearrangement of the smectic mesophase against a higher degree of perfection. Figure 12 indicates that the process follows the Arrhenius equation with an activation energy of about 400-850 kJ/mol. The data points used to determine the activation energy are few and the difference recorded for the two different polymers may simply be due to scatter in the data.

The Avrami Constant. The calorimetric data were analyzed according to the general Avrami equation:

$$1 - \Delta h / \Delta h^{\circ} = exp\ (-Kt^{n}) \qquad\qquad (4)$$

where Δh is the integral heat developed in the process during time t, Δh° is the total heat involved in the process, K and n are the parameters to be determined and t is the time which is set equal to zero at the very start of the exothermal process. Figure 13 show that both processes indeed follow the Avrami equation. The nucleation controlled process is shifted along the ordinate in the Avrami plot with essentially the same slope, i.e. the same Avrami exponent, at the different temperatures. The Avrami exponent for the rapid process is almost independent of temperature and varies for both polymers around an average value between 2.7 and 2.8. For the slower process the exponent is lower with a value near 2.0.

   Some relevant features of nucleation and growth were obtained by polarized light microscopy: (1) The growth rate is constant with time which indicates that the growth is not controlled by diffusion; (2) The growth of the mesophase domains is dependent on direction and is more rapid along the direction of maximum refractive index. On all samples cooled to room temperature, linear cracks oriented parallel to the direction of maximum refractive index were observed. It is postulated that the direction of maximum refractive index is parallel to the domain director for the chain axis. Thus, the growth rate is greatest in the chain axis direction and significantly lower in the perpendicular direction. The growth rate anisotropy is more significant at higher temperatures. There is a significant anisotropy in growth rate approaching almost one order of magnitude (Figure 10). Hence, the growth cannot be regarded as truly three-dimensional; (3) Nucleation is neither perfectly athermal nor perfectly thermal. The nucleation rate was found to decrease markedly with time, so that nucleation seem to be a "mixture" of athermal and thermal types and growth seem to be nucleation controlled (constant growth rate) and less than three-dimensional. This picture is consistent with the n values obtained. Wunderlich (26) reports $n=2$ for athermal nucleation followed by two-dimensional growth and $n=3$ for thermal nucleation followed by two-dimensional growth.

Dielectric Relaxation

The dielectric relaxation was measured for three polymers; HPX-C2, HPX-C11, and HPX-C12 and all samples studied displayed two relatively weak dielectric relaxation processes, here referred to as respectively α and β in order of descending temperature. A typical example (HPX-C12) is shown in Figure 14. The other samples exhibits similar dielectric relaxation behaviour, only different in temperature and strength of the two relaxation processes (Table III).

   The α process appears most clearly at the higher frequencies whereas at the lower frequencies it is hidden by the rapid increase in both dielectric constant and loss due to the Maxwell-Wagner effect. The β process on the other hand is easily revealed in the whole frequency range.

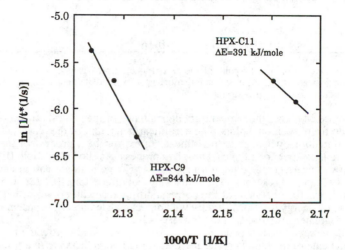

Figure 12. Arrhenius plot $[ln(1/t^*) = f(t)$, where t^* is the time at which the rate of the slower process is at a maximum] of the slow process for both HPX-C9 and HPX-C11#1.

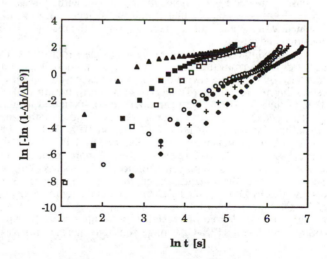

Figure 13. Avrami plots (according to Equation 4) for HPX-C11#1 obtained at different temperatures: 456.9 K (▲); 458.9 K (■); 460.9 K (□); 461.9 K (O); 462.9 K (●); 463.9 K (+); 464.9 K (♦).

Table III. Summary of data for the α and β relaxation processes

Polymer	ΔE (β) [kJ/mol]	T_g [K]	ΔC_p [J/gK][a]	T_{max} (α) [K][b]	Δh_{in} [kJ/kg][c]
HPX-C2	54.4	-	-	340	168
HPX-C11	51.5	292	0.07	295	69
HPX-C12	49.2	297	0.16	302	117

(a) By DSC as the middle point of the glass transition. (b) Temperature for maximum dielectric loss at 10 Hz. (c) Total endothermal heat in all first order transition from the isotropic state down to room temperature.

Figure 14 shows the temperature dependence of the central relaxation time (τ) as obtained from isochronal plots. The non-linear behaviour of the α process is typical for glass transitions whereas the rectilinear curves for the β process of the different samples are indicative for a typical sub-glass process. As shown in Table III, the latter thus follows the Arrhenius equation ($\tau = \tau_0 \, e^{-\Delta E/RT}$) with an activation energy (ΔE) between 49.2 and 54.4 kJ/mol. The latter value obtained for HPX-C2 is very close indeed with what is obtained for the corresponding aromatic polyester (PETP) (28). The activation energy increases slightly with decreasing length of the spacer unit: ΔE is about 10% greater for HPX-C2 than for HPX-C12.

The temperature independent relaxation time (τ_0) is also affected by the spacer length. The data of the liquid crystalline main-chain polyethers are in agreement with the data by Farrow et al. (29) on semicrystalline poly(methylene terephthalate) polymers.

By combination of relaxation strength measurements obtained from isochronal scans and molecular mass data, it could be concluded that the relaxation strength of the β process increases with decreasing molecular mass, i. e. with increasing content of polar end groups. It may thus be postulated that the dielectric β relaxation to some extent is associated with the local main-chain motions of the methylene spacer group adjacent to the polar end groups.

The α process exhibits all the features typical of glass transitions. The temperature dependence deviates from the simple Arrhenius equation (Figure 15). This is further substantiated by DSC data revealing a stepwise increase in specific heat (ΔC_p) at the corresponding temperatures (Table III). It is worth noting that T_g (T_{max}) is lower for HPX-11 than for HPX-C12 (Table III) in contradiction to what could be expected. T_g (T_{max}) should decrease with increasing length of the spacer group. However, crystallinity is higher in HPX-C12 than in HPX-C11 (3) which is also reflected in the significantly higher Δh_{in} value obtained for HPX-C12 (Table III). The non-crystalline component of HPX-C12 should thus be more constrained resulting in a higher value of T_g. HPX-C2 displayed a dielectric α process with a relaxation strength of about 50-70% of the relaxation strength of the other polymers, but no glass transition was revealed by DSC (Table III). This apparently contradictory finding may be explained by the higher overall content of polar end groups in this polymer in combination with the high degree of crystallinity in this material. Thus, dielectric registration of the α process in HPX-C2 is more sensitive than calorimetric.

Conclusions

In this paper it has been shown that the polyethers based on the non-rigid rodlike mesogen bis(4-hydroxyphenoxy)-p-xylene and dibromoalkanes exhibit smectic mesomorphism. Furthermore, the molecular mass of the polyethers follows a pronounced odd-even dependence due to differences in solubility of the different polymers during the polymerization. DSC and polarized light microscopy also revealed an odd-even dependence of the temperatures and enthalpy and entropy changes associated with

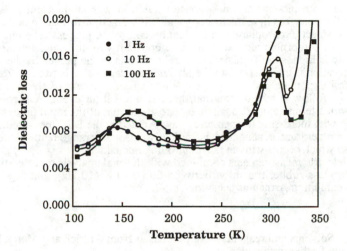

Figure 14. Dielectric loss of HPX-C12/LPE (50/50 blend) plotted versus temperature. (Reproduced with permission from ref. 27. Copyright 1989 Butterworth.)

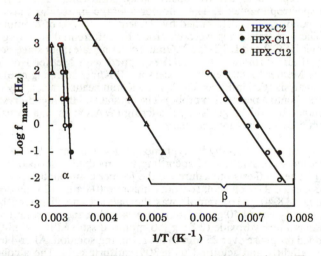

Figure 15. Temperature dependence of the α and β relaxations. The data are obtained from isochronal dielectric loss versus temperature plots. (Reproduced with permission from ref. 27. Copyright 1989 Butterworth.)

crystal melting and isotropization. X-ray diffraction studies confirmed these observations in terms of a higher degree of order in polymers with an even-numbered
spacer group.

The isotropic-smectic mesophase transition was studied under isothermal
conditions by polarized light microscopy and DSC. The transformation of the isotropic
melt to the smectic mesophase is a nucleation controlled process. The linear growth
rate of the mesomorphic mosaic domains is constant with time and anisotropic. The
growth rate is higher in the chain axis direction than in the perpendicular direction.
This observation is consonant with the obtained average value between 2.7 and 2.8 of
the Avrami exponent, indicating a two-dimensional growth of the mesomorphic
domains to follow a mixed athermal/thermal nucleation event. A second slower
process with a temperature dependence opposite to that of the rapid process, the rate
increasing with increasing temperature, was revealed in the DSC measurements. This
reversed temperature dependence is indicative of a rearrangement of the smectic
mesophase which occurs without rate controlling nucleation.

Dielectric measurements combined with thermal analysis revealed two transitions: α, a glass-rubber transition between 295-340 K (10 Hz), and ß, a sub-glass
transition exhibiting Arrhenius behaviour.

Experimental

Materials. Solvents and reagents were purchased from Aldrich and Merck and were
used without further purification.

Characterization. Conformational and energy calculations were performed by using
the MM2 (Molecular Mechanics Program Version 2) in Chem-X from Chemical
Design LTD, Oxford (July 1987 update). 1H-NMR spectra were recorded on a 200
MHz Bruker spectrometer and IR spectra on a Perkin-Elmer 1710 Infrared Fourier
Transform spectrophotometer. The number average molecular mass (M_n) of the
synthesized polymers was determined by elemental analysis of bromine (end-group
analysis). Thermal transitions were recorded by differential scanning calorimetry
(Perkin-Elmer DSC-2 and DSC-7; 10 K/min scanning rate), hot stage polarized light
microscopy (Leitz Ortholux POL BKII equipped with Mettler Hot Stage FP 82
controlled by Mettler FP 80 Central Processor; 10 K/min scanning rate) and thermomechanical analysis (Perkin-Elmer TMA-1; 10 K/min heating rate, applying a pressure
of 0.01 kg/mm2 onto a penetration probe). The crystal and the liquid-crystal structures
were determined by wide-angle X-ray scattering (WAXS) using a focusing Guinier-
Hägg camera with transmission geometry.

Monomer Synthesis. Bis(4-hydroxyphenoxy)-p-xylene and 1,10-bis(4-hydroxy-
phenoxy)decane were synthesized according to a method described by Griffin and
Havens (30). The synthesis procedure for the former compound was as follows: To
35 ml of de-areated 95% ethanol, sodium dithionite (0.04 g, 0.23 mmol) was added.
Hydroquinone (18.40 g, 167 mmol) was then added and dissolved by warming.
α,α'-dibromo-p-xylene (4.40 g, 16.7 mmol) was added and the solution was heated
to reflux. Potassium hydroxide (2.81 g, 50 mmol) dissolved in 20 ml 95% ethanol
was then added dropwise over 25 min to the refluxing solution. After 4 h the solution
were cooled slightly and acidified with 30% sulfuric acid. The alcohol-insoluble
byproducts were filtered off, and the solvent was evaporated from the filtrate. The
solid was washed with 500 ml of water and, after air drying, was recrystallized twice
from 85% ethanol, which gave 1.0 g (19%) of small white crystals. Mp 514-517 K
(slight decomposition).

Polymer Synthesis. A typical polymerization procedure for the α,ω-dibromoalkoxy
polyethers of bis(4-hydroxyphenoxy)-p-xylene (HPX) is as follows: To 7.5 ml of a

50 % by weight sodium hydroxide solution, HPX (0.400 g, 1.24 mmol) was added. The suspension was heated to 85 °C under nitrogen and dibromododecane (0.407 g, 1.24 mmol) dissolved in 7.5 ml of o-dichlorobenzene was added. Finally, under intense stirring, tetrabutylammonium bromide (80 mg, 0.25 mmol) was added. After 7.5 h, the reaction mixture was acidified, washed with 2 M hydrochloric acid, and precipitated into methanol, and the polymer was then filtered off. The polymer was purified twice by dissolution in hot o-dichlorobenzene, followed by precipitation into acetone and methanol respectively. Yield: 0.477 g (79%).

Studies of Kinetics. The number average molecular mass determined by end-group analysis were: M_n=9 700 (HPX-C9), 10 300 (HPX-C11#1), and 6 300 (HPX-C11#2). The molecular mass of HPX-C11#2 was also determined by gel permeation chromatography (Waters 200 GPC) at 408 K, using trichlorobenzene as eluent: M_n=6 300 and M_w=15 600. The GPC data were analysed using the universal calibration procedure with estimated values for the Mark-Houwink parameters (a=1.2) for this semi-rigid polymer. Polarized light microscopy were performed in the hot stage by cooling (20 K/min) the 10 mm thick samples from 488 K (HPX-C9) and 478 K (HPX-C11) to the experimental isothermal temperature (T) followed by photographic recording at different times (t) after the establishment of isothermal conditions. The samples were kept at the maximum temperature (488 K and 478 K resp.) for 1 minute. Samples weighing about 5 mg were cooled in the DSC apparatus at a rate of 80 K/min from 485 K to the experimental isothermal temperature (T) and the exotherm associated with the first order transition was recorded during the isothermal conditions. The samples were kept at the maximum temperature (485 K) for 1 minute.

Studies of Dielectric Relaxation. The number average molecular masses determined by end-group analysis were: M_n=720 (HPX-C2) (dimer), 10 300 (HPX-C11#1), and 7 600 (HPX-C12). HPX-C2 is almost exclusively terminated by hydroxyl end groups. The dielectric work was carried out on 50/50 (w/w) blend samples of the polyethers and linear polyethylene (LPE). The presence of LPE made it possible to produce samples of sufficient mechanical strength and it was verified by dielectric measurements that the LPE component showed no measurable dielectric loss in the temperature and frequency range used in this study. The 200 μm thick samples were coated with gold-palladium in a vacuum sputterer. The dielectric apparatus was a IMASS TDS time domain spectrometer equipped with a Hewlett Packard Series 300 computer. The time domain spectrometer is based on a design by Mopsik (31).

Acknowledgments

These studies have been financed by The National Swedish Board for Technical Development (STU) grant # 86-03476P, and The Swedish Natural Science Research Council (NFR), grant K-KU # 1910-300. The scholarship for HJ from The Royal Institute of Technology, Stockholm, is gratefully acknowledged.

Literature Cited

1. Ober, C.K.; Jin, J.-I.; Lenz, R. W. Adv. in Polym. Sci. 1984, 59, 103.
2. Percec, V.; Yourd, R. Macromolecules 1988, 21, 3379.
3. Jonsson, H.; Werner, P.-E.; Gedde, U. W.; Hult, A. ibid. 1989, 22, 1683.
4. Percec, V.; Yourd, R. ibid. 1989, 22, 524.
5. Blundell, D. J.; Buckingham, K. A. Polymer 1985, 26, 1623.
6. Yoon, H. N.; Jaffe, M. Polym. Prepr., Am. Chem. Soc., Div. Polym. Chem. 1983, 24(1).
7. Alhaj-Mohammed, M. H.; Davies, G. R.; Abdul Jawad, S.; Ward, I. M. J. Polym. Sci., Polym. Phys. Ed. 1988, 26, 1751.

8. Wendorff, J. H.; Frick, G.; Zimmermann, H. Mol. Cryst. Liq. Cryst. Inc. Nonlin. Opt. 1988, 157, 455.
9. Gedde, U. W.; Buerger, D.; Boyd, R. H. Macromolecules 1987, 20, 988.
10. Takase, Y.; Mitchell, G. R.; Odayima, A. Polymer Commun. 1986, 27, 76.
11. Bhattacharya, S. K.; Misra, A; Stein,R. S.; Lenz, R. W.; Hahn, P. E. Polymer Bull. 1986, 16, 465.
12. Liu, X.; Hu, S.; Shi, L.; Xu, M.; Zhou, Q.; Duan, X. Polymer 1989, 30, 273.
13. Hans, K.; Zugenmaier,P. Macromol. Chem. 1988, 189, 1189.
14. Pracella, M.; De Petris, S.; Frosini, V.; Magagnini, P.L. Mol. Cryst. Liq. Cryst. 1984, 113, 225.
15. N'Guyen, T. D.; Boileau, S. Polymer Bulletin. 1979, 1, 817.
16. Cameron, G. G.; Law, K. S. Makromol. Chem., Rapid Commun. 1982, 3, 99.
17. Yamazaki, N.; Imai, Y. Polym. J. 1983, 15, 603.
18. Percec, V.; Auman, B. C. Makromol. Chem. 1984, 185, 617.
19. Yamazaki, N.; Imai, Y. Polym. J. 1985, 17, 377.
20. Percec, V.; Nava, H.; Jonsson, H. J. Polym. Sci., Part A: Polym. Chem. 1987, 25, 1943.
21. Percec, V.; Nava, H. ibid. 1987, 25, 405.
22. Shaffer, T. D.; Jamaludin, M.; Percec, V. J. Polym. Sci., Polym. Chem. Ed. 1985, 23, 2913.
23. Cox, J.D.; Pilcher, G. Thermochemistry of Organic and Organometallic Compounds; Academic Press: London, 1970.
24. Benson, S. J. Chem. Educ. 1965, 42, 502.
25. Demus, D.; Richter, L. Textures of Liquid Crystals; VEB Deutscher Verlag für Grundstoffindustrie: Leipzig, 1978.
26. Wunderlich,B. Macromolecular Physics, Volume 2: Crystal nucleation, growth, annealing; Academic Press: New York, 1978.
27. Gedde, U.W.; Liu, F.; Hult, A.; Gustafsson, A.; Jonsson, H.; Boyd, R.H. Polymer submitted 1989.
28. Coburn, J.; Boyd, R. H. Macromolecules 1986, 19, 2238.
29. Farrow, G.; Mcintosh, J.; Ward, I. M. Macromol. Chem. 1960, 38, 147.
30. Griffin, A. C.; Havens, S. J. J. Polym. Sci., Polym. Phys. Ed. 1981, 19, 951.
31. Mopsik, F.I. Rev. Sci. Instrum. 1984, 55, 79.
32. Jonsson, H.; Wallgren, E.; Hult, A.; Gedde, U. W. Macromolecules 1990, 23, 1041.

RECEIVED March 20, 1990

Chapter 6

Synthesis and Chemical Modification of Chiral Liquid-Crystalline Poly(ester β-sulfide)s

E. Chiellini[1], G. Galli[1], A. S. Angeloni[2], and M. Laus[2]

[1]Dipartimento di Chimica e Chimica Industriale, Centro CNR Macromolecole Stereodinate Otticamente Attive, Università di Pisa, 56100 Pisa, Italy
[2]Dipartimento di Chimica Industriale e dei Materiali, Università di Bologna, 40136 Bologna, Italy

The synthesis and liquid crystalline properties are presented of two classes of chiral (I-n) and prochiral (II-m) thermotropic poly(ester β-sulfide)s. The nematic mesophase behavior of the polymers I-n exhibits distinct even-odd alternations with chemical structure and is compared with that of closely related poly(ester β-sulfide)s III-n and IV-m.
Polymers II-m have been transformed by asymmetric oxidation into chiral poly(ester β-sulfoxide)s V-m. These modified polymers present a cholesteric mesophase of limited persistence. Their optical activity and stability are also discussed.

The design and synthesis of new liquid crystalline polymeric materials endowed with intrinsc chirality deserve attention, as chirality can offer probes of the supermolecular structure and a tool for modulating specific responses of the polymers (1). The chemical transformation of preformed thermotropic polymers can add novel opportunities for the realization of various molecular architectures conventionally unfeasible and best suited for mesophase modification.

In the present paper we describe the mesophase behavior of a new series of *chiral* poly(ester β-sulfide)s I-n and report on the asymmetric transformation of another class of suitably *prochiral* poly(ester β-sulfi-

0097–6156/90/0435–0079$06.00/0

de)s **II-m** to yield modified chiral polymers, such as poly(ester β-sulfoxide)s. While a few diverse chemical or physical modifications of thermotropic polymers have so far been described (2-8), the asymmetric oxidation of functional liquid crystalline polymers has not been performed previously.

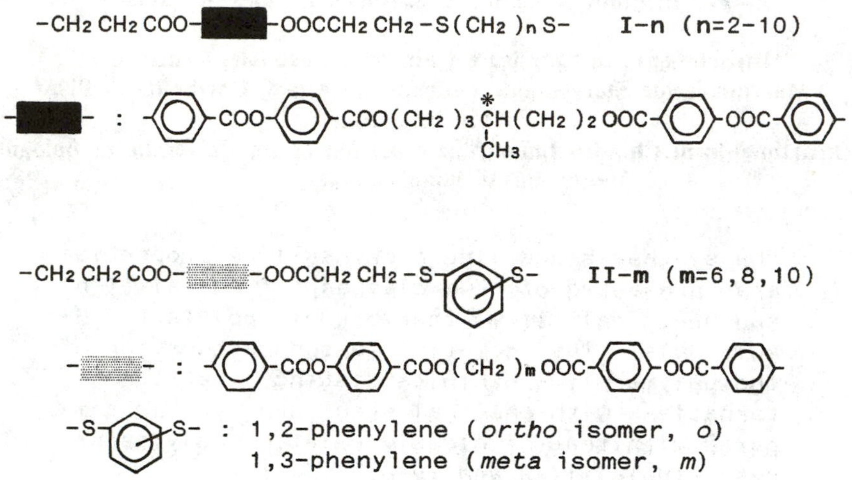
I-n (n=2-10)
II-m (m=6,8,10)
1,2-phenylene (*ortho* isomer, *o*)
1,3-phenylene (*meta* isomer, *m*)

The structurally ordered poly(ester β-sulfide)s are designated according to the number **n** of methylene units in the sulfide segment, or to the number **m** of methylene units in the fully alkylene segment in their repeat unit.

Experimental

The polymers were synthesized according to the base-catalyzed Michael-type polyaddition reaction of α,ω-dimercaptoalkanes to (*R*)-3-methylhexamethylene bis[4-[4-(acryloyloxy)benzoyloxy]benzoate] (9) (polymers **I-n**), as described in detail elsewhere (10), or of 1,2-dimercaptobenzene or 1,3-dimercaptobenzene to α,ω-alkylene bis[4-[4-(acryloyloxy)benzoyloxy]benzoate]s (9) (polymers **II-m**).

(*S,S*)-(−)-2-(*d*-10-camphorsulfonyl)-3-(2-chloro-5-nitrophenyl)oxaziridine (1) was prepared according to literature (11) (optical purity >90%).

In a typical asymmetric oxidation experiment (12), a solution of 1 (2.36mmol) in anhydrous chloroform (20 ml) was added dropwise to a solution of polymer **II-6*m***

(1.18mmol r.u.) in the same solvent (20ml) at -50°C and the mixture was let to react for 1h. The solution was evaporated to small volume and then poured into a large excess of methanol. The polymer was purified by several precipitations from chloroform solution into *n*-hexane. Yield 90%.

The liquid crystalline properties of the polymers were studied by a combination of DSC, optical microscopy and X-ray diffraction analyses.

Results and Discussion

All of samples I-n show one enantiotropic mesophase (Table I). The X-ray diffraction spectra recorded on representative polymers I-3, I-4, I-8, and I-9 indicate a nematic (cholesteric) structure for the mesophase, independent of the length and parity of the sulfide segment. Both melting and isotropization temperatures decrease in a regular alternating manner as the series is ascended (Figure 1). The downward trend is a result of the overall decreasing polarity and molecular rigidity with increasing number n. Polymers with even-numbered spacers show higher transition temperatures than adjacent polymers with odd-numbered spacers. This even-odd effect (13-15) is more pronounced for the lower homologues and

Table I. Mesomorphic properties of chiral poly(ester β-sulfide)s I-n

Sample	n	$\overline{M}n$	$[\Phi]_D^{25a}$	Tm	Ti	ΔHi	ΔSi
		$\cdot 10^{-3}$	deg	K	K	kJ/mol	J/(mol·K)
I-2	2	8.4	+70.4	413	440	3.4	7.7
I-3	3	8.9	+65.3	395	421	3.0	7.2
I-4	4	9.5	+71.9	398	427	4.2	9.8
I-5	5	11.6	+60.0	389	417	3.9	9.4
I-6	6	9.3	+56.6	391	418	4.9	11.7
I-7	7	7.2	+62.9	381	400	4.6	11.5
I-8	8	8.9	+62.0	391	406	6.2	15.3
I-9	9	9.3	+62.0	386	395	4.9	12.4
I-10	10	10.1	+56.5	386	394	7.0	17.7

[a]Molar optical rotation = $[\alpha]_D^{25} \cdot MWr.u./100$ (in chloroform solution).

is attenuated in the higher homologues, for which the mesophasic range is also narrowed.

The dependence of the isotropization enthalpies and entropies upon the length of the sulfide spacer segment is also characterized by a distinct even-odd alternation (Table I). The cholesteric-isotropic transition entropies (Figure 2) for even and odd members lie on two smooth separate curves, which rise with increasing spacer length, even members laying on the upper curve. This increase in isotropization entropies is connected with the increased conformational entropy of the molecules in the isotropic phase. Consistently, the spacer plays an integral role in determining the degree of organization in the liquid crystalline state. The basic difference between the configurational character of the extended conformers of the spacer (16-18) should be responsible for the large oscillations of the phase transition parameters even in the present series of polyesters containing a chiral spacer segment.
The mesophase properties of polymers I-n are compared with those of previously investigated achiral counterparts III-n (15) and chiral structural isomers IV-m (19).

$-CH_2 CH_2 COO-\boxed{}-OOCCH_2 CH_2 -S(CH_2)_n S-$ III-n (n=2-10)

$-\boxed{}-$: $-\langle O \rangle-COO-\langle O \rangle-COO(CH_2)_6 OOC-\langle O \rangle-OOC-\langle O \rangle-$

$-CH_2 CH_2 COO-\blacksquare\blacksquare-OOCCH_2 CH_2 -S(CH_2)_3 \overset{*}{C}H(CH_2)_2 S-$ IV-m (m=6-10)
CH_3

$-\blacksquare\blacksquare-$: $-\langle O \rangle-COO-\langle O \rangle-COO(CH_2)_m OOC-\langle O \rangle-OOC-\langle O \rangle-$

The presence of a methyl substituent in the alkylene segment of polymers I-n decreases the thermal stability of the mesophase by 30-40 K relative to the corresponding unsubstituted samples III-n, whereas the range of persistence remains essentially the same (Figure 1). However, for polymers I-n the isotropization entropies are much lowered and the even-odd alternations are

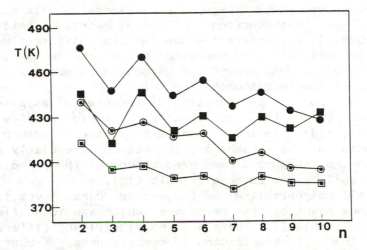

Figure 1. Trends of the melting (squares) and isotropization (circles) temperatures of poly(ester β-sulfide)s I-n (▣, ☉) and III-n (■, ●) with varying number n.

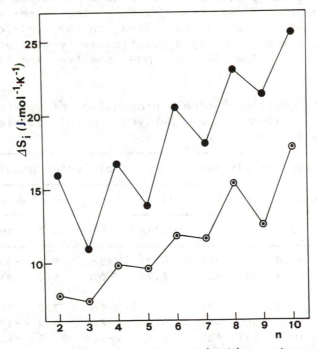

Figure 2. Trend of the isotropization entropy of poly(ester β-sulfide)s I-n (cholesteric-isotropic, ☉) and III-n (nematic-isotropic, ●) with varying number n.

sensibly decreased (Figure 2). These results suggest that the introduction of the methyl substituent (chirality) in polymers I-n can enforce different mutual arrangements of the aromatic diads resulting in an attenuated dependence of the mesophase parameters on the parity of the repeat unit.

We also note the striking differences of samples I-n (n=6-10) from the positionally isomeric samples IV-m (m=6-10) (19). In the latter system the structure and character of the mesophases vary rather irregularly along the series (Figure 3) and predictions of their incidence and stability are quite difficult (19).

All polymers II-m of the second class investigated show one enantiotropic mesophase which was identified as nematic (Table II). There are no significant differences in the isotropization temperatures, but the isotropization entropies increase with increasing number m of methylene groups in the alkylene segment particularly for polymers containing the 1,3-phenylene residue (*meta* isomers, *m*).

Poly(ester β-sulfide)s II-m were oxidized by using stoichiometric amounts of the chiral 2-sulfonyloxaziridine 1 to give chiral poly(ester β-sulfoxide)s V-m in nearly a quantitative yield, according to Scheme 1. The IR and NMR spectra and elemental

Table II. Physicochemical properties of poly(ester β-sulfide)s II-m[a] and derived chiral poly(ester β-sulfoxide)s V-m[a]

m	Poly(ester β-sulfide)				Poly(ester β-sulfoxide)			
	Type	$\bar{M}n$	Tm	Ti	ΔSi	Type	$\bar{M}n$	$[\Phi]_D^{25}$[b]	Tm
		$\cdot 10^{-3}$	K	K	J/(mol·K)		$\cdot 10^{-3}$	deg	K
6	II-6*o*	8.6	377	438	5.3	V-6*o*	8.5	-366	398
8	II-8*o*	5.4	387	420	7.7	V-8*o*	6.3	-316	392
6	II-6*m*	6.9	387	425	6.6	V-6*m*	7.2	-162	408
8	II-8*m*	5.4	389	417	9.7	V-8*m*	6.4	-136	403
10	II-10*m*	10.1	397	420	24.4	V-10*m*	11.2	-154	408

[a] Based on the 1,2-phenylene (*ortho*, *o*) isomer or 1,3-phenylene (*meta*, *m*) isomer. [b] In chloroform solution.

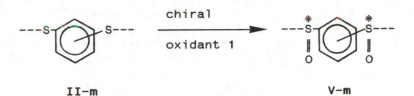

II-m   **V-m**

Scheme 1. Schematic representation of the asymmetric oxidation performed

analyses of samples **V-m** are in full agreement with the proposed structures, while no evidences of residual sulfide or of overoxidized sulfone moieties were detected. The GPC traces of the corresponding parent and modified polymers are practically superimposable, thus indicating that the oxidation reaction takes place without appreciable chain degradation. Polymers **V-m** are optically active (Table II), with those based on the 1,2-phenylene residue (*ortho* isomers, *o*) having greater optical rotatory powers. This clearly indicates that the sulfide groups undergo asymmetric oxidation to chiral sulfoxide groups with a significant prevalent chirality.

The circular dichroism spectra in dilute solution of the poly(ester β-sulfoxide)s are characterized by the same strong absorption profiles for all the prepared samples in correspondence of the electronic transitions of the alkyl aryl sulfinyl chromophore ([20]) (Figure 4). This suggests that the overall optical activity of the polymers is dominated by the chirality of the sulfoxide moieties, which are characterized by the same prevalent configuration at the sulfur atoms. It is not known to what extent the sign and rotatory strength of the absorption bands depend on the local conformation of the macromolecular chain. In principle, such an asymmetric reaction should give rise to two chiral centers per repeat unit resulting in a mixture of a pair of (*R,R*) and (*S,S*) enantiomeric forms and a (*R,S*) meso form (Scheme 2). However, due to the complexity and strong overlaps of the signals of the ^{13}C-NMR spectra of the oxidized samples, the degree of asymmetric induction could not be fully investigated in polymers **V-m**.

To gain a better insight into the enantioselectivity of the asymmetric oxidation studied, we submitted to oxidation in the identical experimental conditions a

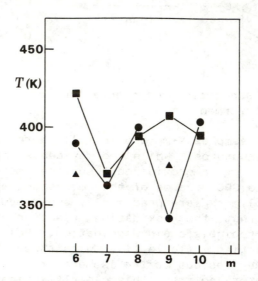

Figure 3. Trends of phase transition temperatures of poly(ester ß-sulfide)s **IV-m** with varying number **m**: (●) melting, (▲) smectic-cholesteric, and (■) isotropization.

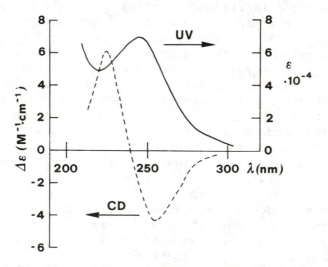

Figure 4. Ultraviolet (UV) and circular dichroism (CD) absorption spectra in dilute solution of poly(ester ß-sulfoxide) **V-6m**.

poly(ester β-sulfide) model compound in which the aromatic-alkylene-aromatic sequence of polymers II-m had been replaced by a 1,4-cyclohexanediyl unit. In the ^{13}C-NMR spectra of the derived poly(ester β-sulfoxide) the signals of the various carbon atoms experiencing the different stereochemical environments of the (R,R)+(S,S) and (R,S) diastereomeric forms are split into individual components of different intensities. An asymmetric induction of 25 ± 5% and an enantiomeric excess of 49 ± 10% of either (R,R) or (S,S) form could, therefore, be evaluated (12). We expect the reaction to proceed with a quite similar enantioselectivity with structurally analogous polymers II-m.

In one parallel oxidation experiment of sample III-4 with the chiral oxidant 1, a poly(ester β-sulfoxide) was obtained which does not show any appreciable optical activity. From this result we infer that even in the case of macromolecular substrates the difference in effective size of the groups directly linked to the sulfur atom plays a role in assisting the discrimination between the two enantiogroups and in determining the magnitude of the asymmetric bias (11).

The mechanism for oxygen transfer by 2-sulfonyloxaziridines is known to be a S_N2 nucleophilic attack by the substrate on the electrophilic oxaziridine oxygen atom (11,21). Therefore, according to the chiral recognition model proposed (11), the preferred diastereomeric transition state for sulfide oxidation should be the one in which the enantiotopic electron pair on sulfur attacks the oxaziridine oxygen atom in such a way that the large (R$_L$) and small (R$_S$) groups of the substrate face the small and large regions of the oxaziridine ring, respectively (Figure 5). That is where the R$_L$ group is as far away as possible from the bulky camphorsulfonyl group. Thus, (S,S)-2-sulfonyloxaziridines

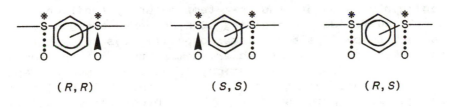

        (R,R)   (S,S)   (R,S)

Scheme 2.   Schematic illustration of the diastereomeric forms obtained will afford sulfoxides of the (S) configuration at the sulfur atom (11). Conversely, the similarity of the Cotton effects observed for the poly(ester β-sulfoxide)s (Figure 4) with those of low molecular weight model compounds of established (S) configuration (20b) is also an argument in favor of the prevalence of the (S,S) configuration in the repeat unit of these polymers. Direct extrapolation of the model to the asymmetric oxidation of systems involving polymeric substrates must, however, be treated with caution, and the absolute configuration of the chiral centers in the poly(ester β-sulfoxide)s prepared should be better assessed.

Examples are known (22-25) of the oxidation of macromolecular substrates containing sulfide functional groups, with chiral polysulfoxides being prepared by different synthetic approaches (26,27). However, the asymmetric oxidation of polymeric precursors was only performed with very limited degrees of chemoselectivity and enantioselectivity (27). The asymmetric oxidation reaction employed in this work is accomplished with a moderate enantioselectivity, which is nevertheless on the order of magnitude as those obtained with low molecular weight substrates (11,28,29).

Poly(ester β-sulfoxide)s **V-m** do exhibit liquid crystalline properties, but in no case could the mesophase behavior be investigated in detail up to the isotropization temperature due to extensive degradation of the samples (Table II). Typically the mesophase extends over 10-20 degrees. Furthermore, the polymers are optically unstable and their optical rotatory power becomes vanishingly small after short permanence times at the elevated mesophase temperatures (Figure 6). It is well documented that chiral sulfoxides can undergo thermal racemization (30,31) involving most frequently the interconversion of enantiomers by pyramidal inversion (30). The thermal stereomutation rate is, however, negligibly slow at the low temperatures. Additionally, alkyl sulfoxides containing β-hydrogens are generally subject to pyrolysis by a stereospecific cis elimination to olefins, even at lower temperatures than those required to effect racemization (32,33). Preliminary NMR investigations point to the occurrence of the latter degradation process for the investigated chiral poly(ester β-sulfoxide)s, but the two concurrent factors may well be responsible for the observed optical instability.

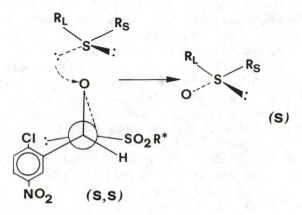

(S)

(S,S)

Figure 5. Representation of the chiral recognition pathway in the sulfide oxidation by chiral oxaziridine 1.

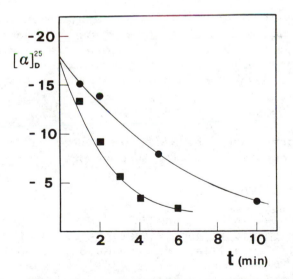

Figure 6. Decrease of the optical rotatory power of chiral poly(ester β-sulfoxide) V-6*m* with permanence time at 110°C (●) and 120°C (■).

Conclusions

Chiral liquid crystalline poly(ester β-sulfide)s can be prepared based on a sequence of rigid-flexible-rigid-flexible segments, in which the mesomorphic properties are tuned in a predictable way by variations in the sulfide spacer structure. The isomeric interchange of the two different flexible segments produces dramatic variations in the thermal behavior of the polymers, highlighting the importance of even subtle changes in affecting the properties of semiflexible thermotropic liquid crystalline polyesters (17,34).

The asymmetric oxidation reaction of prochiral poly(ester β-sulfide)s to optically active poly(ester β-sulfoxide)s can be accomplished with almost theoretical chemoselectivity and moderate to high enantioselectivity degrees. While the asymmetric oxidation of prochiral sulfides should not be a preparative method for chiral sulfoxides, we expect that the structure of the parent polymers might be specifically designed for the preparation of chiral thermotropic poly(ester β-sulfoxide)s.

As a final point, we may speculate on the interest of both performing the same type of reactions under mesophase conditions to address the stereochemical pathway, and using these polymeric materials as chiral catalysts or synthons in organic synthesis (35).

Acknowledgment

The authors thank the *Ministero Pubblica Istruzione of Italy* for partial financial support of the work.

Literature Cited

1. (a) Chiellini, E. and Galli, G., In <u>Recent Advances in Mechanistic and Synthetic Aspects of Polymerization</u>, Fontanille, M. and Guyot, A. eds., Plenum Press: New York, 1987; p 425; (b) <u>Faraday Discuss. Chem.Soc.</u> 1985, <u>79</u>, 241.
2. Finkelmann, H.; Kock, H.J.; Rehage, G. <u>Makromol.Chem., Rapid Commun.</u> 1981, <u>2</u>, 317.
3. Zentel, R.; Reckert, G. <u>Makromol.Chem.</u> 1986, <u>187</u>, 1919.
4. Eich, M.; Wendorff, J.H.; Reck, B.; Ringsdorf, H. <u>Makromol.Chem., Rapid Commun.</u> 1987, <u>8</u>, 59.

5. Creed, D.; Griffin, A.C.; Gross, J.R.D.; Hoyle, C.E.; Venkataraman, K. Mol.Cryst.Liq.Cryst. 1988, 155, 57.

6. Laus, M.; Angeloni, A.S.; Galli, G.; Chiellini, E. Makromol.Chem. 1988, 189, 743.

7. Ebert, M.; Ringsdorf, H.; Wendorff, J.H.; Wustfeld, R. ACS Polym.Div., Preprints 1989, 30(2), 479.

8. Noonan, J.M.; Caccamo, A.F. ACS Polym.Div., Preprints 1989, 30(2), 501.

9. Galli, G.; Laus, M.; Angeloni, A.S. Makromol.Chem. 1986, 187, 289.

10. Chiellini, E.; Galli, G.; Angeloni, A.S.; Laus, M. J.Polym.Sci., Polym.Chem.Ed. in press.

11. Davis, F.A.; Jenkins, R.H., Jr.; Awad, S.B.; Stringer, O.D.; Watson, W.H.; Galloy, J. J.Am.Chem. Soc. 1982, 104, 5412.

12. Chiellini, E.; Galli, G.; Angeloni, A.S.; Laus, M. Macromolecules submitted.

13. Blumstein, A.; Thomas, O. Macromolecules 1982, 15, 1264.

14. Roviello, A.; Sirigu, A. Makromol.Chem. 1982, 183, 895.

15. Galli, G.; Chiellini, E.; Angeloni, A.S.; Laus, M. Macromolecules 1989, 22, 1120.

16. Samulski, E.T.; Gauthier, M.M.; Blumstein, R.B.; Blumstein, A. Macromolecules 1984, 17, 479.

17. Yoon, D.Y.; Bruckner, S.; Volksen, W.; Scott, J.C.; Griffin, A.C. Faraday Discuss.Chem.Soc. 1985, 79, 41.

18. Abe, A.; Furuya, H. Macromolecules 1989, 22, 2982 and references therein.

19. Chiellini, E.; Galli, G.; Angeloni, A.S.; Laus, M.; Pellegrini, R. Liq.Cryst. 1987, 2, 529.

20. (a) Jaffé, H.H.; Orchin, M. Theory and Applications of Ultraviolet Spectroscopy, Wiley: New York, 1962; p 491; (b) Mislow, K.; Green, M.M.; Laur, P.; Melillo, J.T.; Simmons, T.; Ternay, A.L., Jr. J.Am.Chem.Soc. 1965, 87, 1958.

21. (a) Davis, F.A.; Billmers, J.M., Gosciniak, D.J.; Towson, J.C.; Bach, R.D. J.Org.Chem. 1986, 51, 4240; (b) Davis, F.A.; Lal, S.G.; Durst, H.D. J.Org.Chem. 1988, 53, 5004.

22. Janout, V.; Cefelin, P. Coll.Czechoslovak Chem. Commun. 1982, 47, 1818.

23. Marco, C.; Fatou, J.G.; Bello, A.; Perena, J.M. Makromol.Chem. 1984, 185, 1255.

24. Janout, V.; Hrudkova, H.; Cefelin, P. Coll.
     Czechoslovak Chem. Commun. 1984, 49, 1563.
25. Lazcano, S.; Marco, C.; Fatou, J.G.; Bello, A.
     Makromol.Chem. 1988, 189, 2229.
26. Mulvaney, J.E.; Ottaviani, R.A. J.Polym.Sci. A-1
     1970, 8, 2293.
27. Yamaguchi, H.; Minoura, Y. J.Appl.Polym.Sci. 1971,
     15, 1869.
28. Pitchen, P.; Dunanch, E.; Deshmukh, M.N.; Kagan,
     H.B. J.Am.Chem.Soc. 1984, 106, 8188.
29. Di Furia, F.; Modena, G.; Seraglia, R. Synthesis
     1984, 325.
30. Rayner,   D.R.;   Gordon,   A.J.;   Mislow,   K.
     J.Am.Chem.Soc. 1968, 90, 4854.
31. Miller, E.G.; Rayner, D.R.; Thomas, H.T.; Mislow, K.
     J.Am.Chem.Soc. 1968, 90, 4861.
32. Kingsbury, C.A.; Cram, D.J. J.Am.Chem.Soc. 1960, 82,
     1810.
33. Kwart, H.; George, T.J.; Louw, R.; Ultee, W.
     J.Am.Chem.Soc. 1978, 100, 3927.
34. Angeloni, A.S.; Caretti, D.; Laus, M.; Chiellini,
     E.; Galli, G. Polym.J. 1988, 20, 1157.
35. Solladié, G. Synthesis 1981, 185.

RECEIVED April 10, 1990

Chapter 7

Liquid-Crystalline Character of Novel Main-Chain Oligophosphates Bearing Lipophilic and/or Mesogenic Moieties

C. M. Paleos[1], A. Kokkinia[1], and P. Dais[2]

[1]NRC "Demokritos," 15310 Aghia Paraskevi, Attiki, Greece
[2]Department of Chemistry, University of Crete, 71110 Iraklion, Greece

In the present study it is investigated whether certain oligophosphates prepared by the reaction of phosphorous oxychloride with long-chain aliphatic diols and the mesogenic 4,4 diphenyldiol exhibit liquid crystalline character. Specifically, it was found that the oligophosphates resulting from aliphatic diols show liquid crystalline behavior while that originating from the diphenyldiol does not exhibit liquid crystalline character. It has however been established that mesomorphic phases are induced with oligophosphates bearing, in random, mesogenic and lipophilic groups

Recent reports on monomeric and polymerized bolaamphiphiles[1] provide evidence for their potential application in the broader field of molecular organizates (1,2). Thus monomeric bolaamphiphiles have been employed in the formation of monolayer lipid membranes or vesicles (1–3), formation of micelles (4–5) and also for spanning bilayer membranes (1–6). The latter process has resulted in the stabilization of membranes.

Extensive recent studies has established that liquid crystallinity is not only induced by the presence of mesogenic groups (7), but also induced in certain molecules by the existence of distinct polar and non-polar moieties which segregate forming lamellar structures. These amphiphilic molecules (1,7–18), both ionic and non-ionic, exhibit liquid crystalline phases in the melt, and form aggregates in solution which generate molecular organizates (19).

In a recent article we have established the liquid crystalline behavior exhibited by α,ω diphosphate amphiphiles (20). Prompted by this investigation we decided to study the liquid crystalline character of their polymeric counterparts, i.e. of polyphosphates or rather oligophosphates (Scheme I) prepared by the reaction of phosphorous oxychloride with 1,12 dodecanediol and 1,16 hexadecanediol respectively. In addition, the mesomorphic behavior of oligophosphates bearing the biphenyl mesogenic moiety was compared with the liquid crystalline behavior of the purely aliphatic oligophosphates.

0097–6156/90/0435–0093$06.00/0

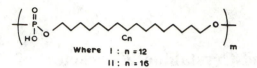

Where I : n = 12
II : n = 16

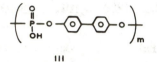

III

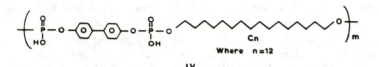

Where n = 12

IV

Scheme I

Experimental

Preparation of Poly (1,12 dodecanediol phosphate), (I). The oligoph-
osphate (I) was prepared by a modification of the method recently
described in a preliminary report (21). Specifically in 0.01 mole of
POCl₃ dissolved in dry dioxane, 0.01 mole of 1,12 dodecanediol, in
dioxane, was slowly added and allowed to stir for several hours.
Polymerization was completed by heating the reaction mixture for a
few hours at 70 °C. Dioxane was distilled off under vacuum, and the
residual oil was hydrolysed by dilute hydrochloric acid in cracked
ice. The precipitated material was centrifuged and washed several
times with water. It was precipitated with an ethanol-ether mixture
and dried over phosphorous pentoxide.

It should be noted that the polymer produced is soluble in dioxane. The very low molecular weight polymer obtained by the previously
employed method (21) could be attributed to its early precipitation
in benzene. Also, the duration of polymerization was increased
because pyridine was not employed in this experiment which was difficult to remove from the precipitated polymer.

Preparation of poly (1,16 hexadecanediol phosphate), (II). This oligomer was prepared by an analogous procedure to the one employed previously.

Preparation of poly(4,4 diphenyldiol phosphate), (III). The procedure
employed for the preparation of III was analogous to the one for
aliphatic polyphosphates.

Preparation of poly(1,12 dodecanediol-4,4 diphenyldiol phosphate)
(IV). In this case, an equimolar mixture of the diols was used which
was the same to the total diol concentration employed for the preparation of the homopolymers.

Proton NMR spectra were obtained on a Varian XL-200 spectrometer operating at 200 MHz, at room temperature. Chemical shifts were measured in DMSO-d₆ and referenced to TMS. The error of integration of the various peaks was estimated ca ±10%.

Optical microscopy studies were performed with a Reichert Thermopan microscope with crossed polarizers. The magnification used was 80X.

DSC thermal studies were performed with a DSC-4 Perkin Elmer thermal analyzer coupled with a System-4 programmer at a scanning rate 10 or 20 °C/min.

Number average molecular weights were determined using a Knauer Vapor Pressure Osmometer.

Results and Discussion

Formation of branched polymers was kept at a low level by an extremely slow addition of the diols to phosphorous oxychloride. By this technique phosphorous oxychloride concentration was always kept at a great excess, thus favoring the formation of linear polymers. The predominantly linear structure of the polymers was established by elemental analysis and proton NMR spectroscopy. The peak assignment was facilitated through homodecoupling experiments and integration of the various resonances. The α- and β-CH₂ protons of the aliphatic chains relative to the phosphate group in polymers I and II absorb at 4.03 and 1.72 ppm respectively, whereas those of the internal methylene protons of the chain form a broad envelope at 1.27ppm. The aromatic protons of polymer III show the characteristic pattern of an AB subspectrum with δ_A=7.61ppm, δ_B=7.28 ppm and J_{AB}=8.75 Hz, indicating that these protons are magnetically nonequivalent. This is due to the low molecular weight of the polymer whose terminal groups represent a significant portion to the polymer chain. The same AB pattern is shown in the NMR spectrum of copolymer IV, and in addition depicts the characteristic absorptions of the methylene protons of the aliphatic chain. Minor peaks at 3.79 and 3.67 ppm may indicate a very small proportion of non-linear polymer. Integration of the aromatic (7.51 – 7.14 ppm) and saturated (1.24ppm) protons indicate that the ratio of aromatic to aliphatic moieties in the various copolymers of IV(a–c), Scheme II, is on average 3.5:1. Since equimolar quantities of diols were employed, one may conclude that the reactivity of 4,4 diphenyldiol is higher than that of 1,12 dodecanediol.

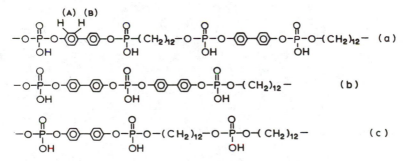

Scheme II

Figure 1. Liquid crystalline texture of polymer I.

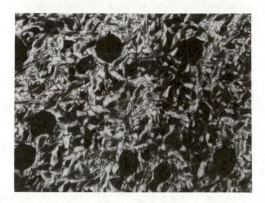

Figure 2. Liquid crystalline texture of polymer II.

Number average molecular weights of the polymers were found to be 2000 for polymer I, 3600 for II and 2100 for polymer III. For the copolymer IV molecular weight was not determined because of its insolubility in all common solvents.

The liquid crystalline behavior of oligophosphates was investigated by optical microscopy and DSC studies. Thus polymers I and II melt at about 60 °C and 75 °C to anisotropic melts which become isotropic at about 85 °C and 95 °C respectively. On cooling from their isotropic melts they form spherulites at about 70 °C and 85 °C respectively as shown in Figure 1 and Figure 2. For comparison the mesomorphic texture of monomeric dodecane-1,12-diphosphate is shown in Figure 3. It is interesting to note that the mesomorphic phase of the C_{12} monomeric diphosphate derivative is in a transient state and quickly reverts to a crystalline phase. In Figure 3 the texture is quite characteristic since both mesomorphic and crystalline phases coexist. The monomeric C_{16} diphosphate derivative melts at 120 °C to a mesomorphic phase, Figure 4, and becomes isotropic at about 170 °C. On cooling from its isotropic melt it supercools and is transformed directly to its crystalline phase.

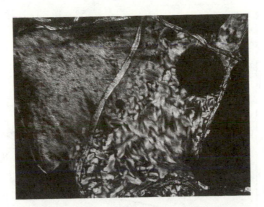

Figure 3. Coexistence of liquid crystalline and crystalline phases obtained on cooling of C_{12} diphosphate derivative.

Figure 4. Liquid crystalline texture of C_{16} diphosphate derivative.

Polymers I and II are thermally stable as evidenced by their DSC traces, Figure 5 and Figure 6, in which the same clearing points were observed both in the first and second heating runs.

The introduction of the biphenyl mesogenic group in polymer III raises the melting and clearing points, and thus depolymerization and decomposition occured before the mesomorphic phase was completely transformed to the isotropic phase. This behavior is depicted in DSC diagram, Figure 7, in which depolymerization is observed after the first transition. A copolymer was prepared in order to lower the melting point. As already mentioned in the preparation of polymer IV, equimolar quantities of 4,4 biphenol and 1,12 dodecanediol were used. But, as determined by NMR the ratio of the aromatic to the aliphatic moieties was not 1:1 but 3.5:1. However, what really counts in this case is that even with this ratio the lowering of the melting

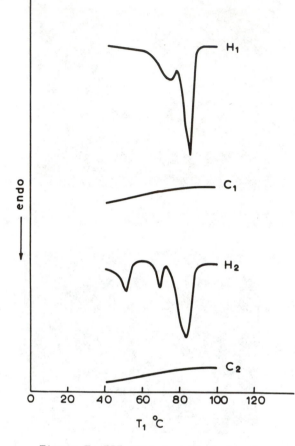

Figure 5   DSC diagram of polymer I.

point is significant.   By exercising pressure on the cover slip the
material melts at about 60 °C and becomes isotropic at about 140 ° C.
On cooling the melt reverts to a liquid crystalline phase at about
140 °C while almost simultaneously crystallizing. This two-phase
system is shown in Figure 8 The DSC trace of the copolymer is shown
in Figure 9.

Concluding Remarks

These first results establish that the presence of distinct polar and
lipophilic groups in phosphate oligomers induces the appearance of
mesomophic phases. Furthermore, it was found that mesomorphic phases
are observed with phosphate copolymers, containing random mesogenic
and   lipophilic   moieties.   Copolymers   with   alternating
mesogenic-lipophilic moieties will be further prepared in an attempt
to enhance the thermal stability and liquid crystalline range of
these polymers.

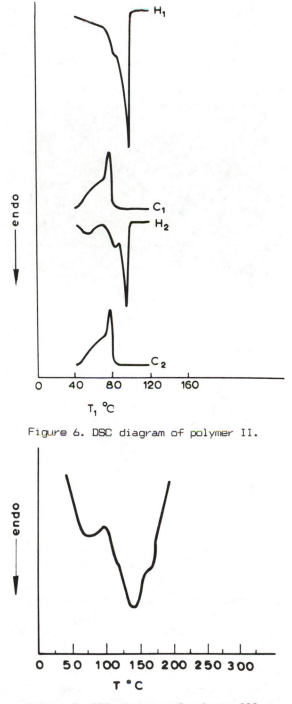

Figure 6. DSC diagram of polymer II.

Figure 7. DSC diagram of polymer III.

Figure 8. Coexistence of mesomorphic and crystalline phases during cooling of polymer IV.

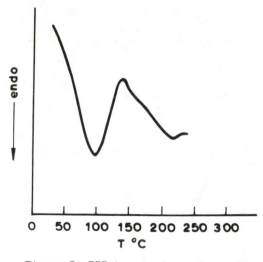

Figure 9. DSC trace of copolymer IV.

Literature Cited

1. H. Fuhrhop, J. H.; Fritsch, D. Acc. Chem. Res. 1986,
   19, 130 and references cited therein.
2. Ringsdorf, H.; Schlarb, B. Angew. Chem. 1988, 27,
   113, and references cited therein.
3. Bader, H,; Ringsdorf, H. J. Polym. Sci., Polym. Chem.
   Edit. 1982, 20, 1623.
4. Menger, F. M. ; Wrenn, S. J. Phys. Chem. 1974, 78, 1387.
5. Yiv, S.; Kale, K. L,; Lang, J.; Zana, R. J. Phys. Chem.
   1976, 80, 2651.
6. Bader, H.; Ringsdorf, H. Faraday Discuss. Chem. Soc. 1986,
   81, 329.
7. Kelker, H.; Hatz, R. Handbook of Liquid Crystals; Verlag
   Chemie, Weinheim, 1980; and references cited therein.
8. Skoulios, A.; Guillon, D. Mol. Cryst. Liq. Cryst. 1988,
   165, 317.
9. Geffrey, G. A. Acc. Chem. Res. 1986, 19, 168,
   and references cited therein.
10. Margomenou-Leonidopoulou, G.; Malliaris, A.; Paleos,
    C. M. Thermochimica Acta, 1985, 85, 157.
11. Busico, V.; Ferraro, A.; Vacatello, M. Mol. Cryst.
    Liq. Cryst. 1985, 128, 243.
12. Michas, J.; Paleos C. M.; Dais, P. Liquid Crystals
    1989, 5, 1737.
13. Gault, J. D.; Gallardo, H. A. Muller, H. J. Mol. Cryst.
    Liq. Cryst. 1985, 130, 163.
14. Needham, G. H.; Willet R. D.; Franzen, H. F. J. Phys. Chem.
    1984, 88, 1984.
15. Iwamoto, K.; Ohnuki. K.; Sawada, K.; Seno, M. Mol. Cryst.
    Liq. Cryst. 1981, 73, 95.
16. Pfannemuller, B.; Welte, W.; Chin E.; Goodby, R. W.
    Liquid Crystals 1986, 1, 357.
17. Busico, V.; Ferraro, A.; Vacatello, M. Mol. Cryst.
    Liq. Cryst. 1985, 128, 243.
18. Bruce, D. W.; Dunmur, D. A.; Lalinde, E.; Maitlis P. M.;
    Styring, P. Nature 1986, 323, 791.
19. Fendler, J. H. Membrane Mimetic Chemistry; Wiley
    Interscience, New York, 1982.
20. Kokkinia, A.; Paleos C. M.; Dais, P. Mol. Cryst.
    Liq. Cryst In Press.
21. Kokkinia, A.; Paleos C. M.; Dais, P. Polymer Preprints
    1989, 30, 448.

RECEIVED March 6, 1990

Chapter 8

Thermotropic Poly(ester-*co*-carbonate)

Yu-Chin Lai[1,2,3], Bruce T. DeBona[1], and Dusan C. Prevorsek[1]

[1]Corporate Research and Development, Allied Signal, Inc.,
Morristown, NJ 07960

A number of thermotropic polyester-carbonates
were prepared through melt-polymerization of
substituted hydroquinones and diphenyl terephthalate and diphenyl carbonate to have high
molecular weight, with reduced viscosity in the
range of 2-3. The molecular weights of the
polymers can be advanced further by solid state
heat-treatment, with the rate of postpolymeri-
zation depending on temperature and Concentration of catalyst. Samples of some compositions
can be spun into high performance fibers and
processed into self-reinforced plastics. The
properties of thermotropic polyester-carbonates
and polyesters were compared as fibers and
plastics.

Although studies concerning main chain liquid
crystalline polymers were originated by Onsager (1a) and
Ishihara (1b) in the late 1940's, extensive work in this
field did not really begin until the early 1970's.
Jackson and Kuhfuss (2) reported the first thermotropic
polyester by modifying polyethylene terephthalate with
various amounts of p-hydroxybenzoic acid (HBA). They
found that the copolyester with HBA content of at least
35 mole % have opaque melts. Subsequent studies in the
area of aromatic polyesters by various authors resulted
in a large number of patents and publication.(3) These
polymers were all derived from unsubstituted and

[2]Current address: Bausch and Lomb, Inc., Rochester, NY 14534
[3]Address correspondence to this author.

0097–6156/90/0435–0102$06.00/0
© 1990 American Chemical Society

substituted aromatic diols and diacids and hydroxy acids with and without flexible spacers.(3) The reasons for the broad interest in this area, among others, are to create melt-processable high strength, high modulus fibers and self-reinforced plastics.

Thermotropic polyesters derived from unsubstituted aromatic diols and diacids usually have melting points which approach or exceed the thermal decomposition point. Thus it is reasonable to expect that some modification in molecular structure would be required to render them melt-processable, even though some adverse effects on liquid crystallinity and mechanical properties of the polymers would result.

The preparations of thermotropic polyester-amides from comparable monomers as those of thermotropic polyesters were a logical extension of a series of studies in thermotropic polyesters and lyotropic polyamides.(4) However the inclusion of carbonates had rarely been explored.(5) Because of the flexibility of carbonate compared to substituted aromatic rings, it should be an even more effective approach in lowering the melting temperatures of the unmodified all aromatic polyesters into the easily processable range.

The solid-state heat-treatments of thermotropic polyester fibers to obtain fibers with high tenacity were reported before.(6) However, the nature of this process was not clear when we commenced this study. The objectives of this paper are to disclose the less explored thermotropic polyester-carbonates and to demonstrate the nature of heat-treatment of the thermotropic polymers using polyester-carbonates as examples. In addition, the properties of thermotropic polyesters and polyester-carbonates as fibers and plastics are compared.

Experimental

Monomers and catalysts. The monomers methylhydroquinone (MHQ), t-butyl-hydroquinone (BHQ), resorcinol (RO), hydroquinone (HQ), diphenyl carbonate and catalyst tetrabutyl titanate (TBT) were purchased (Aldirch) and used as received.

Diphenyl terephthalate (DPT) was prepared by interfacial reaction of sodium phenoxide in water and terephthaloyl chloride (TPC) in methylene chloride, using benzyl triethylammonium chloride as catalyst, followed by washing with hot water and recrystallization from toluene. The yield was usually 90% or higher.

General Procedure for the preparation of polyester-carbonates. A resin kettle was charged with the aromatic diols, DPT and DPC of a selected molar ratio. TBT in the amount of 30-1000 ppm of the combined weights of monomers was also added. The contents were heated under constant nitrogen flow and stirred mechanically

when melted. The polymeri-zation started to occur at
about 215°C as phenol started to distill out. The
temperature was slowly raised as the reaction progressed
and most of the phenol was distilled at 260-270°C.
Vacuum was then applied gradually to reach 0.2-0.5 torr
in about 5 minutes. The reaction inter-mediate, after
half an hour under vacuum, changed from fluid to a very
viscous, golden, fibrous material. The reaction was
stopped when the temperature reached 290°C and the
product was removed from the kettle while still hot. The
product was first broken into pieces and then crushed
into fine pieces. It was then extracted with hot toluene
and dried in vacuo. The yield was usually over 90%.
    When the polymerization was done on a large scale,
a thermocouple was used for detecting the temperature of
the contents and heating medium such as Therminol-66
(Monsanto) was added into the reaction contents.

Polymer characterization. The thermotropic behaviors of
the polyester-carbonates prepared were examined under a
polarized hot stage microscope (Sybron Co.). Normally
the sample became softer at 120-150°C and melted with
birefringence afterwards, depending on the molecular
weight and composition of the polymer. The samples
stayed bire-fringent until 350°C or higher.
    Dilute solution viscosities of most polymers were
measured at room temperature in p-chlorophenol/1,2-
dichloroethane (50/50 by weight), using an Ubbelhode
viscometer. For samples containing HQ/RO, the viscosities were measured at 50°C in pentafluorophenol. In
any case the molecular weight or viscosity of each
sample was reported as n_{sp}/c, or reduced viscosity, at
0.5 g/dl in the units of dl/g.
    The DSC scans for all samples were done using a
DuPont 990 differential scanning calorimeter at a
heating rate of 20°C/minute in argon. In all cases, the
samples were quenched from 350°C and reheated to check
for changes in the transition temperatures.

Heat-treatments of thermotropic polyester-carbonates.
The solid-state heat-treatments of thermotropic
polyester-carbonates in fine pieces were conducted in a
Blue M high temperature inert gas oven (Model
AGC170EMTI). These samples were treated at different
temperatures, for different period of time. The treated
samples were characterized further by DSC and viscosity
measurements. The treated samples were used in fiber-
spining and plastic processing studies.(7,8)

Results and Discussion

Polymer Synthesis. Though solution polymerization is a
feasible process in making polycarbonates and
copolyester-carbonates, (9) it is not acceptable for
making thermotropic polymers because the low solubility

of thermotropic polyesters in inert solvent limits the
growth of molecular weight of product in solution. (3a)
Thus melt-polymerization was applied here to prepare
polyester-carbonates with molecular weight high enough
for processing studies.
Equation 1 shows the reaction scheme which leads to the

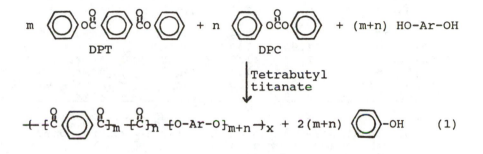

formation of high molecular weight polyester-carbonates
derived from diphenyl terephthalate (DPT), diphenyl
carbonate (DPC), and aromatic diols (Ar in the equation)
such as t-butyl hydroiquinone (BHQ), methyl-hydroquinone
(MHQ), hydroquinone (HQ) and resorcinol (RO).
     For the polymerization to proceed at a reasonable
rate, the use of a transesterification catalyst is
needed.   Compounds which are usually used as a catalyst
for the preparation of polyesters through transesterification can be used here. These include lithium,
sodium, zinc, magnesium, calcium, titanium, maganese,
cobalt, tin, antimony, etc. in the form of a hydride,
hydroxide, oxide, halide, alcoholate, or phenolate or in
the form of salts of organic or mineral acids, complex
salts, or mixed salts.(10) In this study, tetrabutyl
titanate (TBT) in the amount of 1000 ppm was used
normally.
     In order to keep the contents more fluid-like for
proper heat exchange and smooth stirring during the
course of reaction, a high temperature-stable heating
medium such as Therminol-66 was used.

<u>Polymer Composition-Property Relationship</u>.   Tables I
summarized the representative thermotropic polyester-
carbonate samples prepared from BHQ/MHQ/DPT/DPC. The
first series (entry 1-3) of compositions under
investigation was those containing BHQ. Higher DPT/DPC
ratio gave polymer with higher melting temperature.
These samples showed birefringence only under shear when
melted.
     The inclusion of MHQ into the polyester-carbonate
compositions gave polymers with strong birefringence

Table I. Thermotropic Polyester-carbonates Derived
From BHQ/MHQ/DPT/DPC

Feed molar ratio BHQ/MHQ/DPT/DPC	n_{sp}/c	Tg ($^{\circ}$C)*	Tm ($^{\circ}$C)**	Thermotropic Behavior
100/0/50/50	0.58	119	270	***
100/0/40/60	0.76	142	170	***
100/0/30/70	0.80	149	160	***
50/50/50/50	2.98	135	180	****
50/50/52.5/47.5	2.07	130	210	****
50/50/55/45	1.79	133	240	****
50/50/57.5/42.5	1.98	126	260	****
50/50/60/40	2.88	135	290	****

    *   By DSC.
   **   By visual observation under polarized microscope.
  ***   Birefringence under shear only.
 ****   Birefringence without shear.

without shear when melted. In this series of study, the
feed molar ratio of DPT/PTC was changed from 50/50 to
60/40 while the molar ratio of BHQ/MHQ was kept at
50/50. The melting temperature of the polymer was
changed from 180°C to 290°C when the molar ratio of
DPT/DPC of samples was changed from 50/50 to 60/40.
    It must be stressed here that the molar ratios
mentioned in this article always referred to the feed
molar ratio. The true molar ratios of BHQ/MHQ and
DPT/DPC of products might be different from the feed
ratios. The true molar ratios of DPT/DPC in two
polyester-carbonate samples derived from BHQ/MHQ/DPT/DPC
in feed molar ratios of 50/50/50/50 and 50/50/60/40 were
checked by Attenuated total reflection Fourier transfer
infrared spectroscopy. They were found to be 42/58 and
50/50 respectively, indicating that DPC is more
reactive. This was expected in view of the flexibility
of DPC relative to DPT. This identi-fication was made
possible by comparing the IR absorption of the
polyester-carbonate product at 1777.5 cm^{-1} (for the
carbonyl groups of carbonate) and at 1738.8 cm^{-1} (for
the carbonyl groups of ester), using the IR absorptions
of DPT and DPC as controls (to calculate the respective
extinction coefficients of carbonyl groups). However,
the same IR method could not give the ratio of BHQ/MHQ,
because of the overlap of absorption bands at 2925-2975
cm^{-1} attributed to methyl and t-butyl groups of the
product. Because of solubility problem in common organic
solvent, no NMR spectrum of the product was taken. The
ratio of BHQ/MHQ can be obtained by solid state NMR if
available. In view of the structures of BHQ and MHQ, no
difference in reactivity was expected.

Thermotropic polyester-carbonates containing only HQ and RO have melting points near thermal decomposition. However when MHQ and BHQ were included in the composi-tions, the melting temperatures were depressed somewhat close to 300°C as shown in Table II.

Table II. Thermotropic Polyester-carbonates Derived from HQ/RO/BHQ/MHQ and DPT/DPC (molar ratio 50/50)

Feed molar ratio	As is		After annealed*		
HQ/RO/BHQ/MHQ	n_{sp}/c	Tg (°C)	n_{sp}/c	Tg (°C)	Tm (°C)
40/35/15/10	1.18	118	1.11	115	288
40/30/20/10**	0.92	119	insol	127	327
35/35/15/15**	1.49	122	insol	130	330
35/30/25/10	1.65	135	1.52	128	285
35/30/20/15	1.61	122	1.55	119	290
30/30/25/15	0.87	112	0.92	112	285

*   At 240°C for 30 minutes.
** Annealed at 250°C for 6 hours.

Because of the very amorphous nature of these polymers due to the synthetic procedure employed (melt-polymeri-zation and quenching of products) and the compositions studied, true melting temperatures can be obtained from the DSC scans only after the as-prepared samples were annealed at temperature below melting for a short period of time. Almost all polyester-carbonate samples derived from HQ/RO/BHQ/MHQ and listed in Table II were annealed at 240°C (more than 40°C below Tm) for 30 minutes to give samples with more crystallinity, yet they were not over treated to cause changes in molecular weight as demonstrated by constant viscosities and Tg's before and after annealing. Melting temperatures were then measured by DSC. Two samples were treated at higher temperature (250°C) for longer period of time (6 hours) to get the ultimate melting temperature for samples with extremely high molecular weights (samples insoluble in the same solvent used for measuring viscosity of low viscosity samples). Both of them were close to 330°C.

In general, high molecular weight polymers can be obtained routinely with the synthetic procedure as described earlier unless when the stirring of reaction contents were not efficient during the vacuum distillation stage. Normally samples of reduced viscosity of 2-3 can be obtained as long as the melting temperature of the polymer was 290°C or below. The molecular weight of thermotropic polymers prepared by melt-polymerization is much higher than those prepared by solution polymerization.(3a) In those cases the inherent viscosity were normally in the range of 0.3 to 1.0.

The Effect of Solid-State Heat-Treatment on Molecular
weight and Crystallinity. Solid-state heat-treatment
had been done before on thermotropic polyester fiber to
obtain advancement in tenacity (6) and on polyester-
carbonate samples to get advancement in molecular
weight.(11) It was understood that post polymerization
and thermal annealing occurred on thermotropic polymers
at solid state, as the term post-polymerization has been
used interchangeably for solid-state heat-treatment.
        The objectives of carrying out solid-state heat-
treatments of thermotropic polyester-carbonates were
twofold.   One, to study the heat-treatment phenomena.
Two, to prepare samples of higher molecular weight for
processing studies into high performance fibers and
plastics if heat-treatment did cause advances in
molecular weight.
        Selected samples (in fine-pieces) of thermotropic
polyester-carbonates derived from BHQ/MHQ/DPT/DPC,
(contained 1000 ppm of TBT) and also those containing
HQ/RO as described in the proceding section, were heat-
treated in the solid state at temperatures between 40°
below Tm and Tm for different periods of time. It was
found that the crystallinity was increased in all
samples, as suggested by the DSC scans after the sample
was heat-treated. It was also found that the molecular
weights of all samples increased with the advances
depending on the time and temperature of heat-treatment.
Table III summarized some typical examples in this
study. As expected from the general chemical reation
kinetics, the higher the heat-treatment temperature, the
faster the advances in molecular weight.
        Because high molecular weight thermotropic
polyester-carbonate products can be obtained by this
process, the difficulty in obtaining high molecular
weight materials directly from melt-polymerization is
circumvented.
        It was observed that the efficiency in heat-
treatment is irrevelent to the powder size in our
experiments.   However it was found that samples of the
same molecular weight can be treated more efficiently in
the fiber form (particularly in fine denier fiber) than
in powder form
        Because of differences in surface area. For
example, a sample in powder form derived from
BHQ/MHQ/DPT/DPC (50/50/55/45) with reduced viscosity of
1.97, can be treated to give a product with reduced
viscosity of 4.19 at 240°C for 6 hours, while a sample
in fiber form, spun from a sample of the same viscosity,
became insoluble in the same solvent after it was
treated under the same conditions.

The Effect of Catalyst Concentration on Melt-
polymerization and Solid-state Heat-treatment.   In the
preparation of polyester-carbonates of different

Table III. Effect of Heat-treatment on Molecular Weight
(n_{sp}/c) for Thermotropic Polymers Derived
from BHQ/MHQ/DPT/DPC (BHQ/MHQ 50/50)

Heat Treatment		DPT/DPC Molar Ratio		
T (°C)	t (hours)	55/45	57.5/42.5	60/40
Original		1.97	2.07	2.88
240	3	3.66	--	--
	6	4.19	--	--
	12	5.86	--	--
	18	9.45	--	--
	24	12.7	--	--
250	3	--	3.67	6.01
	6	--	6.63	8.76
	9	--	10.9	--
	12	--	16.3	insol
	24	--	insol	--
260	1	--	--	5.28
	3	--	--	insol

compositions as described in the previous sections, TBT
in the amount of 1000 ppm was used universally. This
melt-polymerization did need catalyst. In some cases,
the addition of TBT was withheld intentionally, the
reaction contents stayed fluid at 250-270°C, with no
phenol distilled. However, once a few drops of TBT was
added, phenol started to distill over.

To study the effect of TBT on polymerization
kinetics, the melt-polymerization of BHQ/MHQ/DPT/DPC
(50/50/57.5/42.5) was carried out with TBT in the amount
of 1000, 300, 100 and 30 ppm's respectively. Attempts
were made to carry these polymerizations under the same
reaction conditions, particularly the same distillation
time under the same vacuum and the same reaction
temperature. However some variations in reaction
conditions were still possible due to the equipment
limitations. Table IV summarized some representative
data in this study. As can be seen, the molecular weight
of the product achieved was about the same (2.07 vs
1.81) when the amount of TBT used was 1000 ppm or 300
ppm. However, when the amount of TBT was dropped down to
100 ppm or 30 ppm, the molecular weight of the product
(reduced viscosity was 1.11 and 0.87 respectively)
achieved was only half of those obtained with a high
amount of TBT.

The effect of TBT concentration on solid-state
heat-treatment was also studied. It was assumed here
that the loss of TBT during purification was in

Table IV. Effect of TBT Concentration on Melt-
polymerization and Solid State Heat-treatment
for BHQ/MHQ/DPT/DPC (50/50/57.5/42.5)

Reaction Condition	TBT Concentration (ppm)			
	1000	300	100	30
Melt-polymerization Distillation time (hr)				
under nitrogen	1.8	1.0	2.2	3.0
under vacuo	2.0	1.8	2.0	2.0
n_{sp}/c	2.07	1.81	1.11	0.87
Heat-treatment at 250°C, N_2 for (hr)	n_{sp}/c			
2	--	2.71	--	--
3	3.67	--	--	--
4	--	3.03	--	--
6	6.63	--	--	1.41
9	10.9	--	--	--
12	16.3	--	1.79	--
24	insol	--	--	--
48	--	--	4.80	--
60	--	--	5.63	--
72	--	--	insol	--

proportion to the amount of TBT used. The loss of TBT
for a sample pre-pared with 300 ppm of TBT was analyzed
by ICP analysis. It was found that the purified product
contained 272 ppm of TBT, representing a 50% loss of
TBT, after adjusting for the loss of phenol. This series
of heat-treatment studies were carried out at 250°C for
different periods of time with the results summarized in
Table IV. As expected, the molecular weight of the
sample advanced faster when more TBT was used. When the
TBT concentra-tion used was very low (such as 100 ppm
or 30 ppm), the heat-treatment process became very
inefficient in gaining an advancement in molecular
weight.

Correlation of Transition Temperatures with Molecular
Weight. Table V shows the Tm's and Tg's of thermotropic
polyester-carbonates of higher molecular weight after
heat-treatments. In all three series, the Tm advanced as
the molecular weight increased, but it reached a
asymptotic figure. The Tg's stayed relatively constant.
The Tm' vs reduced viscosity were plotted in Figure 1.

Thermotropic Polyester-carbonates as Fibers and Self-
reinforced Plastics. The major objective of our study
in thermotropic polyester-carbonates was to evaluate

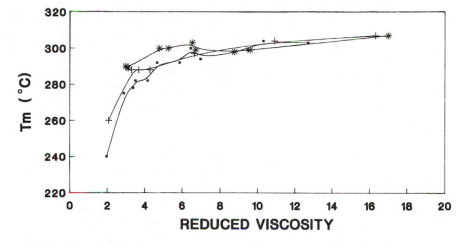

Figure 1. Melting temperature versus reduced viscosity for polyester-carbonates. (■) DPT/DPC 55:45; (+) DPT/DPC 57.5:42.5; (*) DPT/DPC 60:40

Table V. Tm and Tg vs Molecular Weight (in n_{sp}/c) for Thermotropic Polyester-carbonates Derived from BHQ/MHQ/DPT/DPC (with BHQ/MHQ at 50/50)

Composition DPT/DPC		n_{sp}/c,	Tm	and	Tg			
55/45	n_{sp}/c	1.96	2.88	3.38	4.15	6.95	9.45	12.7
	Tm	240*	275	278	282	294	299	303
	Tg	128	129	130	131	134	134	137
57.5/42.5	n_{sp}/c		2.07	3.29	4.27	6.63	10.9	Insol
	Tm		260*	288	288	297	304	307
	Tg		128	132	133	133	137	138
60/40	n_{sp}/c		2.97	3.10	4.77	6.52	8.76	Insol
	Tm		290*	289	300	303	306	308
	Tg		136	137	138	138	137	140

* By visual observation on a polarized microscope.

these polymers as fibers and plastics, in hope that high tenacity fiber and self-reinforced engineering plastics can be obtained. The fiber-spinning and plastic processing studies were described separately.([7,8]) During the course of our study, we had the opportunity to evaluate the only available all aromatic, thermotropic polyester, LCP2000 from Celanese. Presumably it was a copolymer containing hydroxybenzoic (HBA) and hydroxy naphtholic acid (HNA) moieties. The sample was subjected to fiber spinnings and plastic moldings as were the thermotropic polyester-carbonates, using the same equipment and under comparable conditions. Table VI compares the key properties of these two types of thermotropic polymers category by category. The samples compared had the same melting ranges, but were very different in reduced viscosities and solubility characteristics. The data compared were those processed under the most favorable conditions. Interestingly enough, the as-spun fibers from the polyester-carbonate can be heat-treated more efficiently than those fibers (of same tenacity) spun from the polyester. Both of them gave fiber properties far superior to those of nylons and polyethylene terephthalate. These two classes of polymers also had comparative properties (such as tensile strength, tensile modulus, flex modulus, notched Izod impact strength) as plastics; and their properties were far superior to most plastics without any reinforcement.

Table VI. Comparative Properties of Thermotropic
Polyester and Polyester-carbonates

Class of polymer	Polyester	Polyester-carbonate
Composition	HBA/HNA-containing	BHQ/MHQ/DPT/DPC (50/50/57.5/42.5)
n_{sp}/c	> 17 in pentafluoro-phenol at 50°C	2.07 in p-chlorophenol/1,2-dichloroethane at RT
Tm (°C)	272 & 292	250

Properties as fiber

Spun at (°C)	299	260
Properties: tenacity/modulus/denier (in grams/denier)		
as-spun	8.0/405/6	5.7/360/7
heat-treated for 6 hours at	250°C 14.3/445/8	240°C 18.2/410/6

Properties as plastics

Sample viscosity	same	1.81
Molded at (°C)	304	260
Notched Izod ft-lb/in	4.1	3.0
Tensile strength (10^4 psi)	2.07	1.98
Tensile modulus (10^5 psi)	5.21	6.86
Flex modulus (105 psi)	7.50	8.11

Conclusion

High molecular weight thermotropic polyester-carbonates
can be prepared through transesterification of monomers
in the melt, using tetramethyl titanate as catalyst. The
products can be heat-treated in the solid-state to gain
advances in molecular weight and crystallinity, with the
rate of post-polymerization depending on temperature.
Thermotropic polyester-carbonates can be spun and heat-
treated into high performance fibers and processed into
self-reinforced plastics, with mechanical properties
equivalent to those of some all aromatic thermotropic
polyesters.

Literature Cited

1. a) Onsager, L. New York Acad. Sci. 1949. 51, 627.
    b) Isihara, A. J. Chem. Phys. 1951, 19, 1142.
2. a) Kuhfuss, H. F.; Jackson, Jr., W. J. U.S. Patent
       3 778 410, 1974.
    b) Jackson, Jr., W. J.; Kuhfuss, H. F. J. Polym.
       Sci. -Chem. 1976, 14, 2043.
3. Too much to be cited completely. Some examples are:
    a) Jin, J. I.; Antoun, S.; Ober, C.; Lenz, R. W.
       Brit. Polym. J. 1980, 12, 132.
    b) Jackson, Jr., W. J. ibid. 1980, 12, 154.
4. a) Jackson, W. J.; Kuhfuss, H. F. J. Polym. Sci. -
       Chem. 1980, 18, 1685.
    b) Calundann, G. W.; Charbonneau, L. F.; Ease, A. J.
       U.S. Patents 4 339 375, 4 341 688, 4 351 917, 4
       351 918, 4 355 132. 1982
    c) McIntyre, J. E.; Milburn, A. H. Brit. Polym. J.
       1981, 13, 5.
5. a) Inata, H.; Morinaga, T. U.S. Patent 4 107 143,
       1978.
    b) Fayelle, B. EPO 15 856, 1980. U.S. Patent 4 284
       757, 1981.
6. Luise, R. R. U.S. Patent 4 183 895, 1980.
7. Lai, Y. C.; DeBona, B. T.; Prevorsek, D. C. J. Appl.
    Polym. Sci. 1980, 36, 805.
8. Lai, Y. C.; DeBona, B. T.; Prevorsek, D. C. ibid.
    1988, 36, 819.
9. Prevorsek, D. C.; DeBona. B. T.; Kesten, Y. J.
    Polym. Sci. -Chem, 1980, 18, 75.
10. Mark, H. F.; Gaylord, N. G.; Bikales, N. M. Ency.
    Polym. Sci. Tech. 1969, 10, 722.
11. Akkapeddi, M. K.; DeBona, B. T.; Lai, Y. C.;
    Prevorsek, D. C. U.S. Patent 4 398 018, 1983.

RECEIVED March 26, 1990

Chapter 9

Structures and Thermal Properties of Liquid-Crystalline Poly(ester-*co*-carbonate)

M. Kawabe, I. Yamaoka, and M. Kimura

Research and Development Laboratories—1, Nippon Steel Corporation, 1618 Ida, Nakahara-ku, Kawasaki 211, Japan

The thermotropic liquid crystalline copoly(estercarbonates) (PECs) were synthesized from p-hydroxybenzoic acid (HBA), 4,4'-dihydroxybiphenyl (DHBP), and diphenylcarbonate (DPC). Threaded textures, a characteristic of nematic phase, were observed over the investigated composition range. The liquid crystalline transition temperatures (LCTT) of PECs were found significantly low, compared with those of copolyesters based on biphenylene terephthalate and HBA. Density measurement revealed that the structural orders in injection-molded specimens were well developed. The degrees of preferred orientation were increased and the fibrous structures were significantly developed by heat treatment. These results suggested development in lateral order and overall perfection of the structure that occurred upon annealing.

Thermotropic liquid crystalline polymers have recently been a topic of scientific investigation (1-3). Because of their thermotropic liquid crystalline character from rigid structural moieties incorporated in the polymer backbone, these polymers possess outstanding melt processability and excellent mechanical properties.

Jackson and Kuhfuss reported a thermotropic polymers system comprised of poly(ethylene terephthalate) modified with HBA (4). They demonstrated not only the thermotropic liquid crystalline behavior but also the excellent melt processability of these materials to yield specimens with high modulus values. However these copolymers had very low use temperatures as reflected by low heat deflection temperatures.

Economy and co-workers reported the copolyesters based on biphenylene terephthalate and HBA (5-6). These polymers exhibited excellent mechanical properties up to 250-300℃, but these copolyesters had poor processability because of their high LCTTs.

The purpose of this research is to synthesize a series of thermotropic liquid crystalline PECs consisting of HBA, DHBP, and carbonyl units with good melt processability and high heat deflection temperatures. The carbonate linkage is expected to reduce LCTT without reduction of the chain rigidity and heat resistance.

The thermal properties and solid state structures of PECs are elucidated. The effects of heat treatment on the thermal properties and the structures are also discussed.

Experimental

Nomenclature

Copolymer compositions were expressed as a mole percentage of HBA moiety with the remainder being biphenylene carbonate units. For example, PEC having a 40 mole% of HBA moiety was coded as PEC-40HBA.

0097–6156/90/0435–0115$06.00/0

Polymer Preparation

PECs were prepared by monobutyltin oxide-catalyzed, melt transesterification of mixtures of HBA, DHBP, and DPC.

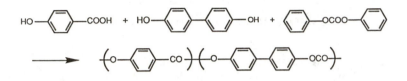

A mixture of the monomers plus catalyst was heated (200 °C) and phenol by-product was removed by atmospheric distillations. The stirred reaction mass was gradually heated to 300-320°C, and finally, the pressure was reduced below 1 Torr.

A series of thermotropic PECs having HBA and biphenylene carbonate units were synthesized in a range of HBA moiety varying from 40 to 80 mole%. Polymer compositions were based on the ratios of starting monomers. They were checked by the yields of the polymers.

The similar dilute solution viscosity data in Table I suggested that the differences of molecular weight among them were not large, so the comparisons of structures and thermal properties of these copolymers would be regarded as significant.

Table I.   Inherent Viscosities of PECs

Polymer	As-molded IV.[a] (dL/g)	Annealed Ta[b] (°C)	IV. (dL/g)
PEC-40HBA	2.77	290	insol.
PEC-50HBA	2.86	270	insol.
PEC-60HBA	3.07	230	insol.
PEC-70HBA	2.84	270	insol.
PEC-80HBA	3.74	290	insol.

a) Inherent viscosity ; 0.1g/dL, pentafluorophenol/chloroform =80/20(w/w).

b) Annealing temperature.

The copolymers were injection molded into miniature dumbbell specimens at the temperature above LCTT with a Custom Scientific Instruments (CSI) Mini-Max injection molding machine. The specimens for thermal treatment were annealed at 200°C for 4 hr and then at a few tens degree below LCTT for 40 hr under vacuum to develop crystalline order.

Measurement Methods

Inherent viscosities (IVs) were measured at 0.1 g/dl concentration in 80/20 pentafluorophenol/chloroform solvent at 30 °C.

Liquid crystal textures were observed by optical microscopy between crossed polars. The instrument used was an Olympus BH-2 polarizing microscope equipped with a Linkam TH-600RH hot stage.

The transition temperatures of the polymers were measured with a Mettler DSC-30 differential scanning calorimeter under a constant flow of nitrogen and a heating rate of 10 °C/min.

Dynamic mechanical analyses were performed on an Orientec Rheovibron DDV-III-EP. The storage modulus spectra were obtained from -150 °C to 250 °C at a heating rate of 2 °C/min and a frequency of 110 Hz.

Densities were measured using a density gradient column (toluene/carbon tetrachloride solvent).

Wide-angle X-ray scattering (WAXS) was done using a Rigaku-denki Geigerflex RAD-3B X-ray diffractometer with nickel-filtered CuKα radiation. The degree of orientation for the crystallites (H) was determined from azimuthal scans of the strongest equatorial reflection using the equation $H=100(1-\phi/180°)$, where ϕ is width (degrees) at half maximum of the peak on the intensity profile.

Microstructures were characterized using a JEOL JSM-T330A scanning electron microscope with an accelerating voltage of 30kV. The samples were fractured under liquid nitrogen after immersion for about five minutes. They were then mounted on aluminum stubs and sputtered with gold using an Eiko Engineering IB-3 sputter coater for enhanced conductivity.

Results and Discussion

Thermal Properties

Typical polarized optical micrographs of the molten samples were shown in Figure 1. Microscopic analyses were carried out at temperatures in the range 25 - 390 °C, and threaded textures, a characteristic optical texture of nematic mesophase, were observed above LCTTs over the investigated composition range. The nematic to isotropic transitions were not observed up to 390 °C.

Figure 2 indicated the influences of HBA content and annealing on the two types of transition temperatures, LCTT and Tg.

The incorporation of biphenylene carbonate moiety into poly(p-oxybenzoate) appeared to reduce the LCTTs in a quasi-eutectic manner, and the LCTT curve had a minimum at 60 mole% HBA content. Arrows on Figure 2 showed LCTTs of some commercial copolyesters based on biphenylene terephthalate and HBA. The LCTT of PEC-60HBA was significantly low, compared with those of the copolyesters. The introduction of flexible carbonate linkage would lower the LCTTs of PECs.

The Tg curve declined gently as HBA content increased because the introduction of biphenylene moiety and carbonate linkage would effectively increase Tg.

The LCTTs and the Tgs of annealed samples were higher than those of as-molded samples over the investigated composition range (40-80 mole% of HBA units).

To check the influences of annealing on the transition behaviors, the as-molded and the annealed specimens of PECs were reheated. Figure 3 showed typical DSC curves for these samples.

The heat treatment affected the thermal transition behaviors and the LCTTs of PECs. The as-molded sample of PEC-70HBA exhibited two transitions at 240 °C and 270 °C. After annealing, the endotherm at lower temperature disappeared; the endothermic peak at 270 °C was shifted to the higher temperature.

The effects of thermal treatment of PECs on Tgs were dynamic-mechanically examined. Figure 4 showed the storage modulus curves of the as-molded specimens and the annealed ones.

The degrees of the storage modulus drops at about 130 °C, assigned to the glass transition, were decreased by heat treatment. These diminutions of storage modulus drops and the increases of the storage modulus over the investigated temperature range suggested that some of less-ordered regions would be changed to ordered regions by annealing.

(a)

(b)

Figure 1. Polarized optical micrographs of the PECs: (a) PEC-60HBA 265 ℃ (magnification X200); (b) PEC-70HBA 300 ℃ (magnification X200).

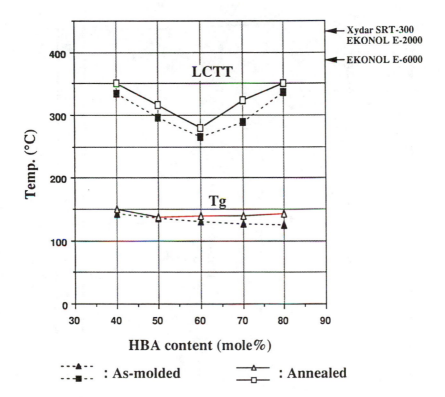

Figure 2. Effects of HBA content and annealing on transition temperatures.

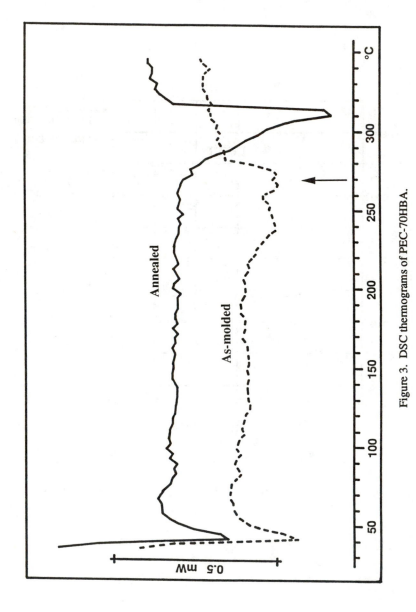

Figure 3. DSC thermograms of PEC-70HBA.

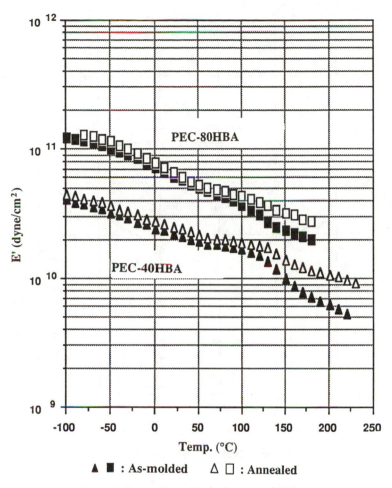

Figure 4. Dynamic mechanical spectra of PECs.

Density Measurement

If the densities of molten PECs were reflected by density-column measurement on the quenched samples, further evidence for the cause of the lowering of LCTT could be elucidated by varying the densities of the molten state. The density curves of the quenched samples in Figure 5 showed a similar pattern with the LCTT curve depicted in Figure 2 as HBA content of PEC was changed. These results would suggest that the lowering of LCTT on PECs was attributable to the reduction of density of the melt which would imply a reduction of the degree of molecular packing.

The large density differences between the as-molded specimens and the quenched ones were observed. On the other hand, the densities of the as-molded specimens are similar with those of the annealed ones. These results indicated that orders in as-molded specimens were well-developed, but quenched specimens were less ordered.

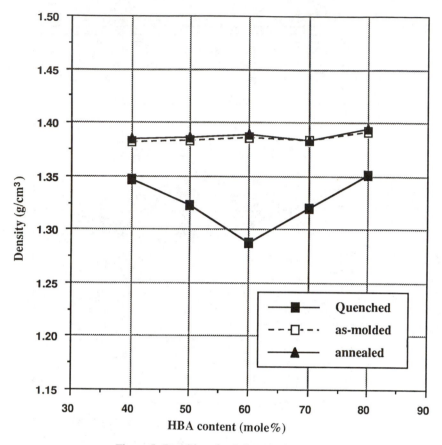

Figure 5. Densities of poly(estercarbonates).

Degree of Crystallite Orientation

The degrees of orientation for the crystallites in the injection-molded specimens of PECs were determined from the azimuthal breadth at half-maximum intensity in the strongest equatorial reflection by X-ray diffraction. Figure 6 showed the degrees of crystallite orientation of the as-molded and the annealed specimens.

It could be seen that the degrees of crystallite orientation increased by thermal treatment. This would result from a development in the overall order of crystallite along the flow direction.

SEM Analysis

SEM micrographs and visual appearances of the fracture surfaces revealed the presences of a hierarchical organization and microstructures in the macrolayers. Starting from the macroscopic level, five macrolayers were observed: two outer skins, two mid layers with a core in between. Due to differences in color, the macrolayers were readily visible to the naked eye. This skin-core morphology is a characteristic of many injection-molded LCPs (7-8).

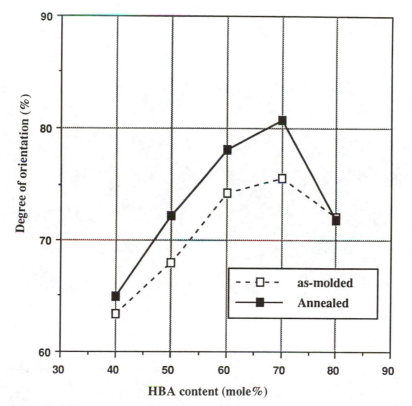

Figure 6. Degrees of orientation of PECs.

Figure 7 and 8 showed the typical fracture surfaces of the as-molded PEC-70HBA and PEC-40HBA by pulling along transverse directions respectively.

As indicated in Figure 7b and 8b, the outer skin layers were "fibrillar" microstructures in nature and were oriented parallel to the flow direction. The high orientation observed could be attributed to the elongational flows that seem to predominate near the surface of the mold. The skin layers were approximately 20-50 μm in thickness.

Observations of fracture surfaces in lower magnification (Figure 7a and 8a) suggested that these levels of organization showed a "mica-like" structure.

No microstructural differences between the mid and the core layers could be detected from comparison of the fracture surfaces (Figure 7c, d and 8c, d). The microstructures in these layers were better described as strips, and the average molecular orientations of the mid and the core layers were still parallel to the flow direction, in contrast with a relatively featureless core with little or no molecular orientation observed by other investigators (7). The variations in the layered structures and the skin core phenomena observed in PECs could be correlated to the process history, the shapes of the specimens, the injection molding parameters, and the molecular structures of the polymers.

The fracture surfaces of annealed samples were shown in Figure 9 and 10. Internal microstructures appeared more fibrous after annealing. These fibrous structures were observed not only in the outer skin layers but also in the mid and the core layers after annealing (Figure 9c, d and 10c, d). The aspect ratios of the macrofibrils in the mid and the core layers were significantly increased by thermal treatment. These might result from more lateral packing of the molecules in the macrofibril and perhaps from development of extended-chain crystals because of the increase of molecular orientation by annealing (Figure 6).

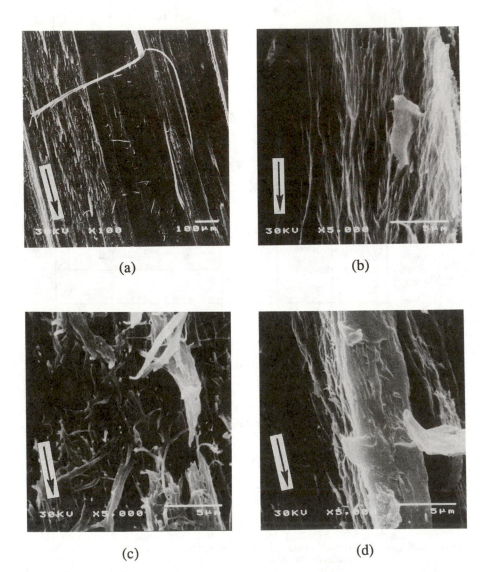

(a)   (b)

(c)   (d)

Figure 7. SEM micrographs facing fracture surfaces of PEC-70HBA as-molded: (a) whole surface (×100); (b) top skin layer (×5000); (c) mid layer (×5000); (d) core layer (×5000).

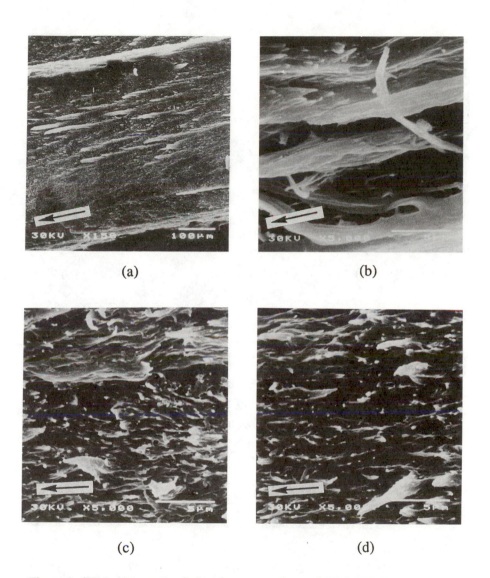

(a)

(b)

(c)

(d)

Figure 8. SEM micrographs facing fracture surfaces of PEC-40HBA as-molded: (a) whole surface (×150); (b) top skin layer (×5000); (c) mid layer (×5000); (d) core layer (×5000).

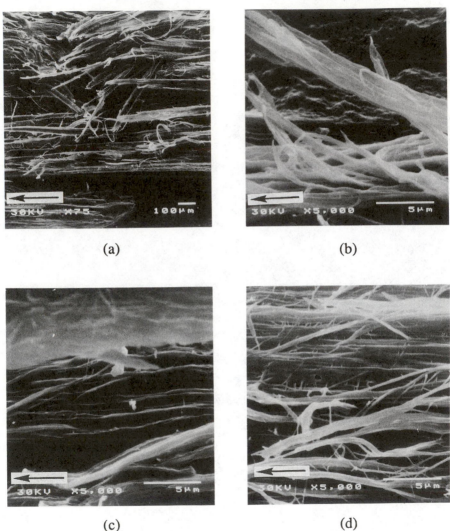

(a)   (b)

(c)   (d)

Figure 9. SEM micrographs facing fracture surfaces of PEC-70HBA annealed: (a) whole surface (×75); (b) top skin layer (×5000); (c) mid layer (×5000); (d) core layer (×5000).

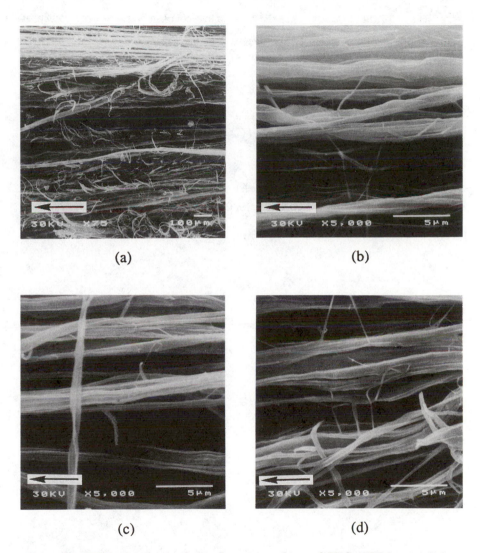

(a)

(b)

(c)

(d)

Figure 10. SEM micrographs facing fracture surfaces of PEC-40HBA annealed: (a) whole surface (×75); (b) top skin layer (×5000); (c) mid layer (×5000); (d) core layer (×5000).

Conclusions

(1)   The thermotropic liquid crystalline PECs could be synthesized from HBA, DHBP, DPC. Threaded textures of PECs, a characteristic of the nematic phase, were observed over the investigated composition range.
(2)   The LCTTs of PECs were found significantly low, compared with those of commercial copolyesters based on biphenylene terephthalate and HBA.
(3)   Microstructures in as-molded specimens were found well ordered.
(4)   The degrees of preferred orientation were increased, and fibrous structures were significantly developed by heat treatment.

Literature Cited

1) Krigbaum, W.R. Polymer Liquid Crystals; Academic Press; New York,1982.
2) Blumstein,A. Polymeric Liquid Crystals; Plenum Press; New York, 1985.
3) Chapoy,L.L. Recent Advances in Liquid Crystalline Polymers; Elsevier Appl.Sci.Publ.; New York, 1985.
4) Jackson,Jr.,W.J.; Kuhfuss,H.F. J.Polym.Sci.,Polym.Chem.Ed. 1976, 14, 2043.
5) Economy,J. U.S.Patent. 3 637 595, 1972; 3 980 749, 1976.
6) Economy,J. et al. J.Polym.Sci.,Polym.Chem.Ed. 1983, 21, 2249.
7) Sawyer,L.C.;Jaffe,M. J.Mater.Sci. 1986, 21, 1897.
8) Weng,T.; Hilter,A.;Baer,E. J.Mater.Sci. 1986, 21, 744.

RECEIVED March 26, 1990

Chapter 10

Synthesis and Microstructure of Aromatic Copolyesters

J. Economy[1], R. D. Johnson[2], J. R. Lyerla[2], and A. Mühlebach[3]

[1]University of Illinois, Urbana, IL 61801
[2]IBM Research Division, Almaden Research Center, 650 Harry Road, San Jose, CA 95120–6099
[3]CIBA-Geigy AG, Forchungszentrum, 180.053, CH–1701, Fribourg, Switzerland

A comprehensive interpretation of the microstructure of the liquid crystalline aromatic copolyesters is presented. The role of the synthetic route and of high temperature processing on the microstructure are clearly defined. As a result of this study a predictive model now exists which permits interpretation of the very subtle chemical processes which can occur at elevated temperatures leading to either randomization or ordering of the microstructure.

The preparation of the homopolymers of p-hydroxybenzoic acid (HBA) and of 2-hydroxy-6-naphthoic acid (HNA) have been described in the literature (1,2) and both have been shown to polymerize as single crystals. However, reports on the synthesis of copolyesters of HBA with either HNA, polyethylene terephthalate (PET), or biphenol terephthalate (BPT) are not well documented in the published literature. Considering the commercial importance of these melt-processible copolyesters of PHBA it is surprising how little progress has been reported on relating the synthesis to the microstructure of these copolymers. Some of the problems associated with the study of these systems arise from their low solubility in most solvents, which greatly limits use of techniques aimed at relating microstructure to the synthetic pathway. An additional complicating factor is the potential for transesterification reactions during polymerization and subsequent processing which may also influence the microstructure. In fact, there appears to be some confusion in the case of liquid crystalline polyesters as to whether they are stable in the nematic melt (3), randomize(4), or undergo crystallization induced ordering (5).

In this paper, some of our recent studies in this area are combined to provide a reasonably comprehensive picture of the reactions which occur during polymerization, and subsequent processing(2,6,7,8). The HBA/HNA 50/50 copolymer system was selected for this particular study because of its solubility and relative ease of synthesis.

Transesterification Processes during Polymerization

The HBA/HNA system is perhaps the simplest system to study among the various commercial systems, since the monomers can either be polymerized in solution or in the melt to low molecular weight oligomers which can be further characterized by NMR techniques. In contrast, preparation of the HBA-PET system proceeds via the transesterification of a melt of PET and the acetoxy benzoic acid monomer(9). The complexity of this reaction has been described recently in a study(10) showing significant variations in composition and microstructure of the 60/40 PHBA/PET copolyester. In the case of the HBA-BPT polymerization, the initial reaction is heterogeneous suggesting that the oligomers would tend to be blocky(11)

In our study of the HBA-HNA copolymer we chose to carry out the reaction in an inert high boiling hydrocarbon solvent (Therminol 66). The degree of reaction could be followed directly by measuring the evolution of acetic acid. As a starting point, we examined the kinetics of homopolymerization of the two monomers i.e. p-acetoxy benzoic acid and 6-acetoxy-2- naphthoic acid. There already were published data(12) on the kinetics of polymerization of HBA showing a zero order rate process, as one might predict for a polymerization which is carried out almost completely in the solid state. We were surprised to find that the HNA behaved almost identically, precipitating early in the polymerization at a DP ~ 4-5 in the form of single crystals, so that most of the reaction was carried out in the solid state(2). As shown in Figure 1, the HNA polymerizes in the solid state two times faster than the HBA suggesting that the acetate end group in the growing HNA polymer is better positioned to react and or to diffuse out through the openings in the structure. On the other hand, the rate of conversion of the 1/1 HBA-HNA mixture proceeds in solution at a rate slightly lower than that for the HBA. A plot of 1/monomer concentration (shown in Figure 2) indicates that the kinetics for copolymerization are best described as 2nd order kinetics.

A study of the reactivity ratios of the two comonomers was also undertaken; however, analysis of the oligomer with DP < 6 was complicated by solubility problems which made it difficult to isolate samples representative of the actual composition at these low DP's. To get around this problem the polymerization was run to 90% completion (based on acetic acid). C^{13} NMR analysis of the copolymer showed the same ratio of monomers as in the starting mixture indicating in the absence of transesterification that the reactivity ratios were the same. End group analysis showed an M_n value of 2000. In addition, analysis of the diad sequences in this polymer showed a distribution of the four possible diads identical to what one might predict for a random copolymer.

The above conclusion on the role of reactivity ratios on microstructure assumes the absence of rapid transesterification reactions between chains. Since such processes might also tend to randomize the microstructure, it seemed important to isolate the role of interchain transesterification. A unique experiment was designed in which a 13_C labeled carbonyl in acetoxy benzoic acid monomer (B*) was reacted with the dimer of HBA-HNA. At 99% ^{13}C enrichment, the only resonances in the carbonyl region of the spectrum will arise from the enriched benzoic acid carbonyl. In the absence of any interchain transesterification the microstructure of the polymer would consist only of B*-B dyads (see scheme 1). On the other hand presence of B*-N dyad in the polymer could only arise from a transesterification reaction where the B* was inserted between the -BN-. As shown in Figure 3, ^{13}C NMR displays a small but distinct peak (ca.14%) at the resonance position corresponding to B*-N diad. Since the polymerization was run at the same temperature of 245°C for170 min and to the same degree of polymerization (M_n~2252) as the earlier experiment on the reactivity ratios, one can conclude that the role of interchain transesterification is relatively small and that the monomers have approximately equivalent reactivity ratios. The reaction of B* with the HBA-HNA dimer was also examined at 225° and 285°C to determine the possibility of a temperature effect. The times of reaction were 20 hrs and one hour, respectively,

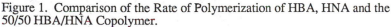

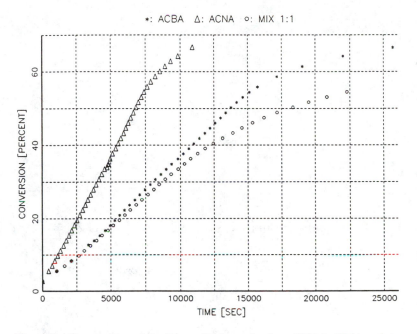

Figure 1. Comparison of the Rate of Polymerization of HBA, HNA and the 50/50 HBA/HNA Copolymer.

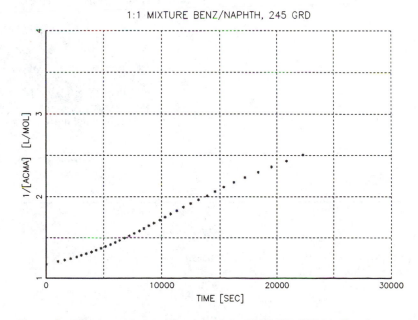

Figure 2. Kinetics of Polymerization of the 50/50 HBA/HNA Copolymers.

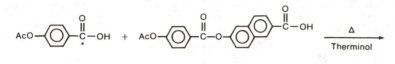

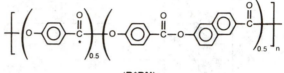

(B*BN)$_n$

Scheme 1. Partially Ordered B* + BN.

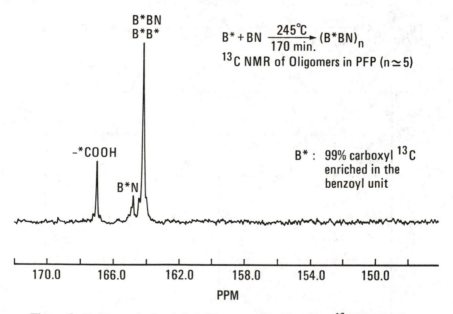

Figure 3. Evidence for Interchain Transesterification from ^{13}C NMR using 99% Carboxyl ^{13}C Enriched Benzoyl.

and were designed to arrive at the same degree of polymerization as the 245° experiment. It was found that the amount of B*-N diad was the same in all three cases. This should not be unexpected if the rate of transesterification is determined by the concentration of carboxylic acid end groups, which would be the same in all three cases.

Some additional insights into the nature of the p-HBA polymerization reaction were obtained by monitoring with IR the change in acetoxy and carboxylic acid end groups. As can be seen in Figure 4, the acetoxy and carboxylic acid peak disappear at approximately the same rate although the growing polyester peak masks the acetoxy peak after 10 minutes. No other peaks such as anhydrides could be detected in the IR spectrum indicating that the mechanism of polymerization proceeds via direct reaction of the carboxylic acid with the ester.

Transesterification during Annealing

In an earlier study, Butzbach et al(13) had reported a sharp increase in crystallinity during extended annealing of the PHBA/HNA(58/42). In our work a sample of the 50/50(HBA/HNA) copolymer was annealed at 210°C for 24 hours and examined by DSC and NMR. As can be seen in figure 5, the DSC thermogram showed an approximate 4x increase in transition enthalpy indicating an increased crystallinity. However, there was no detectable change in the diad ratios in the 13C-NMR spectrum (see Figure 6). One can conclude from these results that the increased ordering in the DSC was primarily the result of slow chain diffusion to provide a higher concentration of the crystalline ordering present prior to annealing. On the other hand, a significant increase in M_n was observed during annealing, i.e., from 10,000 to 30,000. Hence some transesterification reactions are occurring in the solid state but are limited to the end groups. It should be noted that although the 50/50 copolymer is random, because of the rod-like nature of the chains there is a tendency for segregation of like sequences into poorly formed three dimensional crystallites(14). The fact that annealing results in an increase in the transition enthalpy but with no upward shift in temperature is additional evidence along with the 13C NMR dyad analysis that the microstructure has not changed. Comparison of the X-Ray diffraction patterns before and after annealing show practically no change in the relatively broad diffuse pattern.

Transesterification at $T_{c \leftrightarrow n}$

A sample of the 50/50 HBA/HNA copolyester was heated for 16 hours at 250°C. DSC analysis showed a completely altered spectrum with the peak at 250°C completely disappearing and being replaced by three new peaks, namely at 240°C, 320°C and 370°C (see Figure 7). In addition, the sample was completely insoluble in pentafluoro phenol (PFP). End group analysis showed the polymer to have increased in M_N >30,000 which could explain the insolubility. Hence the heating at 250°C was repeated for 72 hrs. in the presence of a capping agent, 2-naphthoic acid, to control the M_N at about 7,000. Again the three new peaks were observed but in this case the sample swelled 20 X in PFP. This experiment was repeated with even more 2-naphthoic acid to bring the M_N down to 3,800. The DSC showed the same three new peaks but now the sample was soluble in PFP. Analysis by ^{13}C-NMR showed a spectrum typical of the random diad distribution of the 50/50 copolyester.

Others have interpreted this behavior in this class of copolyesters as a possible physical ordering process(15,16). A different interpretation and one more

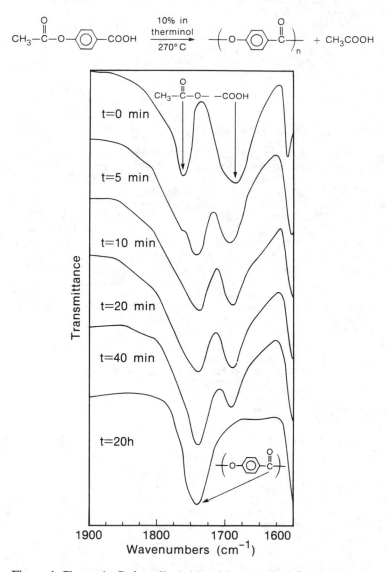

Figure 4. Change in Carboxylic Acid and Acetoxy End Group in the Polymerization of p-Acetoxybenzoic Acid.

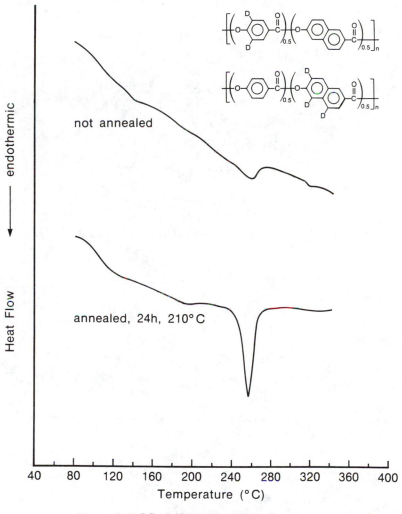

Figure 5. DSC of 48/52 HBA/HNA Copolymer.

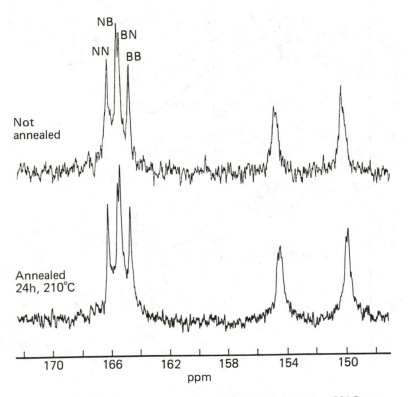

Figure 6. ^{13}C NMR of HBA/HNA (48/52) in PFP at 80°C.

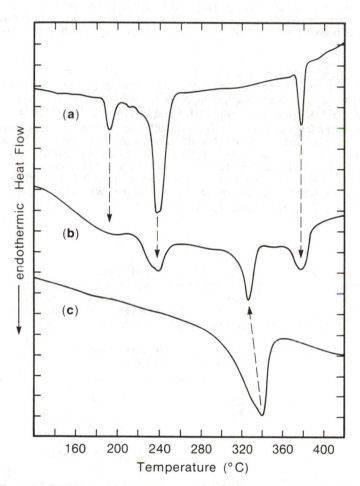

Figure 7. Comparison of DSC Spectra: a) HBA/HNA (50/50) Copolymer Annealed for 16 hrs at 250°C, b) Single Crystals of the Alternating (HBA-HNA) (50/50) Copolymer, c) Homopolymer of PHBA or PHNA.

consistent with the observed data is that interchain transesterification has occurred in the polycrystalline regions to produce the more highly ordered structures of the alternating HBA/HNA copolymer and of the HBA and HNA homopolymers. In fact, the new endotherms at 240°C, 320°C and 370°C are typical of these structures. The insolubility in PFP undoubtedly is due to the better developed crystallinity in the reordered domains. The 13_C NMR data which show a random sequence distribution for the low M_N copolymer can be explained by the fact that the majority of the sample is indeed random and the ordering which has occurred in the original polycrystalline regions (< 10-15%) is beyond the limits of detection by 13_C NMR.

Transesterification Above $T_{c \leftrightarrow n}$
To test the potential for transesterification in this temperature range, an experiment was designed to heat under compression molding conditions a 50/50 mixture of the relatively low M_N homopolymers of PHBA and PHNA at 460°C[8]. This temperature is a few degrees above $T_{c \leftrightarrow n}$ of both homopolymers. Surprisingly the entire sample extruded into the annulus within less than a minute after a pressure of 2000 psi was applied. Analysis by 13_C NMR, DSC and X-ray diffraction proved conclusively that the two homopolymers had undergone a very rapid transesterification to the random 50/50 copolyester of PHBA/HNA (see figure 8). This extremely high rate of ester interchange is consistent with what one would calculate using the known kinetics for transesterification reactions in PET at 275°C-300°C[17].

General Interpretation

The picture that emerges from the results of this work indicates that both the microstructure and resultant morphology are dependent on subtle chemical and physical processes that to a considerable degree are determined by the thermal history of the material. As illustrated in Table I one can clearly distinguish the chemical and physical processes which are operative in these systems.

Table I. Chemical vs Physical Processes
in the LC Polyesters

Temperature	Dominant Processes	Changes in Microstructure or Morphology
$T_{c \leftrightarrow n}$	Chemical	Randomization
$T_{c \leftrightarrow n}$	Physical	Crystallization
Annealing (near $T_{c \leftrightarrow n}$)	Chemical	Crystal Ordering
Annealing (below $T_{c \leftrightarrow n}$)	Physical	Further Crystallization
~T_{ng}	Physical	Nematic Glass

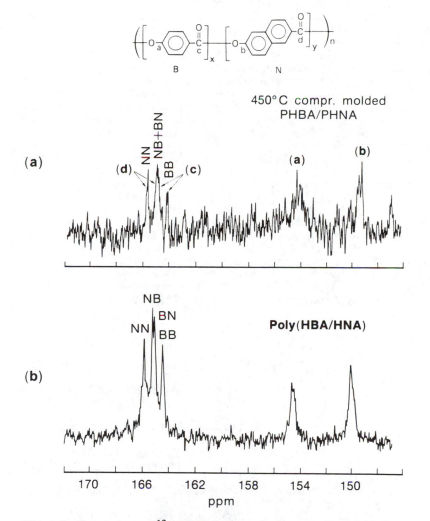

Figure 8. Comparison of ^{13}C NMR Spectra of a Compression Molded Mixture of the Homopolymers of PHBA + PHNA with that of The Random HBA/HNA (50/50) Copolymer.

Thus at temperatures well above $T_{c \leftrightarrow n}$ chemical processes dominate. An extrapolation from ester interchange kinetics at temperatures of 290°C to 450°C would indicate interchain transesterification with approximately 1000 interchanges/chain/10 seconds(8). At the same time M_N increases rapidly resulting in a sharp reduction in carboxylic acid end group concentration and thus reducing the potential incidence of these reactions. The propensity for randomization in the nematic phase will continue with decreasing temperature even below $T_{c \leftrightarrow n}$ but with correspondingly slower rates until the nematic phase freezes into a glass, e.g. in the HBA-HNA system the nematic glass transitions range from 80-130°C depending on composition.

As one goes below the crystal-nematic transition physical processes dominate and the rod-like random chains match up to produce relatively small concentrations of defective three dimensional crystallites. If one anneals near the $T_{c \leftrightarrow n}$ chemical processes dominate and the relatively poorly ordered structures undergo interchain transesterification to yield bimodal domains of highly ordered sequences. As might be expected further polymerization is observed during these annealing experiments. The fact that the DSC scan shows endotherms almost identical to the scans observed with single crystals of the respective homopolymers and of the alternating copolymer indicate that a sufficient degree of aggregation of the respective -B-B-B-, -N-N-N-, and -B-N-B-N- units has occurred to a level of order comparable to that of the pure materials. The driving force to these processes is kinetic and arises from the stronger dipole interactions and higher densities achieved with the improved packing. It is likely that the carboxylic acid concentration is higher at the interface between the nematic phase and the poorly formed crystalline phase thus facilitating the process.

Finally, if we anneal at 50-70°C below the $T_{c \leftrightarrow n}$ physical diffusion processes dominate to yield a significant increase in concentration of the defective crystallites along with increased molecular weight. The potential for "chemical" ordering is sharply reduced at these lower temperatures because of the greatly reduced mobility of end groups within the polycrystalline phase.

Literature Cited

1.   Economy, J.; Storm, R.S.; Matkovich, V.I.; Cottis, S.G. Polym. Sci. Polym. Chem. Ed. 1976, 14, 2207.
2.   Mühlebach, A.; Lyerla, J.; Economy, J. Macromolecules 1989, 22, 3741.
3.   De Meuse, M.T.; Jaffe, M. Mol. Cryst. Liq. Cryst. Inc. Nonlin. Opt. 1988, 157, 535.
4.   Mühlebach, A.; Johnson, R.; Lyerla, J.; Economy, J. Macromolecules 1988, 21, 3115.
5.   Lenz, R.W.; Jin, J-II; Feichtinger, K.A. Polymer 1983, 24, 327.
6.   Mühlebach, A.; Johnson, R.D.; Lyerla, J.; Economy, J. Macromolecules 1988, 21, 3115.
7.   Economy., J.; Volksen, W.; Viney, C.; Geiss, R.; Siemens, R.; Karis, T. Macromolecules 1988, 21, 2777.
8.   Mühlebach, A.; Economy, J.; Johnson, R.; Karis, T.; Lyerla, J. Macromolecules, in Press.
9.   Jackson, W.J. Jr.; Kuhfuss, H.F. J. Polym. Sci. 1976, Polym. Chem. Ed., 14, 2043.
10.   Quack, L,; Hornbogen, E.; Volksen, W.; Economy, J. J. Pol. Chem. 1989, Part A, 27, 775.

11.   Johnson, R.D.; Economy, J.; Lyerla, J.; Mühlebach, A. <u>Amer. Phys. Soc.</u> 1989, St. Louis, MO.
12.   Kricheldorf, H.R.; Schwarz, G. <u>Makromol. Chem.</u> 1983, <u>184</u>, 475.
13.   Butzbach, G.D.; Wendorff, J.H.; Zimmerman, H.J. <u>Polymer</u> 1986, <u>27</u>, 1337.
14.   Hanna, S.; Windle, A.H. <u>Polymers</u> 1988, <u>29</u>, 207.
15.   Lin, Y.G.; Winter, H.H. <u>Macromolecules</u> 1988, <u>21</u>, 2439.
16.   Cheng, S.Z.D. <u>Macromolecules</u> 1988, <u>21</u>, 2475.
17.   Kugler, J.; Gilmer, J.W.; Wiswe, D.W.; Zachmann, H.G.; Hahn, K.; Fischer, E.W. <u>Macromolecules</u> 1987, <u>20</u>, 1116.

RECEIVED April 16, 1990

SYNTHESIS OF SIDE-CHAIN LIQUID-CRYSTALLINE POLYMERS

Chapter 11

Synthesis, Characterization, and Photochemistry of a Cinnamate-Containing Liquid-Crystalline Side-Chain Polymer

John M. Noonan and A. F. Caccamo

Photographic Research Laboratories, Eastman Kodak Company, Rochester, NY 14650–2109

Novel vinyl liquid crystalline (l.c.) polymers were synthesized with the UV-sensitive p-methoxycinnamate chromophore incorporated into the side chain of the polymers. The objective of this synthesis was to determine if a molecularly organized environment could influence the yield of a chemical reaction in the solid state. The investigation into the photochemical and physical processes of these thin films revealed that the photodimerization of the p-methoxycinnamate moieties was very sensitive to their geometrical arrangement in the polymer matrix. The relative quantum yield of cyclobutane formation increased by a factor of approximately 8 for the l.c. p-methoxycinnamate film compared to its amorphous analog. This quantum yield approaches the theoretical limit for this system.

Dramatic synergistic effects in certain chemical reactions have taken place in organized environments, e.g., the crystalline state (1), monolayer assemblies (2), and bilayer assemblies (3). The objective of this research investigation was to determine if synergistic effects in photochemical reactions could be observed in the macromolecular organization of thermotropic-sidechain-l.c. polymers. The chemical reaction chosen for study was the photodimerization of cinnamate moieties contained in the sidechains of certain l.c. polymers. Because the dimerization requires parallel alignment of the double bonds within approximately 4 Å before cyclobutane ring formation will occur and because the lifetime of the triplet excited state is only 10 nanoseconds, the overall quantum efficiency for cyclobutane ring formation in amorphous polymers containing the cinnamate group was only about 4% (4). Therefore, we felt that in a molecularly organized environment, the efficiency of dimerization could be dramatically improved.

Experimental

General Procedure for Vinyl Polymerizations: All polymers (except X) were generated from their corresponding methacrylate monomers by solubilization in solvents to a particular total solids concentration as indicated, and initiated with 1% AIBN (by weight), unless otherwise indicated. The solutions were placed in polymerization vials, septa sealed and purged with N_2 for 20 minutes via syringe needles while cooling to eliminate evaporation. The flasks were then immersed in constant temperature baths (CTB) at the prescribed temperatures for 15 hours (overnight). The viscous solutions were diluted with their respective solvents and precipitated by adding dropwise into the indicated solvents with vigorous stirring. Solids were collected by vacuum filtration and dried at room temperature under oil pump vacuum.

0097–6156/90/0435–0144$06.00/0
© 1990 American Chemical Society

Polymer II: 20% solids in dichloroethane; 70°C CTB; ppt. into MeOH I.V. = 0.23; $T_g > 25°C$.

Polymer IV: 20% solids in dichloromethane; 70°C CTB; ppt. into MeOH; I.V. = 0.40; $T_g = 5°C$.

Polymer VIIa: 10% solids in dichloroethane; 60°C CTB; I.V. = 0.20; $T_g = 30°C$, l.c. 59°C. Elemental analysis: $C_{25}H_{36}O_5$, mol. wt. of repeat unit 416.5. Calcd: C 72.1, H 8.7, O 19.2. Found: C 70.8, H 8.6.

Polymer VIIb: 10% solids in dichloroethane; 60°C CTB; I.V. = 0.29; m.p. 10°C. Elemental analysis: $C_{29}H_{44}O_5$, mol. wt of repeat unit 472.6. Calcd: C 73.7, H 9.4, O 16.9. Found: C 71.3, H 8.9, O 16.2.

Polymer	I.V.	%Solids DCE	% Initiator	CTB °C	Ppt Solvent
IX-1	0.31	10	1.0	60	Et$_2$O
IX-2	0.46	15	0.5	55	Et$_2$O
IX-3	0.79	20	0.5	55	Et$_2$O
IX-4	0.84	20	0.5	50	Et$_2$O
IX-5	1.61	20	0.5	40	Et$_2$O

IX-1: T_g 21°n168°i. Elemental analysis: $C_{27}H_{30}O_{71}$, mol. wt. repeat unit 466.5. Calcd: C 69.5, H 6.5, O 24.0. Found: C 68.7, H 6.5, O 23.6.

Polymer XI: 10% solids in dichloroethane; 60°C CTB; ppt. into MeOH; I.V. = 0.31; T_g 95°C. Elemental analysis: $C_{88}H_{115}O_{30}$, mol. wt. 1653. Calcd: C 63.9, H 7.0, O 29.0. Found: C 63.5, H 6.9, O 29.5.

Polymer X: Elvanol 7130-M (100% hydrolyzed PVA) was dried in a vacuum oven at 37°C and at 8 mm Hg for 24 hours. The dried PVA was placed in molecular sieve-dried-pyridine (5% wt/vol) and heated at 70°C for 6 hours. A mixture of 0.35 mole equivalents of p-methoxycinnamoyl chloride, and 0.55 mole equivalents of benzoyl chloride was added to the swollen PVA gel in two portions. First, half the mixture was added and the reaction was stirred at 70°C for 12 hours followed by addition of the other half. After the addition of the acid chloride was complete the reaction was stirred an additional 12 hours. The reaction solution was cooled and the precipitated pyridine hydrochloride was collected by filtration. The clear pyridine solution was slowly poured into distilled water under vigorous agitation and the precipitated polymer was collected and dried. Polymer X displayed a $T_g = 92°C$ and an I.V. = 1.15 (phenol/chlorobenzene).

The Synthesis of Cinnamate-Containing l.c. Polymers

It was well known that certain low molecular weight cinnamate esters displayed l.c. properties (5). Thus, the p-methoxycinnamate group was attached to a methacrylate via —(—CH$_2$—CH$_2$—)— spacer group in an attempt to obtain a l.c. monomer. Reaction Scheme I describes the synthesis.

Monomer 2 was not l.c. However, this is not a prerequisite for obtaining l.c. polymers (6,7). Monomer 2 was homopolymerized, but the resulting polymer II did not display a l.c. texture. Apparently the (—CH$_2$—CH$_2$—) spacer was not long enough to sufficiently decouple the motions of the main chain from those of the side chains. As a result, a monomer with a flexible spacer group of $(-CH_2-)_6$ was synthesized as described in Scheme II. This monomer did not display a l.c. mesophase. Surprisingly, the homopolymer IV derived from monomer *4* was not l.c.

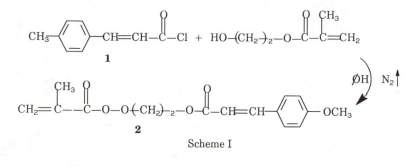

Scheme I

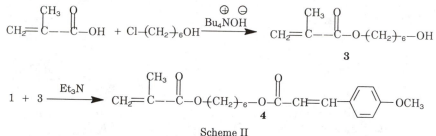

Scheme II

The p-methoxycinnamate group in monomer *4* was replaced with the hexyloxycinnamate and decyloxycinnamate groups, respectively, in an attempt to increase the hydrophobic interactions of the sidechains. The synthesis of these monomers is described in reaction Scheme III. The monomers *7-a* and *7-b* were not l.c. They were each homopolymerized yielding polymers VII-a and VII-b, respectively. Both polymers were stiff viscous glasses with T_g's below room temperature. The DSC scans of polymer VII-a showed a T_g of -30°C and Ti of 59°C with a l.c. texture thought by microscopy to be smectic. Polymer VII-b had a crystalline melting point at 10°C by DSC analysis, but showed no l.c. textures. Photochemical reactions were not initiated on polymer VII-a because characterization of the l.c. texture was not unequivocal and because the films of this polymer were extremely tacky due to its low T_g.

The aspect ratio of the p-methoxycinnamate moiety was extended by synthesizing the phenyl ester of p-methoxycinnamate (see Scheme IV). Microscopy studies did not reveal any l.c. mesophase in monomer *9*. However, the resulting homopolymer IX possessed l.c. order. Microscopy observations on polymer IX revealed a very stable homogeneous nematic texture. The DSC results showed a T_g at 21°C and a nematic to isotropic transition at 168°C.

Photochemistry of the Amorphous p-Methoxy Cinnamate vs the l.c. p-Methoxy Cinnamate Polymeric Films

The photosensitized crosslinking of amorphous polymeric films containing dimerizable cinnamate moieties show quantum yields of dimerization of about 0.04 (4). Sensitization of such films by ketocoumarins proceeds via triplet-triplet energy transfer (9) (see Scheme V). The irradiations were performed at 405 nm at a wavelength where the cinnamate chromophore did not absorb, but at a wavelength where the sensitizer did absorb.

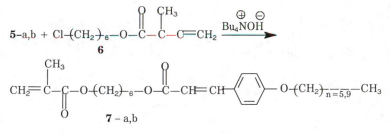

5 a,b

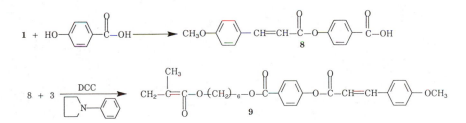

7 – a,b

Scheme III

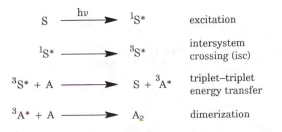

Scheme IV

$$S \xrightarrow{h\nu} {}^1S^* \qquad \text{excitation}$$

$${}^1S^* \longrightarrow {}^3S^* \qquad \text{intersystem crossing (isc)}$$

$${}^3S^* + A \longrightarrow S + {}^3A^* \qquad \text{triplet–triplet energy transfer}$$

$${}^3A^* + A \longrightarrow A_2 \qquad \text{dimerization}$$

Scheme V

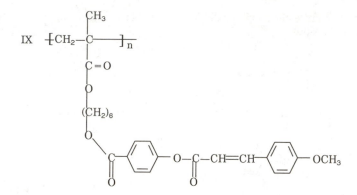

where S = sensitizer; $^1S^*$ = singlet-excited state sensitizer; $^3S^*$ = triplet-excited state sensitizer; A = cinnamate moiety; $^3A^*$ = triplet-excited state cinnamate; A_2 = dimer of A.

There are several energy wastage steps competing with those outlined above and Samir Farid has treated these processes very rigorously (10). Consider Scheme VI. In the case of the amphorous cinnamate-containing polymers the triplet energy migrates stepwise from one chromophore to the next until it decays, or reaches an "appropriate" site. An "appropriate" site has the double bonds of the cinnamate moiety aligned parallel to each other, and approximately 4 Å apart. It visits an average 10 sites before it finds an "appropriate" site (10). Since the lifetime of the triplet excited state is only 10 nanoseconds (10) decay to the ground state dominates over excimer formation.

How might the highly ordered l.c. p-methoxycinnamate polymer affect the energy transfer and migration processes and in turn increase the efficiency of crosslinking? If the chromophores are placed in an ordered micro-domain, e.g., a l.c. polymer matrix, far more "appropriate" reaction sites should exist. Consequently, under the very best conditions, within the lifetime of the triplet excited state, each triplet excited chromophore should find an "appropriate" site for excimer formation. As a result, by eliminating the decay to ground state associated with energy migration, the efficiency of dimerization could theoretically increase from approximately 5% (p-methoxycinnamate) to 50% (see Scheme VI).

Model Polymer Compositions for Photochemical Investigations

The compositions of the polymers for the photochemical investigations were adjusted so that each polymer contained 2.1 molal in the p-methoxycinnamate moiety. The following polymers were used:

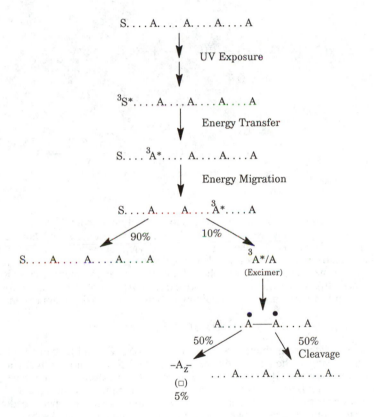

Scheme VI

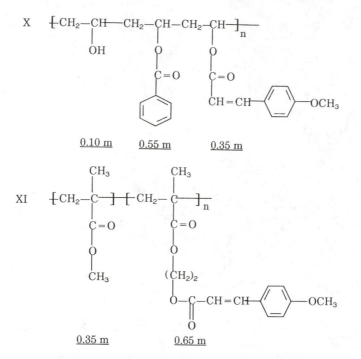

Copolymers X and XI were random-amorphous copolymers with T_g's of 92°C and 95°C, respectively. They were soluble in chlorinated solvents, cyclohexanone, and DMF. It should be noted that the relative quantum yield of cyclobutane formation in the solid state (films) for the polyvinylalcohol (PVA)-p-methoxycinnamate film was about 25% greater than for the amorphous PVA-cinnamate polymer film previously investigated (10). Consequently, the relative quantum yield of dimerization for an amorphous p-methoxycinnamate containing polymer film would be approximately 5%.

Solution Spectroscopy of the Model Polymers

The solution (dichloromethane) spectra of polymers X, XI , and IX can be seen in Fig. 1. Surprisingly, the absorption spectrum of the l.c. p-methoxycinnamate polymer, IX, showed a 10 nm red shift relative to the two non-l.c.-cinnamate polymers, X and XI. We suggest that, in the thermodynamically most stable conformation the p-methoxycinnamate moiety was intramolecularly perturbed by the phenyl ester group. The ester, phenyl p-methoxycinnamate also shows this perturbation, but the alkyl and cycloalkyl esters do not. CPK (Corey-Pauling-Koltun) molecular models sow that the phenyl ester group could assume an orientation that was almost coplanar with the cinnamate moiety which could easily give rise to a 10 nm red shift (see Fig. 2).

Film Spectroscopy. Each polymer (IX, X, XI) was dissolved in cyclohexanone at 3% by weight, and the sensitizer, 3,3'-carbonylbis(5,7-dipropyloxy coumarin) concentration was 3% by weight of the polymer. The three polymer solutions were each cast onto a quartz substrate via spin coating. The thickness of the films were adjusted until an optical density (O.D.) of between 2.0 and 3.0 was obtained at a wavelength of 300 nm. At this film thickness the concentration of the sensitizer was approximately 3×10^{-3} M. Figure 3 shows the absorption characteristics of the *non-*

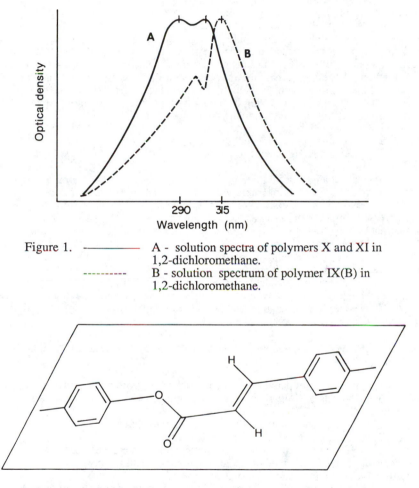

Figure 1. ——————— A - solution spectra of polymers X and XI in 1,2-dichloromethane.

---------- B - solution spectrum of polymer IX(B) in 1,2-dichloromethane.

Figure 2. Schematic representation of the planar conformation of the p-methoxycinnamate moiety.

l.c. p-methoxycinnamate polymer films X and XI before and after UV irradiation. Irradiations were performed on an optical bench with a PEK 200 watt super high pressure mercury light source. The appropriate filters were used ($\lambda \geq 405$ nm) to insure that only the sensitizer absorbed the incident radiation.

As expected, UV irradiation caused a proportionate decrease in the original absorption curves at all wavelengths. However, the film spectrum of the *l.c.*-cinnamate polymer, IX, unexpectedly was different than the solution spectrum, and in addition, UV irradiation of the film did not show a proportionate decrease in the original spectrum at all wavelengths. In fact, there was a large decrease in the absorption band at 290 nm and a slight increase at 315 nm (see Fig. 4). We suggest that the thermodynamically most stable polymer conformation was not achieved in the film-forming process. Instead, the phenyl ester group has been forced into a non-coplanar conformation which does not perturb the π cloud of the p-methoxycinnamate chromophore. We suggest that this was probably due to the constraints that the packing of the l.c. side chains had put on the phenyl ester group.

The spectral changes during UV irradiation of the *l.c.* p-methoxycinnamate film could be attributed to the following three processes:

1.   isomerization
2.   conformational changes
3.   cyclobutane formation

See illustration in Fig. 5.

Spectral changes of all three processes take place in the 290 nm to 310 nm region, and each process would disrupt the order in the side chains of the l.c. cinnamate polymer enormously. We feel that this disruption would allow the residual p-methoxycinnamate groups to assume their thermodynamically most stable conformation which would account for an increase in the O.D. in the 315 nm region. (Upon heating the films above their T_g's a proportionate decrease in the original spectra at all wavelengths took place. Apparently, the side chains reoriented themselves in a way approaching a perpendicular position to the plane of the film.) Since all three processes cause changes between 250-350 nm, the relative quantum yield could not be determined by monitoring the disappearance of the cinnamate double bonds. Consequently, the relative quantum yield was determined on *films* of the polymers using a photographic procedure.

Photographic Speed Determination of the Amorphous and l.c.-p-methoxycinnamate Polymers

The three most important parameters which must be controlled when comparing the photographic speeds of similar negative-working-light-sensitive polymers are the following:

1.   The concentration of the light-sensitive moieties
2.   The light-sensitive polymer film thickness
3.   The molecular weight of the polymers under investigation.

The concentration of the p-methoxycinnamate moiety in all the polymers under investigation was adjusted to 2.1 molal.

Each polymer was first cast onto a quartz plate, via a spin coater, in order to establish the coating conditions which would yield films of equal optical density (O.D.) and therefore thickness. These conditions were then used to coat each polymer formulation onto an aluminum substrate. The films were dried in air at ambient temperature. Typical absorption curves obtained for these films on quartz can be seen in Fig. 6.

In order to confirm that the *film thicknesses* monitored on quartz were the same as those on aluminum under identical coating conditions, the following experiment was performed. Sample films on aluminum, 10 cm^2, were obtained for each polymer under coating conditions established on quartz. The polymer films were dissolved off the aluminum support with methylene chloride into 10 mL volumetric flasks. The

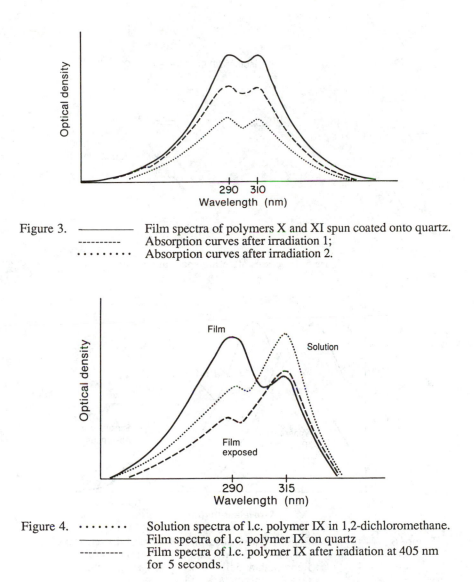

Figure 3. ——————— Film spectra of polymers X and XI spun coated onto quartz.
- - - - - - - - - Absorption curves after irradiation 1;
· · · · · · · · · Absorption curves after irradiation 2.

Figure 4. · · · · · · · · Solution spectra of l.c. polymer IX in 1,2-dichloromethane.
——————— Film spectra of l.c. polymer IX on quartz
- - - - - - - - - Film spectra of l.c. polymer IX after iradiation at 405 nm
for 5 seconds.

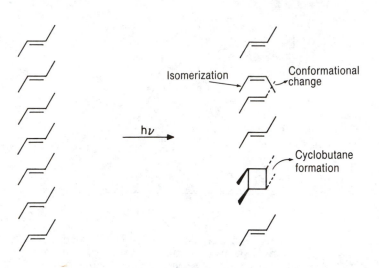

Figure 5. Processes affecting spectral changes.

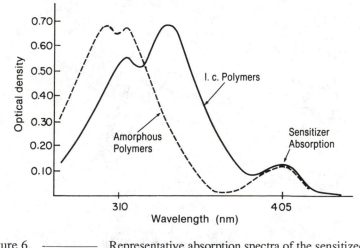

Figure 6. —————— Representative absorption spectra of the sensitized l.c.
                     polymer (IX) films before irradiation at 405 nm.
          - - - - - - - - Representative film absorption spectra of the
                     non-l.c. polymers X, XI.

optical density of these solutions were then monitored at 400 nm and were found to be identical. The molecular weights of the polymers involved in this study were determined by Gel Permeation Chromatography (GPC) and light scattering. The molecular weight of the PVA-p-methoxycinnamate polymer could not be determined by either gel permeation chromatography or light scattering. However, the degree of polymerization (d.p.) of the PVA backbone was known to be approximately 1700 (11) which made the absolute M.W. of the derivatized polymer approximately 268,000.

The formulation used to cast the films in order to determine the photographic speed of the p-methoxycinnamate polymers was the following:

1. The p-methoxycinnamate polymers were dissolved in cyclohexanone at 3% by weight.
2. The coumarin sensitizer, K, concentration was adjusted to 3 x 10⁻³ M.

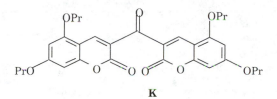

K

The polymer films on the aluminum support were exposed in the same manner as described for the films on quartz. However, in addition the films were irradiated through a photographic step tablet in which the density varies in increments of 0.15 per step. After irradiation, the films were developed by soaking in cyclohexanone for one minute or until no further change was noted in the number of crosslinked polymer steps.

Table I shows the molecular weights and the relative photographic speed results of the *amorphous* p-methoxycinnamate polymer films. The difference in the photographic speed between polymers X and XIA, XIB could be attributed to the difference in the backbone configuration and the effect it has on the orientation of the cinnamate chromophore as well as the M.W. difference.

Table I. Photographic Sensitivity of the
Amorphous-Cinnamate Polymers

| Polymer | P.S.E.W.[a] | | | ABS.[b] | Relative Photo |
	I.V.[d]	Mw	Mn	Mw	Speed[c]
X (PVA-control)	1.12	--	--	268,000	1.0
XI-A	0.31	51,074	21,946	70,700	0.25
XI-B	0.90	299,606	76,029	455,000	2.6

[a]Polystyrene equivalent weights determined by gel permeation chromatography.
[b]Absolute molecular weights determined by light scattering.
[c]Exposure time/exposure time of control.
[d]Inherent viscosity determined at a conc. of 0.5 g/dl in 1:1 phenol-chlorobenzene.

Also, the -C_2H_4- spacer removes some of the orientational influence the polymeric backbone might exert on the cinnamate moiety. The latter effect most certainly would increase the photographic speed of the polymer. The effect that the molecular weight played on the photographic speed of these light-sensitive polymers was best illustrated by comparing polymer XI-A with XI-B (Table I and again in the molecular weight series shown in Table II).

Table II. Photographic Sensitivity of the l.c.—p-Methoxycinnamate Polymers

Polymer	I.V.[d]	P.S.E.W.[a] Mw	Mn	ABS.[b] Mw	Relative Photo Speed[c]
X	1.12	--	--	268,000	1
IX-1	0.31	85,000	34,800	213,000	5.6
IX-2	0.46	164,000	59,800	402,000	13.5
IX-3	0.79	368,000	94,600	821,000	25
IX-4	0.84	428,000	140,900	1,170,000	33.3
IX-5	1.61	883,700	333,636	1,760,000	83.3

NOTE: See notes for Table I.

Table II shows the relative photographic speed results of the anisotropic-l.c. p-methoxycinna-mate polymer films. The photographic speed of the l.c. p-methoxycinnamate polymers exceeded our most optimistic expectations. It was necessary to handle films of polymers IX-3, IX-4, and IX-5 under safe-light conditions to avoid premature fogging. Polymer IX-5 (Table II) had a molecular weight 6X that of the PVA cinnamate control, X, but was 83X faster. The molecular weight difference certainly accounted for some of this increased speed, but obviously could not account for all of it.

In order to eliminate the M.W. differences between our PVA-p-methoxy-cinnamate control, and the l.c.-p-methoxycinnamate polymers, which would give a direct comparison between their relative photographic speeds, and hence their relative quantum yields of cyclobutane formation, a plot of the relative photographic speeds of the l.c. polymers in Table II vs MW was constructed (see Fig. 7). The plot in Fig. 7 was linear up to approximately 1,200,000 M.W. units. Please note that the plotted line should go through the origin. Since the M.W. of the PVA-p-methoxycinnamate control was approximately 268,000, then the M.W. equivalent l.c. p-methoxy-cinnamate polymer would have a relative photographic speed of 8 ± 2 (see Fig. 7). This 8-fold increase in the photographic speed corresponds to a relative quantum yield increase approaching the theoretical limit of approximately 10X, and could be directly attributed to the effect that the microscopic order in the polymer matrix had on the triplet energy transfer process.

Conclusions and Recommendations

The work presented here has demonstrated, for the first time, that the molecular organization of thermotropic-side chain-liquid-crystalline l.c. polymers could drastically influence the yield of a chemical reaction.

The synthetic schemes for the synthesis of novel p-methoxycinnamate-containing vinyl monomers and the properties of the corresponding l.c. polymers is described. The fraction of reaction sites in the l.c. polymer films increased to a level which caused the relative quantum yield of cyclobutane formation to approach the theoretical limit.

The presence of the p-methoxycinnamate moiety allowed the films of these polymers to be probed spectrophotometrically. The surprising difference in the solution vs film spectra of the l.c. polymers was accounted for by the unexpected intramolecular perturbation of the p-methoxycinnamate moiety by the phenyl ester group. The unexpected spectral changes during UV irradiation of the l.c. polymer films could be attributed to conformational changes, isomerization, and cyclobutane formation.

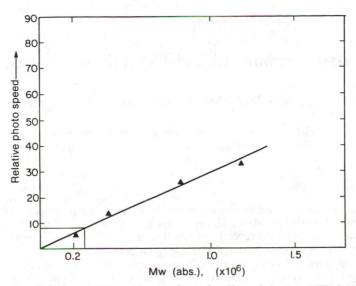

Figure 7.   Relative photographic speed as a function of absolute MW for the l.c. polymer IX.

Acknowledgments

Dr. J.L.R. Williams encouraged us to investigate the area of l.c. polymers at an embryonic stage of its development, and then provided the environment and support which was necessary to carry out this investigation.

Dr. Samir Y. Farid spent many hours discussing the results and interpretation of the many experiments he helped us design. Samir's insight was invaluable to the success of this investigation.

Mr. L. W. Fisher not only determined the MW's, but also displayed the patience in the special handling of these polymers. Mr. J. C. Reiff provided the GPC data. Dr. F. Saeva's discussions were very helpful.

Literature Cited

1.   Wegner, G. Makromol. Chem. 1972, 154, 35.
2.   Day, D. R.; Ringsdorf, J. J. Polym. Sci. Polym. Lett. Ed., 1978, 16, 205.
3.   (a) Johnson, D. S.; Sanghers, S.; Pons, M.; Chapman, D. Biochim. Biophys. Acta., 1980, 602 57. (b) Hub, H. H.; Hupfer, B.; Koch, H.; Ringsdorf, H. H. Angew. Chem., Int. Ed. Engl. 1980, 19, 938.
4.   (a) Shankoff, T. A.; Trozzolo, A. M. Photogr. Sci. Eng., 1974, 19, 173. (b) Reiser, A.; Egerton, P. L. Photogr. Sci. Eng., 1979, 23, 144.
5.   Jones, Jr., F. E.; Ratto, J. In Liquid Crystals and Ordered Fluids; J. F. Johnson, R. S. Porrer, Eds.; Plenum Press: New York, 1973, 2, 723.
6.   Perplies, E.; Ringsdorf, H.; Wendorf, J. H. Makromol. Chem. 1974, 175, 553.
7.   Plate, N. A.; Shibaev, E. P. J. Polym. Sci., Macromol. Rev. 1974, 8, 117.
8.   a) Farid, S. Y. et al. Pure and Appl. Chem. 1979, 51, 241. (b) Williams, J.L.R. et al. Pure and Appl. Chem. 1977, 49, 523.
9.   Herkstroeter, W. G.; Farid, S. Y. J. Photochem. 1986, 35, 71.
10. K. Toyoshima, In Polyvinyl Alcohol, Properties and Applications; C. A. Finch, Ed.; John Wiley and Sons, Ltd.: London, 1973.

RECEIVED May 19, 1990

Chapter 12

All-Hydrocarbon Liquid-Crystalline Polymers

T. C. Sung, J. J. Mallon[1], E. D. T. Atkins[2], and S. W. Kantor

**Department of Polymer Science and Engineering,
University of Massachusetts, Amherst, MA 01003**

A series of thermotropic hydrocarbon liquid crystalline polymers have been prepared based on a biphenyl mesogen. Liquid crystalline vinyl monomers with the general formula $H_{2n+1}C_n$-C_6H_4-C_6H_4-$(CH_2)_m$-$CH=CH_2$ were polymerized with $AlEt_3/TiCl_4$ to form high molecular weight side-chain liquid crystalline polymers. The monomers and polymers with $n \geq 2$ exhibited smectic B and/or smectic E mesophases. The polymers were found to be highly isotactic. All-hydrocarbon main-chain liquid crystalline polymers with the general structure

$\pm C_6H_4$-C_6H_4-$(CH_2)_Y \pm_X$ were prepared by a carbon-carbon coupling reaction. Number average molecular weights have been obtained in the range of 4-6,000. Homopolymers with $Y=6, 8, 10$ and copolymers with $Y=6/8, 6/10, 8/10$ were found to exhibit smectic mesophases.

Thermotropic liquid crystalline (LC) polymers with mesogenic segments and flexible spacers in the main-chain and side-chain have received considerable attention (1-3). The majority of these LC polymers have been polyesters, although other mesomorphic polymers have included polyamides, polyazomethines and polyesteramides. Nearly all of these liquid crystalline polymers contained heteroatoms such as N, O or Si in their structures. These heteroatoms not only provide for simple and versatile synthetic methods but also contribute to mesophase stability through hydrogen bonding or dipole-dipole attractive forces. However, these attractive forces are not believed to be necessary for the formation of the liquid crystalline phase. As early as 1949 (4), Onsager demonstrated that above a critical concentration, a solution of rodlike molecules will spontaneously adopt a nematic order. Intermolecular attractive forces were not required to bring about phase separation. The only interaction assumed in this model is one in which the rodlike molecules cannot interpenetrate each other due to an "excluded volume". Although Onsager's theory predicted the formation of a nematic phase, recent molecular

[1]Current address: Aerospace Corporation, P.O. Box 92957, Los Angeles, CA 90009
[2]Current address: Physics Department, University of Bristol, Bristol, England

0097–6156/90/0435–0158$06.00/0
© 1990 American Chemical Society

dynamic and Monte Carlo studies (5) have also demonstrated the existence of smectic order in the absence of intermolecular attraction. Therefore, it is now recognized that shape anisotropy alone is sufficient to produce liquid crystalline phases, without the need for any attractive forces. All-hydrocarbon LC polymers should be good candidates to test the validity of this theory. Although intermolecular attractive forces are still present in hydrocarbon molecules, the forces present are weaker than those in molecules containing strong dipoles or hydrogen bonds.

Low molecular weight hydrocarbon liquid crystals have been known for some time. A review of the effects of various structural elements on the properties of low molecular weight hydrocarbon liquid crystals may be found in the literature (6). However, with the exception of the work being performed in our laboratory (7-9), all-hydrocarbon LC polymers have received relatively little attention. A recent report by Memeger describes substituted polyphenylene vinylene polymers that were characterized as liquid crystalline (10).

The objectives of our research were the following:
1. Synthesize all-hydrocarbon side-chain and main-chain polymers.
2. Demonstrate the liquid crystalline behavior of the new polymers.
3. Investigate their structure-mesophase property characteristics.
4. Investigate the tacticity-mesophase property relationships for the side-chain LC Polymers.

An interesting extension of our work is the possibility to blend some of the more rod-like hydrocarbon LC polymers with other flexible random coil all-hydrocarbon polymers such as polystyrene or polyethylene. Depending on both the transition temperatures and solubilities, some mixtures may form phase separated immiscible blends and ultimately generate "molecular composites".

Structures [I] and [II] are representative of the general structures of the target polymers synthesized in our studies. Biphenyl was selected as the mesogen not only

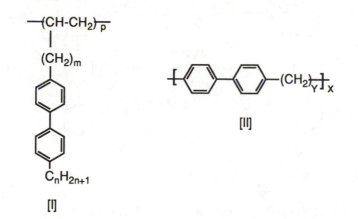

because it may be readily incorporated synthetically but also because the low molecular weight hydrocarbon model compounds listed in Table I were found to exhibit LC properties (11). The biphenyl model compounds that are liquid crystalline are disubstituted in the para position of each ring. This report reviews some of our recent work on the synthesis and characterization of hydrocarbon side-chain [I] and main chain [II] LC polymers.

Table I. Biphenyl LC Model Compound

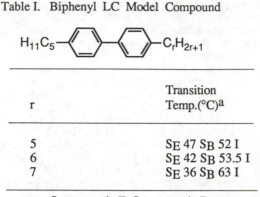

r	Transition Temp.(°C)[a]
5	S_E 47 S_B 52 I
6	S_E 42 S_B 53.5 I
7	S_E 36 S_B 63 I

a. S_E= smectic E; S_B= smectic B;
I=Isotropization or clearing transition.

Hydrocarbon Side-Chain Liquid Crystalline Polymers

Synthesis. The general synthetic route for the hydrocarbon side-chain LC polymers
[I] is shown in Scheme 1 and the detailed synthetic procedures have been reported
previously (8).

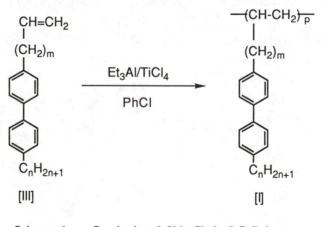

Scheme 1.   Synthesis of Side-Chain LC Polymers

Six polymers were prepared using a $Et_3Al/TiCl_4$ catalyst. The structures of the
monomers and polymers were confirmed by ^1H NMR spectra and elemental
analyses. All the polymers were soluble in chlorobenzene, chloroform and hot
benzene at 60°C. The GPC molecular weights (MW) of these polymers, relative to
polystyrene standards, are shown in Table II. The molecular weights are generally
high with broad molecular weight distributions. All of the polymers were capable of
being drawn into fibers or cast into films.

Table II.   GPC Molecular Weights for Side-Chain LC Polymers [I]

n	m	Mn	Mw	Mw/Mn
0	4	49,000	450,000	9
0	6	43,000	660,000	15
2	4	26,000	250,000	10
2	6	90,000	850,000	9
4	4	14,000	330,000	24
4	6	70,000	710,000	10

The general synthetic route which allows the length of the tail (n) and spacer (m) to be easily varied is shown in Scheme 2 for a representative monomer [III] with n=4 and m=6. The detailed synthetic procedures for the monomers have also been reported previously (7). Control of the tail length may be accomplished by the proper selection of the Friedel-Crafts acylating agent followed by a Wolff-Kishner reduction to give the desired 4-alkyl-4'-bromo-biphenyl derivative. Although only two different tail lengths were used in this work, Gray (12) has prepared numerous bromobiphenyl derivatives with a wide variety of tail lengths. The third step of the reaction sequence shown in Scheme 2 is a dilithium tetrabromocuprate catalyzed coupling reaction of the Grignard reagent with a dibromoalkane in THF solution (13). The coupling reaction is best effected by adding the Grignard reagent and a catalytic amount of Li_2CuBr_4 dissolved in THF to an excess of the desired α,ω–dibromoalkane. An excess of dibromoalkane is used to limit coupling with both ends of the α,ω-dibromoalkane. The spacer length (m) may be easily controlled by selecting the appropriate α,ω-dibromoalkane. The elimination of HBr in the final step is accomplished through a Hoffmann elimination. A total of six monomers were prepared according to Scheme 2.

<u>Characterization.</u> The liquid crystalline properties of the side-chain monomers (III) and polymers (I) have been studied by Differential Scanning Calorimetry (DSC), Polarized Optical Microscopy (POM) and X-ray diffraction. The thermal transition data and phase types for all monomers (III) and polymers (I) are summarized in Table III. A representative DSC scan for the monomer (III) and polymer (I) with a four-carbon tail (n=4) and six-carbon flexible spacer (m=6) are shown in Figures 1 and 2 respectively. The first peak at -24°C shown in Figure 1 is the crystal to smectic B phase transition for the monomer, designated as $K \rightarrow S_B$, whereas the second peak at 42°C is the smectic B to isotropic phase transition, designated by $S_B \rightarrow I$. Figure 2 also shows clearly the presence of two peaks for the polymer. The lower temperature peak at 145 °C represents the transition from smectic E to smectic B and the higher temperature peak at 181 °C is the clearing temperature. The photomicrograph for the same monomer and polymer with a four-carbon tail and six-carbon flexible spacer are shown in Figures 3 and 4 respectively. Figure 3 shows a lancet texture for the monomer that is typical of Smectic B liquid crystals and the fine grain texture more typical of the polymer is shown in Figure 4. Based on both the DSC and POM, it appears that side-chain LC polymers (I) have broader liquid crystalline temperature ranges than their corresponding monomers (III); the POM photomicrographs for the

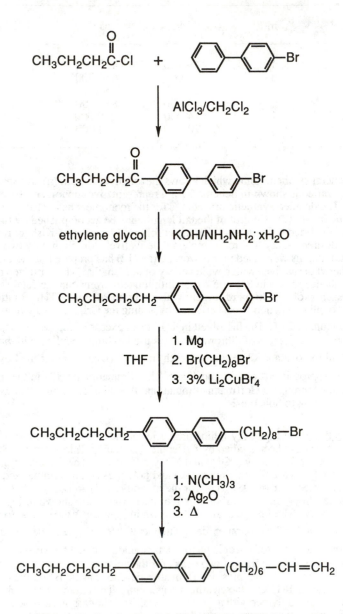

Scheme 2. Synthesis of a Side-Chain LC Monomer

Table III. Thermal Phase Transitions for Side-Chain Monomers [III] and Polymers[I]

Structures	n	m	Transition Temp, °C
[III]	0	4	K[a] 1.3 I
[I]	0	4	K 151 I
[III]	0	6	K 18.3 I
[I]	0	6	K 126 I
[III]	2	4	S_B 26.3 I
[I]	2	4	S_E broad S_B 197 I
[III]	2	6	K 9.4 S_B 28.2 I
[I]	2	6	S_E broad S_B 176 I
[III]	4	4	K 24.4 S_B 38.5 I
[I]	4	4	S_E 158 S_B 206 I
[III]	4	6	K -24.6 S_B 42.4 I
[I]	4	6	S_E 145 S_B 181 I

a. K=Crystal

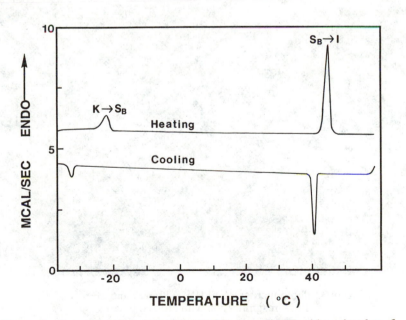

Figure 1. Second heating and cooling DSC scans of [III] with n=4 and m=6.

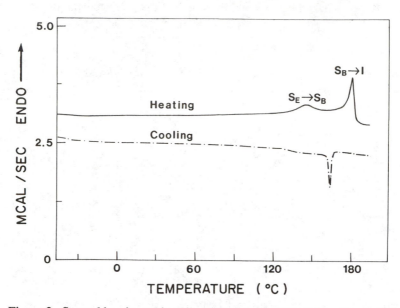

Figure 2. Second heating and cooling DSC scans of [I] with n=4 and m=6.

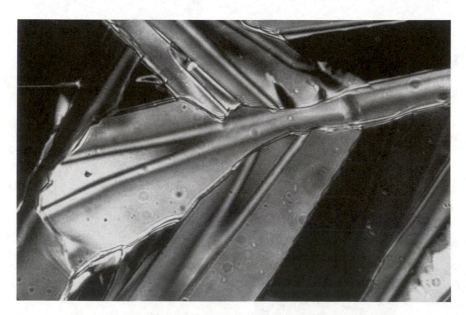

Figure 3. Photomicrograph of [III] with n=4 and m=6 at 35°C.
120X magnification with crossed polarizers.

Figure 4. Photomicrograph of [I] with n=4 and m=6 at 160°C.
120X magnification with crossed polarizers.

polymers are not as distinct as for the monomers. It is also interesting to note that all the side-chain LC polymers and the small molecular weight model compounds listed in Table I exhibited only Smectic E and B mesophases. The X-ray pattern in Figure 5 for the same monomer (III) with n=4 and m=6 shows an inner ring corresponding to a layer spacing of 24.5Å and an outer ring corresponding to an intermolecular spacing of 4.6Å. The X-ray diffraction results are summarized in Table IV. A difference about 3 to 4Å between the observed and calculated layer spacing may be attributed to a combination of the space between the ends of the molecules (14), or uncertainty of the backbone structure in the case of polymers. Detailed fiber diffraction results for the polymer have been published previously (9).

To summarize our results on side-chain all-hydrocarbon LC polymers [I], we found that they displayed Smectic E phases at lower temperatures and Smectic B phases at higher temperatures. The corresponding LC monomers exhibited only Smectic B phases. The presence of tails seems to enhance and stabilize liquid crystallinity. The monomers and polymers that do not have tails are not liquid crystalline. All of the substituted biphenyl hydrocarbon liquid crystals found in the literature are disubstituted. Examples of monosubstituted biphenyl hydrocarbon liquid crystals have not been found. The results obtained from this work support the idea that disubstitution of the biphenyl moiety is necessary to obtain a liquid crystalline phase.

Tacticity. The Ziegler-Natta catalyst chosen for the polymerization reaction is Et3Al/TiCl4, a catalyst that has been reported to give a 50/50 mixture of atactic and isotactic polymer when used to polymerize 1-octadecene or 1-propene. It was reasonable to expect that this catalyst would give similar results with the polymers prepared in this work. However, the isotactic content of our hydrocarbon polymers

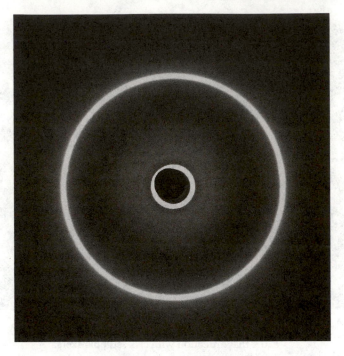

Figure 5. X-ray diffraction pattern of [III] with n=4 and m=6 at 35°C.

ranged from 65% ± 10% to 90% ± 10% as determined by ^{13}C NMR. The experimental details have been published previously (8). One possible explanation for the high isotactic content is that the biphenyl mesogen of the monomers tend to stack together with the biphenyl groups in the growing polymer as shown in Figure 6, thereby resulting in a predominately isotactic polymer.

<u>Hydrocarbon Main-Chain Liquid Crystalline Polymers</u>

<u>Synthesis.</u> Hydrocarbon main-chain LC polymers with the general structure [II] were synthesized as shown in Scheme 3. The monomers (IV) were synthesized by following the literature procedure(15). This polycondensation reaction proceeds through the carbon-carbon (C-C) coupling between the aryl bromide bond and the aryl magnesium bromide group formed in the mixture. In contrast to the low yield generally obtained from uncatalyzed C-C coupling reactions, the transition metal catalyzed C-C coupling reaction proceeds selectively under mild conditions to give the desired product in high yield (16). The nickel catalyst used in this work has been widely utilized to prepare conducting polymers such as polyphenylene, polythiophene etc.(17). The intermediate [V] is insoluble in diethyl ether but soluble in THF. It is also well known that the di-Grignard reagent is soluble in THF. The reaction of the monomer [IV] with an equimolar amount of Mg does not give the sole product [V], as shown in Scheme 3, but gives a mixture of di-Grignard and mono-Grignard in addition to unreacted starting material [IV]. However, if an equimolar ratio of monomer [IV] and Mg is employed, the coupling reaction proceeds without losing the stoichiometric relationship required for the polycondensation.

Table IV. X-ray Diffraction Results for the Side-Chain LC Monomers [III]
and polymers [I]

Structures	n	m	Intermolecular Spacing, Å	Observed Layer Spacing, Å	Calculated Molecular Length, Å
[III]	2	4	4.6	18.8	16.3
[I]	2	4	4.6	32.0	29.0
[III]	2	6	4.6	21.3	18.8
[I]	2	6	4.6	37.0	33.0
[III]	4	4	4.6	21.5	18.8
[I]	4	6	4.6	36.0	33.0
[III]	4	6	4.6	24.5	21.2
[I]	4	6	4.6	40.0	37.0

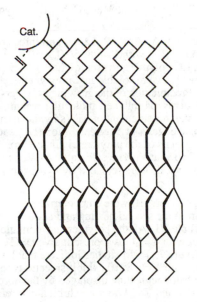

Figure 6. Insertion of LC monomer into growing polymer chain attached to catalyst.

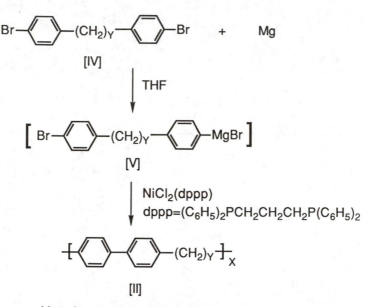

Y=6, 8, 10, 12, 14, 6/8, 6/10, 8/10

Scheme 3. Synthesis of Main-Chain LC Polymers

Five homopolymers with Y=6, 8, 10, 12, 14 and three copolymers with 50 mole% of each comonomer [IV] with Y=6/8, 6/10, 8/10 were prepared. Unlike the hydrocarbon side-chain LC polymers, the main-chain polymers had low solubility in common organic solvents but were soluble in diphenyl ether and biphenyl at high temperature (>150°C). Precipitation of the growing polymers from the polymerization solvent seems to limit their molecular weights. The low solubility of the polymers prevented the measurement of MWs by the usual solution methods. However, the approximate degrees of polymerization (DP) and MW were calculated assuming that both end-groups of the polymer chain are bromine atoms. These results are summarized in Table V. The calculated MW's and DP's would be less if some of the bromine chain ends were replaced by hydrogen atoms.

Characterization. The polymers [II] were characterized by DSC, POM and X-ray diffraction. DSC data from the second heating and cooling scans at a rate of 20°C/min are listed in Table VI. All the polymers have multiple phase transitions except the homopolymer with Y=14. A representative DSC can for the homopolymer with Y=8 is shown in Figure 7. POM studies indicate that both the copolymers and the three homopolymers with Y=6, 8, 10 exhibit smectic-type textures. Figure 8 is a photomicrograph of the batonnet texture displayed by the homopolymer with Y=10. Homopolymers with Y=12 and Y=14 do not exhibit any birefringence. Both the copolymers and homopolymers with Y=6,8,10 gave the same X-ray diffraction patterns as the side-chain LC monomer shown in Figure 5 when the temperature was kept between the first transition temperature (T_1) and the clearing temperature. Therefore, these temperatures (T_1) are interpreted as crystal to smectic phase transition temperatures for the corresponding polymers. Evidence for the existence of nematic phases could not be obtained. The X-ray diffraction data are summarized in Table VII. All the evidence indicates that the hydrocarbon main-chain polymers with Y=6, 8, 10, 6/8, 6/10, 8/10 are liquid crystalline while polymers with Y=12 and Y=14 are not liquid crystalline. This is probably due to the lower mesogenic content, <50%, for polymers with Y=12 and 14. The homopolymer with Y=12 is not liquid crystalline but gives multiple phase transitions by DSC shown in Figure 9. Birefringent texture was not observed by POM even after annealing the sample for 24 hours at 160°C. X-ray scattering experiments also do not show evidence for the existence of liquid crystalline phases. The multiple peaks in Figure 9 probably represent crystal-crystal transitions. The melting and clearing temperatures from the DSC results are summarized in Figures 10 and 11. Based on these data, it appears that lengthening the flexible spacer causes an increase in melting point but a decrease in clearing temperature. In addition , copolymers have broader mesomorphic ranges than the homopolymers and both the melting points (T$_1$) and clearing temperatures are lower than for the homopolymers. These conclusions are tentative since molecular weights have a large effect on transition temperatures, particularly at low or intermediate molecular weights.

Table V. Elemental Analyses

[II]

[II]	C	H	Br	MW	DP
Y=6	88.06	8.52	3.51	4556	18.6
Y=8	87.97	8.75	3.46	4619	16.9
Y=10	87.02	9.52	3.59	4451	14.7
Y=12	87.06	9.39	3.65	4379	13.6
Y=14	87.52	8.94	3.67	4358	12.1
Y=6/8[a]	87.11	8.63	3.32	4819	18.6
Y=6/10[b]	88.97	8.63	2.71	5904	21.8
Y=8/10[c]	88.21	8.51	3.05	5246	18.3

a. copolymer from 50 mole% of two monomers with Y=6 and 8.
b. copolymer from 50 mole% of two monomers with Y=6 and 10.
c. copolymer from 50 mole% of two monomers with Y=8 and 10.

Table VI. Polymer[II] Transition Temperature(°C)

[II]	$T_1{}^c$	$T_2{}^d$	$T_3{}^e$	Tcl[f]
Y=6(H)[a]	104		228	270
Y=6(C)[b]			223	259
Y=8(H)	127	155	184	218
Y=8(C)	104		180	208
Y=10(H)	143			181
Y=10(C)	136			181
Y=12(H)	145	152	158	170
Y=12(C)	137	146	156	165
Y=14(H)				153
Y=14(C)				143
Y=6/8(H)	88			242
Y=6/8(C)				234
Y=6/10(H)	100			211
Y-6/10(C)				204
Y=8/10(H)	114			194
Y=8/10(C)	89			184

a.H=Heating b.C=Cooling c.First Transition Temperature d.Second Transition
Temperature e.Third Transition Temperature f.Clearing Temperature

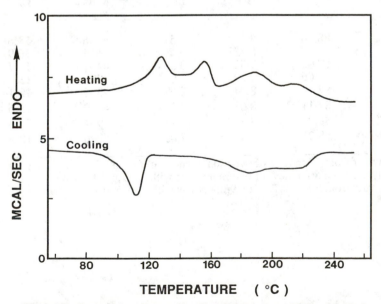

Figure 7. Second heating and cooling DSC scans of [II] with Y=8.

Figure 8. Photomicrograph of [II] with Y=10 at 150°C.

Table VII. X-Ray Diffraction Results for Main-Chain LC Polymers [II]

Y	Intermolecular Spacing, Å	Observed Layer Spacing, Å	Calculated Layer Spacing, Å
6	4.5	14.7	14.8
8	4.5	16.1	16.8
10	4.5	18.3	18.8
6/8	4.5	15.2	-
6/10	4.5	15.7	-
8/10	4.5	16.1	-

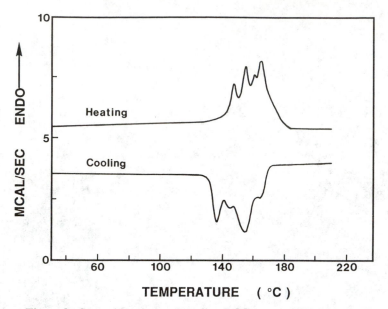

Figure 9. Second heating and cooling DSC scans of [II] with Y=12.

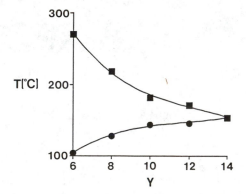

Figure 10. Melting points (●) and clearing temperatures (■) for homopolymers [II] versus methylene spacers Y.

Conclusions

All-hydrocarbon main-chain and side-chain liquid crystalline polymers based on a biphenyl mesogen have been synthesized. Both side-chain and main-chain polymers exhibit smectic phases only. Side-chain LC polymers exhibit the anticipated properties of low transition temperatures and solubility in non-polar solvents. However, hydrocarbon main-chain LC polymers have higher transition temperatures and poorer solubility in common organic solvents. In order to prepare higher molecular weights hydrocarbon main-chain LC polymers, the propagating intermediates have to be kept in solution during the polymerization reaction. A suggested approach to increase solubility might be to place some lateral hydrocarbon groups on the benzene ring. This approach has worked very successfully in the case of LC polyesters and polyamides (18).

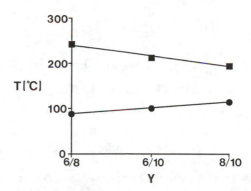

Figure 11. Melting points (●) and clearing temperatures (■) for comopolymers [II] versus methylene spacers Y.

Acknowledgments

The suport of the National Science Foundation for the Materials Research Laboratory and the Center for University of Massachusetts-Industry Research on Polymers (CUMIRP) is gratefully acknowleged.

Literature Cited

1. Ober, C. K.; Lenz, R. W. Adv. Polym. Sci. 1984, 59, 103.
2. Meyer, R. B.; Krigbaum, W. R.; Cifferri, A. Polymer LiquidCrystals; Academic: New York, 1982.
3. McArdle, C. B., Ed. Side Chain Liquid Crystal Polymers; Chapman and Hall: New Youk, 1989.
4. Onsager, L. Ann. N.Y. Acad. Sci. 1949, 51, 627.
5. Frenkel, D. ; Lekkerkerker, H.N.W.; Stroobants, A. Nature. 1988, 322, 822.
6. Toyne, K. J.; In Thermotropic Liquid Crystals; Gray, G.W., Ed.; John Wiley and Sons: New York, 1987, Chapter 2.4 .
7. Mallon, J. J.; Kantor, S. W. Macromolecules. 1989, 22, 2070.
8. Mallon, J. J.; Kantor, S. W. Macromolecules. 1989, 22, 2077.
9. Mallon, J. J.; Kantor, S. W. Macromolecules. 1990, 23, 1249.
10. Memeger.W, Jr. Macromolecules. 1989, 22, 1577.
11. Dabrowski, R.; Zytinski, E. Mol. Cryst. Liq. Cryst. 1982, 87, 109.
12. Gray, G.W.; Harrison, K. J.; Nash, J.A.; Constant, J.; Hulme, D.S.; Kirton, J.; Raynes, E.P. In Liquid Crystals and Ordered Fluids 2; Johnson, J.F.; Porter,R.S.; Eds.; Plenum Press: New York, 1974, p. 617.
13. Friedman, L; Shani, A. J. Am. Chem. Soc. 1974, 96, 7101.
14. Turner Jones, A. Makromol. Chem. 1964, 71, 1.
15. Jaunin, R.; Baer, T. Helv. Chim. Act. 1957, 233, 2245.
16. Tamao, K.; Kodama, S.; Nakajima, I.; Kumada, M.Tetrahedron. 1982, 38, 3347.
17. Yamamoto, T. ; Sanechika, K.; Yamamoto, A. Bull. Chem. Soc. Jpn. 1983, 56, 1497.
18. Ringsdorf, H. Tschirner, P. Hermann-Schonherr, O. Wendorff, J. H. Makromol. Chem. 1987, 188, 1431.

RECEIVED March 26, 1990

Chapter 13

Amorphous—Liquid-Crystalline Side-Chain AB Block Copolymers

Synthesis and Morphology

Joerg Adams and Wolfram Gronski

Institute of Macromolecular Chemistry, University of Freiburg, 7800 Freiburg, Federal Republic of Germany

By a new synthetic method LC-block copolymers were prepared to investigate the effect of a restricted geometry on a liquid crystal. Monodisperse A-styrene-B-cholesteryl block copolymers were synthesized by anionic polymerization of the polymer backbone containing a polystyrene A-block and a 1,2-polybutadiene B-block. The olefin double bonds of the polybutadiene were converted into hydroxyl groups by hydroboration and in a second polymeranalogous reaction cholesteryl was connected to the alcohol group. The resulting polymer shows the same liquid-crystalline mesophase as the correspondent LC-homopolymer. In the case of a weight ratio of 1/1 between the two blocks a lamellar morphology was observed via electron microscopy. By small angle X-ray scattering the orientation of the mesogenes in the ultra thin layers of the liquid-crystal were investigated.

In polymer science two non-crystalline ordered types of polymers are known, liquid-crystalline polymers (1,2) and block copolymers (3). It is obvious to combine the properties of the two polymer classes in an AB-block copolymer in which an amorphous A-block is connected to a liquid-crystalline B-block. If block copolymers with well defined block lengths can be synthesized, microphase separated systems with liquid-crystalline microphases may be realized, possessing long range ordered microdomain morphologies as conventional block copolymers. Two main problems can be investigated with these systems. One is the question how the size and type of the liquid-crystalline microphase, which can be controlled by the molecular weight and composition, affects the liquid-crystalline --> isotropic phase transition. This problem has recently been studied in the case of submicron size droplets of low molecular weight liquid crystals in a polymer matrix (4). The effect of a restricted geometry on the phase transition temperature can be treated by elementary thermodynamics (5). For a liquid crystalline sample enclosed between parallel planes an increase or decrease of the transition temperature with decreasing sample thickness is predicted depending on whether wetting or dewetting of the walls by the nematic phase occurs.

0097–6156/90/0435–0174$06.00/0

A more exact treatment is available in the frame of the Landau - de Gennes theory (6). It was shown that the influence of an ordering contact potential at the boundary is analogous to the effect of a strong magnetic field leading to an increase of the transition temperature until the first order transition turns into a second order transition at a critical thickness. The critical point is predicted to occur at a thickness of the order of 100 Å depending on the strength of the ordering force.

The critical thicknesses are thus in the range of the dimensions of lamellar, cylindrical or spherical mesophases in block copolymers with ordered morphologies. The question is whether the phase boundary between the amorphous and the liquid-crystalline phase in a block copolymer will exert an ordering effect as assumed in the original theory or rather a disordering influence. The latter case and transitions between the two cases have also been treated recently by an extension of the theory (5). Therefore a theoretical framework exists, within which the transition behaviour of amorphous / liquid-crystalline block copolymers can be described.

In addition to basic problems concerning the thermodynamic behaviour another question arises which is connected to the polymeric nature of the system. Unlike droplets of low molecular weight crystals in an amorphous matrix the liquid-crystalline and the amorphous phase at the interphase are coupled through polymer chains. Considering a block copolymer composed of an amorphous and a LC side chain polymer block which are the subject of this investigation, the question arises whether this coupling will have an effect on the director orientation with respect to the interphase by virtue of the fact that the polymer main chain of the LC block is preferentially oriented perpendicular to the interphase (7). If block copolymers with a long range ordered lamellar morphology are prepared, the director orientation with respect to the oriented lamellae may be influenced by the orientation of the polymer chains with respect to the interphase and the coupling of the mesogens to the polymer backbone. Different situations are expected for nematic and smectic systems and for different spacer lengths through which the mesogens are connected to the polymer backbone.

It is not the intention of the present paper to answer all questions concerning the thermodynamic and orientational behaviour, but rather to demonstrate the synthetic methods by which block copolymers of ordered geometry can be prepared and to give a first description of the morphological and structural features of a particular system. To develop ordered morphologies it is necessary to polymerize block copolymers with a narrow molecular weight distribution which is usually realized by living anionic polymerization. For LC side chain polymers this polymerization is not suitable because most mesogens possess groups which react with the living anion. This imposes severe restrictions in the choice of the mesogenic group. In this paper a route will be shown how to synthesize monodisperse LC-homopolymers and block copolymers by two successive polymeranalogous reactions (8) with the possibility to use the broad variety of mesogenic units known up to now. The starting polymer is anionically polymerized polystyrene-block-1,2-polybutadien (**PSPB**), the molecular weight distribution of which will be conserved during the two reactions. In the first step the olefin double bonds of the **PB**-block are converted into alcohol groups by hydroboration with 9-borabicyclo[3.3.1]nonane (**9-BBN**) followed by an oxidation with H_2O_2 (9). In the second step the mesogenic units are connected to the polystyrene-block-polyalcohol polymer (**PSPBOL**) by an oxycarbonyl linkage to cholesteryl.

Reaction Scheme

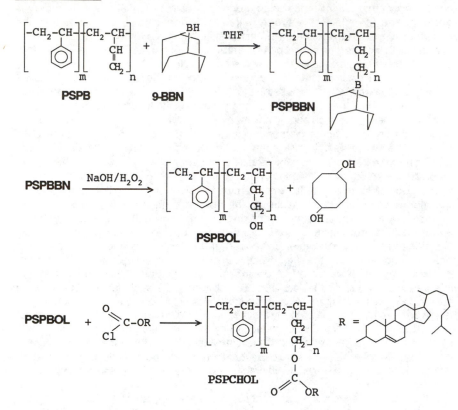

Another possibility to link the mesogen to the polymer backbone is by polymer-analogous esterification with a carbonic acid. This method will be described in a subsequent communication.

Experimental Part

Starting Material. 1,2-polybutadiene (**PB**) and the A-polystyrene-B-1,2-poly-butadiene block copolymer (**PSPB**) were obtained by "living" anionic polymerisation under high-vacuum conditions at -78°C with sec-buthyllithium as initiator.

Poly(2-hydroxyethylethylene) (PBOL).   The hydroboration of **PB** was carried out in the same way as described by Chung et al.(9). An excess of 9-bora-bicyclo[3.3.1.]nonane (0.5 M THF solution)(Aldrich) was allowed to react with the polybutadiene for 4 h at -10°C under nitrogen in a solution of dry THF. Stoichiometric amounts of NaOH and H_2O_2 were added and the resulting polyalcohol was purified by distillation with methanol to remove trace amounts of boric acid, and by precipitation from diethyl ether.

Polystyrene-block-poly(2-hydroxyethylethylene) (PSPBOL). The polymer was obtained from **PSPB** in the same way as the homopolyalcohol except a longer time (12 h) for the hydroboration reaction.

Poly(2-cholesteryloxycarbonyloxyethylethylene) (PCHOL). 0.25 g (3.45mmol) of **PBOL** were dissolved in dry pyridine unter nitrogene and a solution of 1.55 g (3.45 mmol) of cholesteryl chloroformiate (Aldrich) in 20 ml of dry THF was added at 0°C. After 1 h the reaction mixture was allowed to warm up to room temperature and after stirring for 12 h at this temperature the polymer was precipitated into methanol. The product was purified 4 times by precipitation form a THF solution into hot ethanol. Yield: 85%.

$(C_{32}H_{52}O_3)_n(484.7)_n$	Calc	C 79.28	H 10.81	O 9.91
	Found	C 79.54	H 10.86	O 9.60

Polystyrene-block-poly(2-cholesteryloxycarbonyloxyethyl-ethylene) (PSPCHOL). The same procedure as for the synthesis of **PCHOL** was used. The resulting polymer was purified by precipitation into a mixture of Methanol and diethylether (vol. ratio 4:1). Yield: 90%.

$(C_{69.23}H_{89.23}O_3)_n(969.5)_n$	Calc.	C 85.79	H 9.28	O 4.93
	Found	C 85.50	H 9.25	O 5.25

DSC-measurements. The thermal behavior was determined with a Perkin-Elmer DSC instrument and heating rates of 25, 16, 9 and 4 °/min.
Electron-microscopy. The morphology of the block copolymers was investigated with a Zeiss EM 902.
X-ray pattern. A Kiesig camera was used to analyse the mesophase of the LC polymer and the correlation between the block copolymer and the mesogenic groups.
Shear apparatus. A simular apparatus as discribed by Hadziioannou et al. (10) was used to orientate the block copolymer. The polymer was sheared between two heatable metal plates by backward and forward sliding of the plates at 130°C for 4h under nitrogen atmosphere.

Results and Discussion

Synthesis. The main demand for the synthesis of this new type of liquid crystalline polymer is a high conversion and the absence of any side reactions for both reaction steps. For the first investigations described in this communication IR-spectroscopy was used to determine the conversion. The molecular weight distribution obtained from gel permeation chromatography (GPC) is used to monitor side reactions.

The absence of any absorption at 3340 cm^{-1} in the IR-spectra of **PCHOL** (Figure 1) shows that all hydroxyl groups of **PBOL** are converted. With the results of the elemental analysis and the ^{1}H-NMR data the conversion was calculated to be almost 100%. Also no significant change in molecular weight distribution ocurred, shown by the GPC traces of **PB** and **PCHOL** in Figure 2 and the values of M_w/M_n in Table I for the homopolymers **PB**, **PBOL** and **PCHOL**.

The main interest of this work concerns the question whether the reaction conditions can also be applied to the block copolymers. The experiments showed that no modifications of the reaction conditions had to be made. Two polystyrene-block-1,2-polybutadiene with a PS/PB ration of 89/11 and a molecular weight of 55 000 g/mol for **PSPB I** and 118 000 g/mol for **PSPB II** were used as starting materials. High conversion as well as no change in polydispersity is found for both polymers (Table I). This ratio will lead to a lamellar morphology of the block copolymers if phase separation occurs.

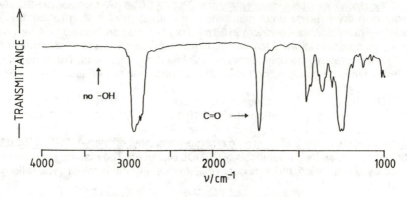

Figure 1.   IR-spectrum of **PCHOL**

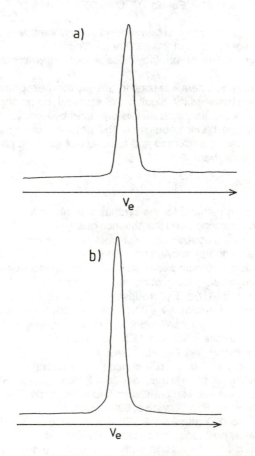

Figure 2.   GPC taken in CHCl$_3$. a: **PB**, b: **PCHOL**

Table I. Molecular weight distribution and molecular weight of the homopolymers and the block copolymers

Polymer	M_w/M_n [a]	M_n [g/mol]
PB	1.10	62 000[a]
PBOL	1.13	----
PBCHOL	1.13	----
PSPB I	1.17	55 000[a]
PSPB II	1.09	118 000[a]
PSPCHOL I	1.19	118 000[a]
PSPCHOL II	1.14	153 000[b]

a: Measured by membrane osmosis

b: Measured by GPC

Thermal and LC Phase Behaviour.   The thermal transitons of the **PCHOL** homopolymer are shown in Figure 3 and Table II. The polymer exhibited a glass transiton at 75°C and a first order LC - isotropic transition at 206°C. The results are typical for a smectic phase of a cholesteryl containing LC polymer and are in agreement with other polymers of this type (11). The analysis of X-ray fiber pattern further proved the existence of a smectic A phase.

The thermal behaviour of the block copolymer **PSPCHOL I** is shown in Figure 4. The phase separation into an amorphous and a liquid crystalline phase is apparent by a T_g of the PS phase and a T_i of the mesophase appearing at the positions of the homopolymer without any significant shift (Table II). T_g of the PCHOL-phase was not clearly resolved.

Table II.   Phase Transition Temperatures and Phase Transition Enthalpies ΔH of the Homopolymer and Block Copolymers

Polymer	Phase Transition Temperature	ΔH [J/g]
PCHOL	g 75°C s 206°C i	1.6
PSPCHOL I	g 98°C s 203°C i	0.3
PSPCHOL II	g 99°C s 201°C i	0.4

g : glassy
s : smectic
i : isotrop

As expected, the phase transition enthalpy ΔH decreased in comparison to the homopolymer. The magnitude of the decrease from 1.6 J/g to 0.3 J/g and 0.4 J/g for **PSPCHOL I** and **PSPCHOL II** respectively cannot be explained only by the fact that 50 weight-% of the polymer is polystyrene. The decrease of the transition enthalpy can be explained by the restricted size of the liquid crystalline microphase (see below). The thermodynamic theory (5,6) predicts a decrease of the discontinuity in the order parameter and the transition enthalpy with decreasing size of the LC microphase. The observed slightly larger ΔH for the block copolymer with higher molecular weight, i. e. larger domain size of the LC phase also conforms to theoretical predictions.

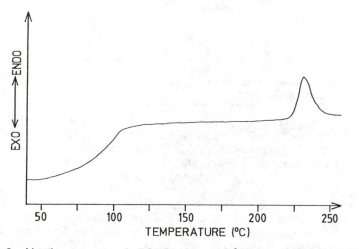

Figure 3. Heating curve of **PCHOL** at 40 °C/min. (Reproduced with permission from reference 8. Copyright 1989 Hüthig and Wepf Verlag, Basel.)

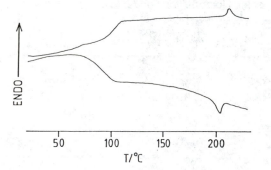

Figure 4. Heating and cooling curve of **PSPCHOL I** with a heating- (cooling-) rate of 20 °C/min. (Reproduced with permission from reference 8. Copyright 1989 Hüthig and Wepf Verlag, Basel.)

Corresponding changes of the transition temperature also predicted by the theory will be difficult to detect because of the small magnitude of the temperature shift (1°C or less). The observed slight decrease of T_i of the block copolymers with respect to the **PCHOL** homopolymer (Table II) is explained by other effects.

The main reason for this decrease is probably due to a small amount of residual cholesterol monomer which was not removed during the purification procedure. Another difficulty arises from the fact that small changes in transition temperature and enthalpy can also be caused by the thermal and mechanical history of the sample, e. g. by presence of internal stress. However, the large decrease of ΔH in the block copolymer with respect to the **PCHOL** homopolymer must have other causes. It is either a thermodynamic effect, as discussed above, or it may also be caused by a disordering of the mesogens at the phase boundary. Since the amorphous / LC phase boundary must be sharp the disordering cannot originate from an intermixture of styrene and mesogenic units in an interphase region. Although the phase boundary is sharp on a local scale the interphase may be of a very irregular structure which could oppose the ordering of the mesogens at the interface. In this explanation the decrease of ΔH occurs because a surface layer of PCHOL at the phase boundary is in the disordered state and only the material in the interior of the LC microphase takes part in the transition. Further systematic investigations are needed to clarify which of the two explanations are correct.

Morpology of the Block Copolymer. In order to get information on the morphology of the system, polymer films were cast from $CHCl_3$-solution and electron micrographs of these films were taken from ultra-thin sections stained with osmium tetroxide. Figure 5a shows an EM micrograph of the block copolymer with the higher molecular weight. The micrograph demonstrates the phase separation into the amorphous PS phase, the LC phase appears dark because of staining with OsO_4.

The basic morphology is the lamellar type as expected from the 1 : 1 composition of the block copolymer. The long period of the lamellar is 510 Å. Beside the lamellar structure a honeycomb structure with a larger periodicity of 820 Å is also present.

The block copolymer with M_n = 118 000 g/mol has been treated by oscillatory shear (10) above the glass transition of PS producing the well ordered lamellar morphology in Figure 5b with a repeat distance of 350 Å. The result shows that at sufficiently low molecular weight thin films can be obtained from the melt by shear, possessing long range ordered lamellae oriented preferentially in the direction of shear. The oriented sample has been used to investigate the question whether a correlation does exist between the supramolecular lamellar order of the block copolymer and the liquid crystalline order in LC lamellar domains. Figure 6 shows the X-ray photograph of this film, the X-ray beam being directed normal to the shear direction. The inner reflection surrounding the beam stop originates from the long periodicity of the lamellar macrolattice. It is deformed into an ellipse with its long axis lying normal to the direction of shear, corresponding to a high degree of orientation of the lamellae as expected from the EM picture in Figure 5b. From the intensity maximum of this first order reflection the same period is obtained as from the EM micrograph. At large angle the reflection of the smectic layers within the LC lamellae can be seen. An enhancement of the intensity perpendicular to the lamellae normal is observed. This proves that the smectic layers are oriented preferentially normal to the polystyrene lamellae. Therefore, for the particular smectic system investigated, a defined correlation between lamellar order and LC order does exist. The situation is schematically shown in Figure 7 showing alternating amorphous and liquid-crystalline

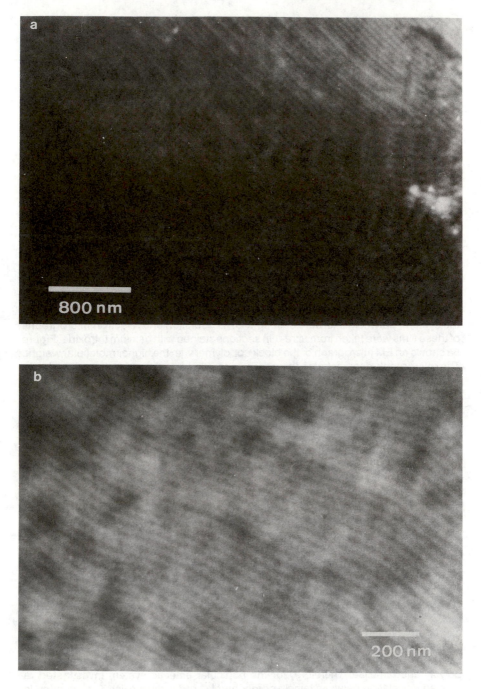

Figure 5.   Electron micrographs of the block copolymers stained with OsO_4,
             a: **PSPCHOL II**, b: **PSPCHOL I**

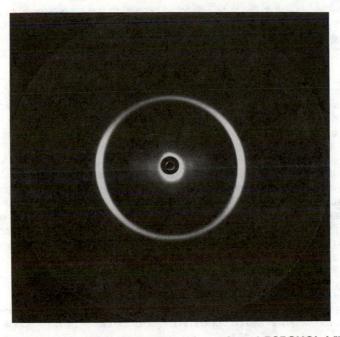

Figure 6. X-ray diffraction pattern of an oriented **PSPCHOL I** film

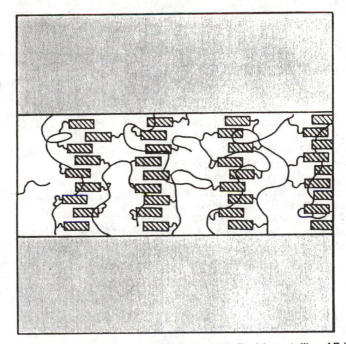

Figure 7.   Model of the orientation in the smectic liquid crystalline AB-block copolymer

lamellae, the layers within the lattice directed normal to the amorphous lamellae. Of course, this is an oversimplified representation. Not shown is the distribution of the smectic director in the lamellar plane.

Conclusion

The synthesis described in this paper renders possible the preparation of block copolymers of uniform molecular weight composed of amorphous and LC side chain blocks. Beside the specific cholesterol mesogen introduced by carbonate linkages leading to a smectic system other mesogens with various spacer lengths can be introduced, e.g. by esterification.

This opens the possibility to tailor block copolymers with a wide variety of LC phases and phase transition temperatures. A interesting possibility is the preparation of thermoplastic LC elastomers of the ABA-type with amorphous A-blocks having a high T_g and an elastomeric LC B-block with low T_g. An uniform director orientation can be achieved in these systems by stress as shown recently for chemically crosslinked elastomers (12). Various applications of these systems in which optical uniaxiality and transparancy are induced by strain can be envisaged.

Acknowledgments

The authors thank Prof. *H. Finkelmann* and Dr. *W. Gleim* for valuable suggestions and discussions. The work was carried out in the SFB 60 of the *Deutsche Forschungs-gemein-schaft*.

Literature cited

1.   Shibaev, V. P.; Platé, N. A. Polym. Sci. USSR (engl. Trans.) 1978, 19, 1065; Vysokomol. Soedin., Ser. A 1977, 19, 923.
2..   Finkelmann, H.; Ringsdorf, H.; Wendorff, J. H. Makromol. Chem. 1978, 179, 273
3.   Gallot, B. Adv. Polym. Sci. 1978, 29, 85
4.   Golemme, A.; Zumer, S.; Allender, D. W.; Doane, J. W. Phys. Rev. Lett. 1988, 61, 2937
5.   Poniewierski, A.; Sluckin, T. J. Liquid Cryst. 1987, 2, 281
6.   Sheng, P. Phys. Rev. A 1982, 26, 1610
7.   Hasegawa, H.; Hashimoto, T.; Kawai, H.; Lodge, T. P.; Amis, E. J.; Glinka, C. J.; Han, C. C. Macromolecules 1985, 18, 67
8.   Adams, J.; Gronski, W. Makromol. Chem., Rapid Commun. 1989, 10, 553
9.   Chung, T. C.; Raate, M.; Berluche, E.; Schulz, D. N. Macromolecules 1988, 21, 1903
10.   Hadziioannou, G.; Mathis, A.; Skoulios, A. Colloid & Polyme. Sci. 1979, 257, 136
11.   Shibaev, V. P.; Platé, N. A.; Freidzon, YA. S. J. Polm. Sci., Polym. Chem. Ed. 1979, 17, 1655
12.   Finkelmann, H.; Kock, H.; Rehage, G. Makromol. Chem., Rapid Commun. 1981, 2, 317

RECEIVED May 9, 1990

Chapter 14

Side-Chain Liquid-Crystalline Polyphosphazenes

R. E. Singler[1], R. A. Willingham[2], C. Noel[3], C. Friedrich[3], L. Bosio[4], E. D. T. Atkins[5], and R. W. Lenz[6]

[1]Army Materials Technology Laboratory, Watertown, MA 02172–0001
[2]Army Natick Research, Development and Engineering Center, Natick, MA 01760–5014
[3]Laboratoire de Physicochimie Structurale et Macromoleculaire, 10 rue Vauquelin, 75231 Paris Cedex 05, France
[4]Laboratoire de Physique Des Liquides et Electrochimie, 10 rue Vauquelin, 75231 Paris Cedex 05, France
[5]H. H. Wills Physics Laboratory, University of Bristol, Tyndall Avenue, Bristol BS8 1TL, United Kingdom
[6]Polymer Science and Engineering Department, University of Massachusetts, Amherst, MA 01003

An overview of the synthesis and characterization of a new class of side-chain liquid crystal polymers with a phosphorus-nitrogen backbone is presented. Using poly-(dichlorophosphazene) as a common reactive high polymer intermediate, low-molar-mass aromatic azo and stilbene mesogens with flexible spacer groups have been attached to the phosphazene (P-N) chain. Single-substituent polymers and mixed-substituent polymers, containing mesogen and trifluoroethoxy cosubstituents, have been prepared and studied by differential scanning calorimetry, optical microscopy, and X-ray diffraction. Both nematic and smectic mesophases have been observed. This synthetic approach offers many opportunities, both with molecular weight control and side-chain chemistry, for tailoring molecular structure to prepare different mesophases and optimize physical properties for non-linear optical applications.

The concept of a side chain liquid crystal polymer has been demonstrated in a number of laboratories and is well documented in the literature (1). Most of the side chain liquid crystalline polymers reported to date contain polysiloxane, polyacrylate or polymethacrylate main chains. More recent studies on the effect of backbone flexibility now include the use of flexible poly(ethylene oxide) or more rigid poly(α-chloroacrylate) chains.

0097–6156/90/0435–0185$06.00/0

Polyphosphazenes represent a new approach to the design and synthesis of side-chain liquid crystal polymers. Polyphosphazenes are inorganic main-chain polymers consisting of alternating phosphorus-nitrogen atoms in the main chain with two substituents attached to each phosphorus atom. The top of Figure 1 shows the general structure for a side chain liquid crystal polymer: a polymer backbone with a side chain comprised of a flexible spacer and a rigid rod (mesogen). The remaining structures in Figure 1 are examples of side-chain liquid crystalline polyphosphazenes. The middle structure represents a phosphazene mixed-substituent "copolymer" which contains a mesogen side chain along with a non-mesogen (R) side chain. Phosphazene homopolymers (Figure 1, bottom) contain two identical substituents attached to phosphorus. Both phosphazene homopolymers and copolymers can be readily synthesized by processes which are described in the next section.

Synthesis

The most commonly used approach (Figure 2) for the synthesis of polyphosphazenes involves a ring-opening polymerization of hexachlorocyclotriphosphazene (trimer) to give open-chain high molecular weight poly(dichlorophosphazene), followed by a substitution process to yield poly(organophosphazenes) (2,3). By using highly purified trimer and by carefully monitoring the polymerization process, one can obtain soluble high molecular weight poly(dichlorophosphazene) and avoid the branching and crosslinking which can lead to a crosslinked matrix, "inorganic rubber". The second step is also critical: poly(dichlorophosphazene) is a reactive inorganic macromolecule, and chlorine substitution is required in order to prepare hydrolytically stable polymers. This also allows one to prepare a variety of different polymers. Shown in Figure 2 are three general examples consisting of alkoxy, aryloxy, and amino substituted phosphazene polymers. Commercial poly(fluoroalkoxy- and aryloxyphosphazene) elastomers have been prepared, where two or more nucleophiles are used in the substitution process (4). Polyphosphazenes which contain organometallic and bioactive side chains have also been reported (4).

The two-step synthesis process shown in Figure 2 affords several possibilities for preparing new side-chain liquid crystal polymers. The polymerization process allows one to vary the molecular weight and molecular weight distribution (MWD), and potentially change the properties of the liquid crystalline state. Most of the polyphosphazenes reported in the literature, including the examples in this paper, are derived from the bulk uncatalyzed process: this generally produces high molecular weight polymer (one million or greater) with a broad MWD. Catalyst systems have been developed which give lower molecular weight poly(dichlorophosphazene) with a narrower MWD (4). One catalyst, boron trichloride, has also been used to prepare poly(dichlorophosphazene) with narrow molecular weight fractions ranging from approximately ten thousand to over one million (5). Thus, it may be possible to prepare side-chain liquid crystal polyphosphazenes with a range of molecular weights and change the properties of the mesophase.

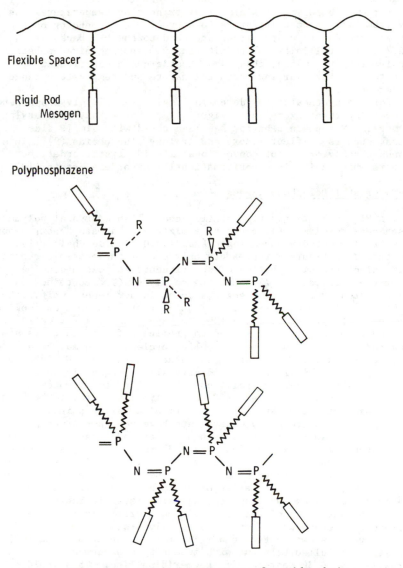

Figure 1. General structures for side-chain liquid crystal polymers.

Polyphosphazene synthesis provides additional possibilities for preparing liquid crystal polymers with different properties. As noted above, the substitution process (Figure 2) enables one to synthesize a wide variety of polymers. The phosphazene inorganic backbone is a highly flexible polymer chain: glass transition temperatures can vary from below room temperature to above 100°C, depending on the size and the nature of the group attached to the P–N backbone (3,6,7). Crystallinity can be altered by preparing mixed-substituent "copolymers" (8). Thus, changing the side-chain substituent affords the potential for varying both the nature and temperature range of the mesophase.

For the synthesis of side chain liquid crystal polyphosphazenes, the most important examples in Figure 2 are the alkoxy- and aryloxy-polymers. Mesophase behavior has been noted with simple side chains, such as trifluoroethoxy and aryloxy side chains (9). This mesophase behavior is not conventional liquid crystal order, but polymers which exist in a conformationally disordered state (10).

Liquid Crystal Polyphosphazenes

In Figure 3 are examples of side-chain liquid crystal polymers. Phosphazene copolymer (I) contains a mixture of an azophenoxy mesogen and a trifluoroethoxy non-mesogen (ca. 1.3:0.7) side chain (11). Structure I in Figure 3 represents only one of the possible repeat units, since the distribution of substituents is presumed to be random (see Figure 1). Although the copolymer (I) contained only slightly greater than one mesogen side chain per repeat unit, it formed a reversible thermotropic liquid crystal phase upon cooling from the isotropic state above 180°C. Analysis by differential scanning calorimetry (DSC) showed no prominent first-order transitions for either the heating or cooling cycles, which may be due to the irregular distribution of substituents along the P–N backbone. Microscopic analysis in the liquid crystalline region is shown in Figure 4. The polymer crystallized upon cooling to room temperature. However, upon heating above 120°C, the polymer transformed back into a mesophase which persisted up to the final clearing point.

High temperature X-ray experiments have been performed on copolymer I in an attempt to obtain a better understanding of the nature of the mesophase. Fibers obtained from I, which were pulled from the melt and quench cooled in water, gave very diffuse patterns. If the fibers were allowed to air cool, the pattern shown in Figure 5 was obtained. It shows a series of sharp arcs centered on the equator and meridian, and the spacings are listed in Table I. More accurate measurements of the low angle equatorial signals were obtained by increasing the specimen to film distance to 170 nm, and the pattern is shown in Figure 6. The reflection at spacing 4.33 Å consists of a full circle but with intensity enhancement at approximately 20-30 degrees off the meridian line. This would approximately place it on a layer line with spacing about 4.9 Å. The equatorial reflections do not index on an obvious lattice. In fact the reflections at 32.4 Å and 25.5 Å are broader than the other reflections, indicating a different origin for these two reflections.

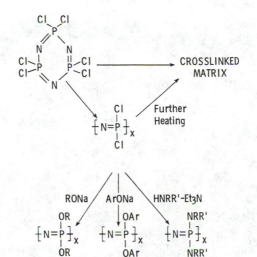

Figure 2. Synthesis of poly(dichlorophosphazene) and poly(organophosphazenes).

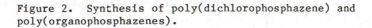

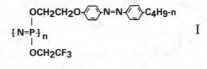

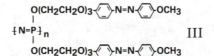

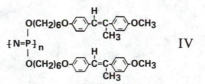

Figure 3. Side-chain liquid crystal polyphosphazenes.

Figure 4. Optical micrograph of polymer I at 128 °C. Crossed polarizers. Magnification 134X. (Reproduced from Ref. 11. Copyright 1987 American Chemical Society.)

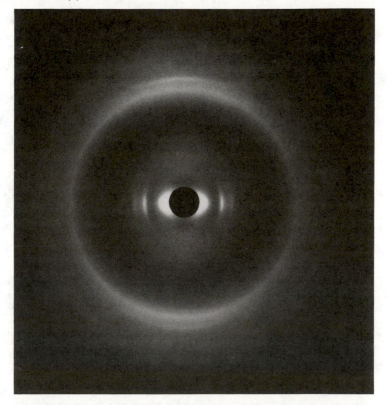

Figure 5. Polymer I. X-ray diffraction pattern obtained from fibers drawn from the melt and cooled in air. Wide-angle pattern.

The sketch of the wide angle pattern is shown in Figure 7. Unit cells taking these two values for the basic parameters $\underline{a}$ and $\underline{b}$ and taking $\underline{c} = 4.9$ Å do not give the calculated densities that related to the average measured density of 1.2 g/mL.

Table I

Spacing Å	Description
32.4	Equatorial Arc
25.5	" "
20.4	" "
12.3	" "
9.55	" "
4.33	Ring With Intensity Enhancement in Angle Region 20° - 30° From Meridian
3.99	Meridional Arc
3.45	" "

For copolymer I, it is reasonable to relate the diffraction ring at spacing 4.33 Å (Figure 6) with the stacking of the mesogenic slabs in the side chain. The observation that this interplanar spacing is spherically averaged and other reflections show a respectable degree of orientation argues for two organizational regimes within the sample which of course frustrates the analysis of the X-ray patterns. The fact this material is a copolymer with an irregular distribution of side chains along the polymer main chain probably accounts for the difficulty in indexing the crystalline state and giving a more well defined explanation for the nature of the mesophase.

The situation is quite different for a phosphazene polymer II (Figure 3) containing solely the n-butyl phenylazophenoxyethoxy side chain. On cooling from the isotropic state (above 185°C) a viscous mesophase is first observed which shows the characteristic focal conic texture typical of low molar mass smectic A phases. A photomicrograph of the smectic A phase at 162.4°C is shown in Figure 8. As the temperature falls below 160–155°C, a smectic C-like mesophase appears. Below 150–145°C, the polymer crystallizes, and in the oriented form can be indexed on an orthorhombic unit cell with $\underline{a} = 35.7$ Å, $\underline{b} = 17.85$ Å, and $\underline{c}$(fiber axis)= 9.85 Å. The doubling of the 4.9 Å chain repeat has been reported for other polyphosphazene structures and can arise from alternating perturbations of the side chains or a change in the backbone conformation (12).

Figure 6. Polymer I. X-ray diffraction pattern obtained from
fibers drawn from the melt and cooled in air. Low-angle pattern
showing broadness of first two reflections.

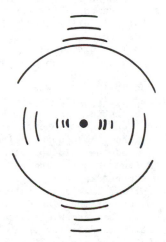

Figure 7. Sketch of wide-angle X-ray pattern for polymer I.

Figure 8. Optical micrograph of polymer II at 162.4 °C. Crossed polarizers. Magnification 200X.

A more extensive series of side-chain liquid crystalline polyphosphazenes and their corresponding cyclic trimers were reported by Allcock and Kim (13,14). The side chain consisted of aromatic azo mesogenic units attached to the phosphorus atoms through flexible oligomeric ethyleneoxy spacer units. Polymers were studied by DSC and optical microscopy. For polymer III, transition from the crystalline to the liquid crystalline state occured at 118°C, and this phase formed an isotropic fluid at 127°C. On cooling, the transition from the isotropic to the liquid crystalline state occured at 126°C, with a final crystallization at 94°C. Polarized optical microscopy of the mesophase at 121°C showed a typical nematic pattern. The corresponding cyclic trimer with three ethyleneoxy spacer units did not form a stable mesophase; however the trimer with two ethyleneoxy spacer units (not shown) displayed a typical nematic texture between 192 and 166°C upon cooling from the isotropic phase. This study showed that the terminal para substituent on the mesogen, the length of the spacer unit, and the cyclic or long-chain polymer character of the phosphazene all played a role in determining whether or not the material was liquid crystalline and also played a role in determining the nature of the liquid crystalline state.

The study by Percec, Tomazos and Willingham (15) looked at the influence of polymer backbone flexibility on the phase transition temperatures of side chain liquid crystalline polymethacrylate, polyacrylate, polymethylsiloxane and polyphosphazene containing a stilbene side chain. Upon cooling from the isotropic state, polymer IV displays a monotropic nematic mesophase between 106 and 64°C. In this study, the polymers with the more rigid backbones displayed enantiotropic liquid crystalline behavior, whereas the polymers with the flexible backbones, including the siloxane and the polyphosphazene, displayed monotropic nematic mesophases. The examples in this study demonstrated how kinetically controlled side chain crystallization influences the thermodynamically controlled mesomorphic phase through the flexibility of the polymer backbone.

Nonlinear Optical Polymers

Part of the rationale for investigating liquid crystal polyphosphazenes is for nonlinear optical (NLO) applications. In general, for NLO activity, polymers must either contain noncentrosymmetric side chains (eg., side-chain liquid crystal polymers) or highly delocalized (conjugated) backbones (16). Alternatively, polymers can be doped with low molar mass compounds to obtain NLO activity.

Recently, a mixed-substituent polyphosphazene (polymer V) was synthesized and the second-order NLO properties were investigated (17). The nitrostilbene/trifluoroethoxy ratio was approximately 36:64. Due to the low glass transition temperature of V ($T_g = 25^{\circ}$C), the second harmonic signal decayed to zero within a few minutes. However, polymer V is a prototype which offers many opportunities for further tailoring the molecular structure of polyphosphazenes to generate an optimum combination of NLO and physical properties (17).

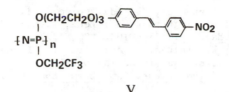

V

Another recent report (18) involves the evaluation of phosphazenes with nonmesogen side chains that exhibit NLO activity. Modification of the intrinsic optical response of the P-N chain was achieved through substitution of side groups including chloro, anilino, dimethylamino, and trifluoroethoxy. Cyclic phosphazenes were also included in this study. Results suggest that phosphazenes possess inherent NLO activity which can be enhanced by the suitable selection of substituent groups.

Conclusion

At the present time, only a few side-chain liquid crystal polyphosphazenes have been synthesized and investigated. Opportunities exist to prepare a wide variety of side-chain liquid crystalline polyphosphazenes, based on the polymerization-substitution process outlined in Figure 2. Alternative approaches, such as side chain modification of polyphosphazenes prepared by the thermal decomposition of N-silylphosphoranimines (19), may provide even further options for preparing liquid crystal polyphosphazenes.

Literature Cited

1. Side-Chain Liquid Crystal Polymers; McArdle, C. B., Ed.; Blackie, Chapman and Hall, New York, 1989.
2. Allcock, H. R. Phosphorus-Nitrogen Compounds, Academic: New York, 1972.
3. Singler, R. E.; Schneider, N. S.; Hagnauer, G. L. Polym. Eng. and Sci. 1975, 15, 321.
4. Inorganic and Organometallic Polymers; Zeldin, M.; Wynne, K. J.; Allcock, H. R., Eds.; ACS Symposium Series No. 360, American Chemical Society, Washington, DC, 1989; Chapters 19-21.
5. Sennett, M. S., U.S. Patent 4 867 957, 1989.
6. Allcock, H. R.; Connolly, M. S.; Sisko, J. T.; Al-Shali, S. Macromolecules 1989, 21, 323.
7. Allcock, H. R.; Mang, M. N.; Dembek, A. A.; Wynne, K. J. Macromolecules 1989, 22, 4179.
8. Beres, J. J.; Schneider, N. S.; Desper, C. R.; Singler, R. E. Macromolecules 1979, 12, 566.

9 . Schneider, N. S.; Desper, C. R.; Beres, J. J. In Liquid
     Crystalline Order in Polymers; Blumstein, A., Ed.; Academic: New
     York, 1978; p 299.
10. Wunderlich, B.; Grebowicz, J. Adv. Polym. Sci. 1984, 60/61, 1.
11. Singler, R. E.; Willingham, R. A.; Lenz, R. W.; Furukawa, A.;
     Finkelmann, H. Macromolecules 1987, 20, 1727.
12. Singler, R. E.; Willingham, R. A.; Noel, C.; Friedrich, C.;
     Bosio, L; Atkins, E., submitted for publication in
     Macromolecules. Polym. Prepr. (Am. Chem. Soc. Div. Polym. Chem.)
     1989, 30(2), 491.
13. Kim, C.; Allcock, H. R. Macromolecules 1987, 20, 1726.
14. Allcock, H. R.; Kim, C. Macromolecules 1989, 22, 2596.
15. Percec, V.; Tomazos, D.; Willingham, R. A. Polym. Bull. Berlin
     1989, 22, 199.
16. Nonlinear Optical Properties of Organic Molecules and Crystals;
     Chemla, D. S.; Zyss, J., Eds.; Academic, New York, 1987.
17. Dembek, A. A.; Kim, C.; Allcock, H. R.; Devine, R. L. S.; Steier,
     W. H.; Spangler, C. W. Chem. Mater. 1990, in press.
18. Exharos, G. J.; Samuels, W. D. Presented at the Meeting of the
     Materials Research Society, Boston, MA, December 1989; paper S2.4
19. Neilson, R. H.; Wisian-Neilson, P. Chem. Rev. 1988, 88, 541.

RECEIVED April 3, 1990

PHYSICS OF LIQUID-CRYSTALLINE POLYMERS: NETWORKS

Chapter 15

Rigid Rod Molecules as Liquid-Crystalline Thermosets

Andrea E. Hoyt[1], Brian C. Benicewicz[1], and Samuel J. Huang[2]

[1]Materials Science and Technology Division, Los Alamos National Laboratory, Los Alamos, NM 87545
[2]Institute of Materials Science, University of Connecticut, Storrs, CT 06269-3136

Rigid rod molecules endcapped with conventional crosslinking groups such as maleimide and nadimide were prepared and studied by differential scanning calorimetry and hot stage polarized light microscopy. Nematic liquid crystalline phases were identified in several of the new monomers. Thermally induced polymerization occurred in the nematic phase region and resulted in retention of the nematic texture in the final crosslinked solid. In many cases, isotropization was not observed at normal heating rates due to crosslinking and solidification in the nematic phase.

In 1975, Roviello and Sirigu reported on the preparation of polyalkanoates that melted into anisotropic phases (1). These mesophases were quite similar to those of conventional low molecular weight liquid crystals. Since this initial report, there has been considerable interest in liquid crystalline polymeric materials from both an academic and industrial viewpoint. The initial motivation for the development of liquid crystalline polymers from the industrial viewpoint was largely due to the pursuit of high tensile property fibers. Polymers that exhibit liquid crystalline order in solution or in the melt can transfer a high degree of molecular orientation to the solid state, which can lead to excellent mechanical properties (2). The technology of spinning rodlike aromatic polyamides from anisotropic solutions led to the first commercial product based on this idea. Subsequently, melt processing of liquid crystalline polyesters was also reported (3). Currently, there are at least three commercial thermotropic liquid crystalline polymers that can be used in a wide variety of thermoplastic polymer applications (4).

The field of thermoset polymers is an area of polymer science that has not yet been widely integrated with liquid crystal polymer research. It is interesting to note that Finkelmann et al. (5) reported on the formation of crosslinked elastomeric liquid crystalline networks in 1981. Although the general concept of a

0097–6156/90/0435–0198$06.00/0

liquid crystal thermoset (LCT) has been addressed, very few physical or mechanical properties have been reported (6–10). It was claimed that these materials exhibited very low shrinkage upon curing but quantitative information was not disclosed. We have initiated an investigation to prepare and evaluate liquid crystal thermosets as high performance composite matrix materials. In this report, we present our preliminary results on the synthesis and characterization of both ester and amide based materials developed from the LCT concept.

Experimental

Differential scanning calorimetry (DSC) was performed using a Perkin-Elmer DSC-2C at heating rates ranging from $20^{\circ}C$ min^{-1} to $40^{\circ}C$ min^{-1} under an argon atmosphere. Infrared (IR) spectra were recorded with a Perkin-Elmer 283 spectrophotometer using KBr pellets. Proton (^1H) nuclear magnetic resonance (NMR) spectra were recorded using a JEOL PMX60SI NMR spectrometer at 60 MHz. All of the monomers synthesized for this study showed satisfactory spectra and elemental analyses.

The maleimide, nadimide, and methyl nadimide endcaps were prepared by reacting the appropriate anhydride and p-aminobenzoic acid in acetone at room temperature to yield the corresponding amic acids. The amic acids were cyclodehydrated using acetic anhydride in the presence of sodium acetate according to the method of Rao (11). The endcaps were then treated with oxalyl chloride according to the method of Adams and Ulich (12) to yield the acid chlorides.

2,2′-Dimethyl-4,4′-diamino-1,1′-biphenyl dihydrochloride was obtained from Professor Lorraine Deck at the University of New Mexico and used as received. Maleic anhydride was obtained from Eastman Kodak and p-aminobenzoic was obtained from National Starch. The other anhydrides, diols, and reagents were purchased from Aldrich Chemical. All reagents were used without further purification. The monomers were synthesized via a Schotten-Baumann type procedure using the appropriate acid chloride endcap and diamines or diols. Triethylamine was used as a scavenger for HCl. The monomers were recrystallized from appropriate solvents in yields of approximately 70%.

Solubilities were determined by placing the monomers in the appropriate solvents at five or ten percent (w/w) increments. Mild heating was used in some cases and the solutions were allowed to cool to room temperature. A positive room temperature solubility was recorded if the monomer did not recrystallize after sitting overnight at room temperature. The maximum room temperature solubilities for the monomers in Table I are within the ranges given in the table.

Results and Discussion

One possible version of the LCT concept involves the design and preparation of new monomers consisting of a rigid rodlike central unit, a characteristic of conventional liquid crystals, capped at both ends with well known crosslinking groups. The crosslinking groups were chosen from the common functionalities used for thermoset materials such as epoxy, maleimide, acetylene, etc. This

concept is shown schematically in Figure 1. In the present study, the results on
maleimide, nadimide, and methyl nadimide endcapped monomers are reported.
In order to design materials that would also possess high thermal stability, it was
decided to construct the rigid rodlike portion of the molecule from aromatic amide
and ester units. This simple design could, of course, lead to rather intractable
materials, particularly in the case of the aromatic amides, unless measures are
taken to improve the tractability of these monomers. There are several known
methods to improve the tractability of wholly aromatic liquid crystalline polymers.
These methods attempt to disrupt the crystalline order of the p-linked chain without affecting it to such a degree that liquid crystallinity is lost. The techniques
used to reduce melting points or improve solubilities have been reviewed (13,14)
and include the use of bent and swivel monomers, flexible spacers, and bulky
ring substituents. One of the structural modifications that produced dramatic
improvements in melting points and solubilities was reported by Gaudiana et al.
(15). The specific structural modification responsible for the improvements was
the 2,2'-disubstituted 4,4'-biphenylene moiety. This substitution pattern forces
noncoplanarity of the phenyl rings while maintaining the rodlike conformation
of the backbone. The intermolecular interactions were also greatly affected. In
the present work, a series of amide monomers employing the 2,2'-disubstituted
4,4'-biphenylene unit were prepared to determine if this modification could
sufficiently reduce the intermolecular attractive forces to produce low molecular
weight liquid crystalline amides. In addition, a series of ester (non-hydrogen
bonded) based monomers was also prepared and characterized. The general
synthetic scheme used to prepare these materials is shown in Figure 2.

Solubilities. Rogers et al. (16–18) have prepared and reported on a large number
of aromatic polyesters and polyamides containing the 2,2'-disubstituted 4,4'-
biphenylene moiety. Some of the polymers were soluble in common solvents such
as tetrahydrofuran and acetone, with solubility as high as 50% in one case. The
solubilities of the amide monomers synthesized in this work are shown in Table I.
These compounds exhibited virtually no solubility in common solvents such as
acetone but displayed fairly high solubilities in several amide solvents, with and
without added salts. Lyotropic liquid crystallinity was not observed although
solubilities were as high as 40% (w/w) in some solvents.

The solubilities of the maleimide endcapped ester based monomers were also
investigated. These monomers contained unsubstituted rigid central cores. The
structures of these monomers are given in Table II. In contrast to the substituted
amide monomers discussed previously, the room temperature solubilities of these
unsubstituted monomers were low, e.g. less than 5% in 1,1,2,2-tetrachloroethane
(TCE), o-dichlorobenzene, tetrahydrofuran, pyridine, p-dioxane, m-cresol, trifluo-
roacetic acid, and 60/40 (v/v) phenol/TCE.

Thermal Behavior. The thermal behavior of the compounds prepared in this study
was investigated using a capillary melting point apparatus, hot stage polarized
light microscopy, and differential scanning calorimetry. Melting was not observed

Table I. Room Temperature Solubilities of Amide Monomers

COMPOUND	X	Y	T_m,°C	SOLUBILITY, % (w/w)			
				DMAc	DMAc/LiCl	DMF	NMP
1		CH$_3$	>350	<20	30–40	<20	20–30
2		CH$_3$	>350	20–30	30–40	<20	25–30
3		CH$_3$	>350	<20	>40	<25	20–30

for Compounds 1–3 in a capillary melting tube under normal conditions. However, the DSC trace of Compound 1 showed a sharp endotherm at approximately 340°C, immediately followed by an exotherm as shown in Figure 3. This exotherm is presumably due to crosslinking reactions. The DSC trace of Compound 2 showed no sharp melting endotherm, but exhibited two broad endotherms in the range of 270–350°C, followed immediately by a crosslinking exotherm as shown in Figure 4. Similar thermal behavior has been attributed to a retro Diels-Alder reaction of the two conformational isomers of the nadimide group (19). The retro Diels-Alder reaction results in the formation of cyclopentadiene and a maleimide endgroup. This reaction is followed by several possible crosslinking or addition reactions. Compound 3 exhibited similar behavior.

These monomers were also examined at very high heating rates. If the capillary melting apparatus or the microscope hot stage was heated to 350°C prior to inserting the sample, Compound 1 melted, flowed, and solidified within a few seconds. Solidification was assumed to be due to polymerization and crosslinking. Liquid crystalline phases were not observed for Compound 1. However, when Compound 2 was subjected to this same experiment the sample melted into a nematic phase. The sample crosslinked in the nematic phase after approximately ten seconds and the nematic texture was preserved in the solid state. Compound 3 showed similar behavior when placed on a preheated microscope hot stage at 340°C. This melting behavior can be explained if reaction occurred in the solid state during heating of the sample. At slow or normal heating rates, partial reaction of the monomer can occur below the crystalline melting point and prevent the material from flowing. At very high heating rates, melting and crosslinking (solidification) are observed in rapid succession.

X = Conventional crosslinking group

Figure 1. Schematic diagram of LCT concept.

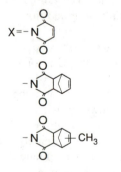

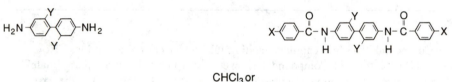

Figure 2. Synthetic scheme for rigid rod monomers.

A series of ester based maleimide endcapped monomers was also synthesized. These monomers are shown in Table II. The DSC traces of the ester based monomers showed sharp endotherms in the range of 270–300°C, followed immediately by a crosslinking exotherm. A representative trace is shown in Figure 5. The monomers melted into a nematic liquid crystalline phase and crosslinked shortly after melting. The nematic texture was retained in the crosslinked solid state. Melting was not always observed at low heating rates but was observed at lower heating rates than the amide based monomers. As described earlier, this may be due to partial reaction in the solid state during the heating cycle.

Table II. Thermal Transitions of Bismaleimide Monomers

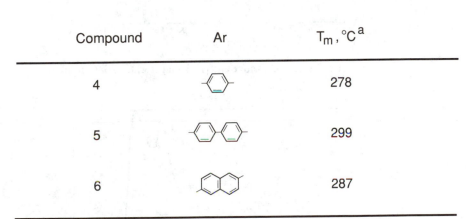

Compound	Ar	T_m, °C [a]
4	—⟨benzene⟩—	278
5	—⟨biphenyl⟩—	299
6	⟨naphthalene⟩	287

[a] *Melting point determined by DSC, reported as extrapolated onset.*

Conclusions

Rigid rod thermoset molecules were prepared by combining rigid, rodlike central units, similar to those found in LCP's, with well known crosslinking functionalities. The initial results have shown that liquid crystalline phases were obtained in these types of materials. The thermal polymerization of these monomers was conducted in the nematic phase and the nematic order was preserved in the crosslinked solid. Melting temperatures for the monomers reported here were high and this limited the time available in the fluid state. Crosslinking usually occurred within seconds after melting. Future efforts will concentrate on complete characterization of the monomers discussed herein and on the synthesis and characterization of additional rigid, rodlike thermosetting monomers with improved processability.

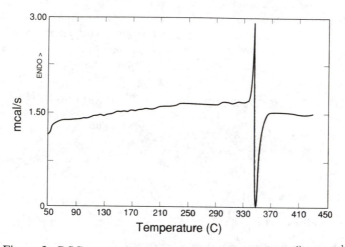

Figure 3. DSC trace of Compound 1. Heating rate, $20^{\circ}C\ min^{-1}$.

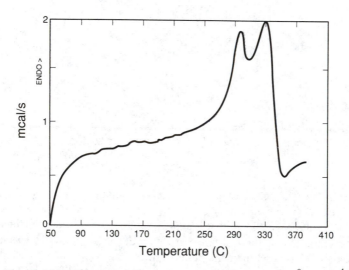

Figure 4. DSC trace of Compound 2. Heating rate, $40^{\circ}C\ min^{-1}$.

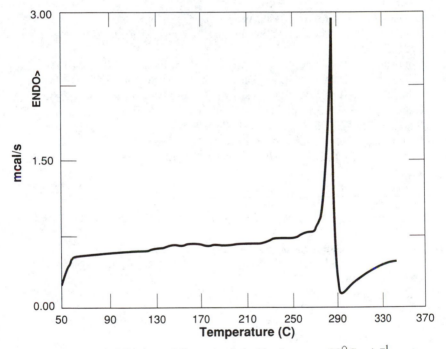

Figure 5. DSC trace of Compound 4. Heating rate, 20°C min^{-1}.

Acknowledgments

We are grateful to the Materials Science and Technology Division of Los Alamos
National Laboratory and the Department of Defense (DARPA contract No.
N00014-86-K-072) for support of this work.

Literature Cited

1. Roviello, A.; Sirigu, A. J. *Polym. Sci., Polym. Lett. Ed.*, **1975**, *13*, 455.
2. Calundann, G. W.; Jaffe, M. *Proc. Robert A. Welch Foundation*, **1983**, p. 247.
3. Jackson, Jr., W. J.; Kuhfuss, H. F. *J. Polym. Sci., Polym. Chem. Ed.*, **1976**, *14*, 2043.
4. *Chemical Business*, April **1989**, pp. 44–6.
5. Finkelmann, H.; Kock, H. J.; Rehage, G. *Makromol. Chem., Rapid Commun.* **1981**, *2*, 317.
6. Conciatori, A. B.; Choe, E. W.; Farrow, G. U.S. Patent 4 440 945, **1984**.
7. Conciatori, A. B.; Choe, E. W.; Farrow, G. U.S. Patent 4 452 993, **1984**.
8. Conciatori, A. B.; Choe, E. W.; Farrow, G. U.S. Patent 4 514 553, **1985**.
9. Calundann, G. W.; Rasoul, H. A. A., Hall, H. K. U.S. Patent 4 654 412, **1987**.
10. Stackman, R. W. U.S. Patent 4 683 327, **1987**.
11. Rao, B. S. *J. Polym. Sci., Part C: Polym. Lett.* **1988**, *26*, 3.
12. Adams, R.; Ulich, L. H. *J. Am. Chem. Soc.* **1920**, *42*, 599.
13. Jackson, Jr., W. *J. Br. Polym. J.* **1980**, *12*, 154.
14. Kwolek, S. L; Morgan, P. W.; Schaefgen, J. R. In *Encyl. Polym. Sci. Eng.*; Wiley: New York, **1987**; Vol. 9, p. 1.
15. Gaudiana, R. A.; Minns, R. A.; Sinta, R.; Weeks, N.; Rogers, H. G. *Prog. Polym. Sci.* **1989**, *14*, 47.
16. Rogers, H. G.; Gaudiana, R. A.; Hollinsed, W. C.; Kalyanaraman, P. S.; Manello, J. S.; McGowan, C.; Minns, R. A.; Sahatjian, R. *Macromolecules* **1985**, *18*, 1058.
17. Sinta, R.; Gaudiana, R. A.; Minns, R. A.; Rogers, H. G. *Macromolecules* **1987**, *20*, 2374.
18. Sinta, R.; Minns, R. A.; Gaudiana, R. A.; Rogers, H. G. *J. Polym. Sci., Part C: Polym. Lett.* **1987**, *25*, 11.
19. Wong, A. C.; Ritchey, W. H. *Macromolecules* **1981**, *14*, 825 and Wong, A. C.; Garroway, A. N.; Ritchey, W. H. *Macromolecules* **1981**, *14*, 832.

RECEIVED April 25, 1990

Chapter 16

Synthesis, Structure, and Properties of Functionalized Liquid-Crystalline Polymers

Rudolf Zentel[1], Heinrich Kapitza[1], Friedrich Kremer[2], and Sven Uwe Vallerien[2]

[1]Institut für Organische Chemie, Universität Mainz, J.-J.-Becher-Weg 18–20, D–5600 Mainz, Federal Republic of Germany
[2]Max-Plank-Institut für Polymerforschung, Postfach 3148, D–6500 Mainz, Federal Republic of Germany

Chiral lc-polymers can be prepared by a proper functionalization of lc-polymers with chiral and reactive groups. These elastomers are interesting, because they combine the mechanical orientability of achiral lc-elastomers with the properties of chiral lc-phases, e.g. the ferroelectric properties of the chiral smectic C* phase. The synthesis of these elastomers was very complicated so far, but the use of lc-polymers, which are functionalized with hydroxyl-groups, has opened an easy access to these systems. Also photocrosslinkable chiral lc-polymers can be prepared via this route.

During the last years our interest was focused on structure-property relations in various types of functionalized liquid crystalline (lc) polymers. At the beginning these were lc-polymers (main chain, side group and combined polymers) functionalized with reactive groups, which allow a network formation via a crosslinking reaction. The schematic representation of the lc elastomers prepared in this way is given in Figure 1. For a discussion of the synthesis and properties of these materials see Refs. (1–2). The most obvious property of the lc elastomers is their good mechanical orientability that means strains as small as 20% are enough to obtain a perfect orientation of achiral liquid crystalline phases.

The mechanical orientability made it interesting to look for a combination of this elastomer property with the properties of chiral liquid crystalline phases (cholesteric and chiral smectic C*, see Figure 2), which result from a proper functionalization of low molar mass liquid crystals with chiral groups. These phases show very special properties, which are: selective reflection of light (cholesteric phase) and ferroelectric properties (chiral smectic C* phase). The goal of this work was not to prepare densely crosslinked thermosets, in which the structure of these phases is locked in unchanged up to the decomposition temperatures (3–4), but to prepare slightly

0097–6156/90/0435–0207$06.00/0
© 1990 American Chemical Society

Crosslinked Side-Group Polymers

Crosslinked
Main-Chain Polymers

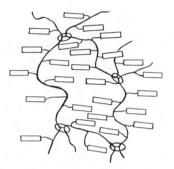

Crosslinked Combined Main-Chain/
Side-Group Polymers

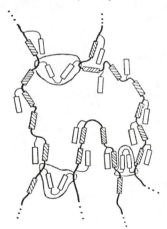

Figure 1. Schematic representation of different types of
lc elastomers.

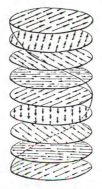

Cholesteric Phase   Chiral Smectic C* Phase

Figure 2. Schematic representation of the cholesteric and chiral smectic C* phase. The repeating distance along the helical axis (pitch) is between 200 nm to some μm.

crosslinked elastomers and to study the influence of mechanical stretching on the helical superstructure of the cholesteric and the chiral smectic C* phase (see Figure 2). In this case a change of the selective reflection (cholesteric) or a piezo-electric response (chiral smectic C*) should result from a mechanically induced change of the helical superstructure (see Ref. (5) for a detailed discussion of this topic).

Combined lc polymers (6) were selected as starting materials for this approach, because they form broad lc phases and they often show liquid crystalline polymorphism. Hence the chance to observe the desired chiral phases in the newly prepared polymers was great. In addition half of the mesogens in these polymers are incorporated into the polymer chain, while the other half (the mesogenic side groups) orients parallel to the polymer chains (7). Therefore a strong influence of the stretching of the sample, which leads primarily to an orientation of the polymer chains, on the orientation of the mesogenic groups can be expected.

The first chiral combined lc polymers prepared for this purpose showed the desired cholesteric and chiral smectic C* phases only at high temperatures (8) (the melting point was always above 100°C). By using lateral substituents (see Figure 3) it is possible however to suppress the melting temperature and to obtain polymers with a glass transition temperature of about room temperature, without losing the cholesteric and chiral smectic C* phases (9).

The synthesis of chiral elastomers of the schematic structure shown in Figure 4 can be accomplished by crosslinking chiral copolymers (5) according to Scheme I (for one chiral elastomer based on lc side group polymers see Ref. (10)). The phase behavior of some of these lc elastomers is summarized in Table I. Cholesteric and chiral smectic C* phases are found depending on the temperature. The assignment of these phases was primarily done on the basis of X-ray measurements and polarizing microscopy (5). The ferroelectric properties of some of these polymers in the chiral smectic C* phase were later confirmed by dielectric spectroscopy (frequency range: 10^{-1} – 10^9 Hz) (11) (see Figure 5). The observation of both the Goldstone and the Soft mode is a direct proof of the ferroelectric properties.

First X-ray measurements show that the helical superstructure of the cholesteric and chiral smectic C* phase can be untwisted by stretching the elastomer (5). High strains of 300% are necessary for this purpose (compared to 20% for the achiral elastomers). Nevertheless these results show that the chiral lc elastomers have the potential to act as mechano-optical couplers (cholesteric phase) or as piezo-elements (chiral smectic C* phase) (5), because the mechanically induced change of the helical superstructure has to change the optical transmission or reflection properties or the spontaneous polarization. Both effects however have not yet been measured directly.

The uncrosslinked and the crosslinked polymers described in Table I still have some drawbacks. To begin with, the synthesis of polymers with strong lateral dipole moments (see polymer 3a, b in Table I (5)) is rather complicated, because the chiral groups have to be introduced prior to the polycondensation reaction (9), which they must survive unchanged. This limits the number of useful chiral groups and excludes e.g. chiral esters, which are well known from low molar mass liquid crystals (12). In addition the crosslinking has to

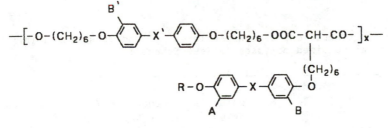

$$A, B, B' : Br, H$$

$$X, X' : -, -N=N-, -N{\overset{+}{\underset{O^-}{=}}}N-$$

R: $-CH_2-CH-\overset{*}{C}H_2-CH_3 \equiv R_1$
$\quad\quad\quad |$
$\quad\quad\quad CH_3$

$\quad-CH_2-\overset{*}{C}H-CH_3 \equiv R_2$
$\quad\quad\quad\quad |$
$\quad\quad\quad\quad Cl$

$\quad-CH_2-\overset{*}{C}H-\overset{*}{C}H-CH_2-CH_3 \equiv R_3$
$\quad\quad\quad\quad |\quad\quad|$
$\quad\quad\quad\quad Cl\quad CH_3$

$\quad-\overset{*}{C}H-(CH_2)_5-CH_3 \equiv R_4$
$\quad\quad |$
$\quad\quad CH_3$

Figure 3. Combined lc polymers with lateral substituents [9].

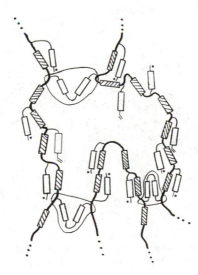

Figure 4. Schematic representation of a network prepared from chiral combined polymers [5].

Table I: Phase transitions of uncrosslinked (<u>a</u>) and crosslinked (b)
chiral combined polymers (<u>5</u>)(see Scheme I)

No.	Y	X_1	X_2	R	molecular weight(GPC)	Cross-linker (mol%)	Phase transitions / °C a)
<u>1a</u>	-N=N-	-N=N-	-N=N-	R_1	60 000	--	c109sc*124n*149i
<u>1b</u>	-N=N-	-N=N-	-N=N-	R_1	--	10	c 99sc*114n*141i
<u>2a</u>	-N(O)=N-	-N=N-	-N=N-	R_1	47 000	--	g24sc*115n*149i
<u>2b</u>	-N(O)=N-	-N=N-	-N=N-	R_1	--	10	g25sc*110n*147i
<u>3a</u>	-N(O)=N-	--	--	R_2	17 000	--	g20sc*124n*130i
<u>3b</u>	-N(O)=N-	--	--	R_2	--	20	g19sc*110n*123i

a) c:crystalline or highly ordered smectic phase, g:glassy frozen
phase, sc*: chiral smectic C*, n*: cholesteric phase,
i: isotropic melt

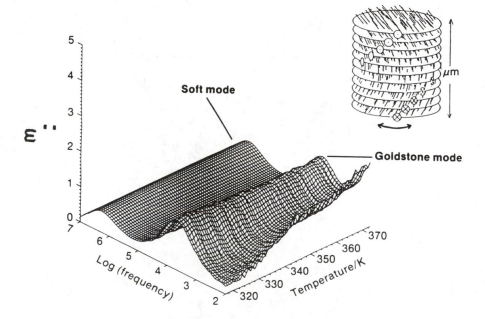

Figure 5. Dielectric loss ε" vs temperature and vs logarithm 10 of
frequency for a thin (10 µm) and aligned sample of polymer <u>3a</u> (see
Tab. 1).

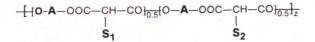

1a–3a

A: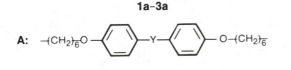

S₁: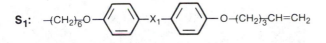

S₂: $-(CH_2)_6O-\langle benzene \rangle-X_2-\langle benzene \rangle-O-R_1$ or R_2

R_1: 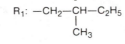 R_2: $-CH_2-\overset{*}{C}H-\overset{*}{C}H-C_2H_5$ with Cl and CH_3

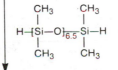

Crosslinked chiral elastomers

1b–3b

Scheme I

be done in isotropic solution, because the lc polymer and the siloxane (see Scheme I) are not miscible. It would, however, be most interesting to crosslink in substance in the different liquid crystalline phases. This would make it possible to crosslink oriented lc samples and to investigate the influence of different liquid crystalline phases on the crosslinking reaction and on the properties of the resulting network. A photo-crosslinking of groups covalently bound to the polymer chain would make this possible.

In order to achieve this goal a synthetic route to functionalized lc polymers was developed (13) (see Scheme II and Table II). Due to the much higher reactivity of hydroxyl groups compared to phenolic groups, it is possible to prepare linear polymers of the structure presented in Scheme II.

Starting from these polymers it is possible to introduce the chiral acids known from low molar mass liquid crystals (12) and to obtain the chiral homopolymers presented in Scheme III and Table III. These polymers show a high spontaneous polarization in the chiral smectic C* phase (14) (see polymer 7, Table III) and selective reflection of visible light in the cholesteric phase (see polymer 9, Table III) (13).

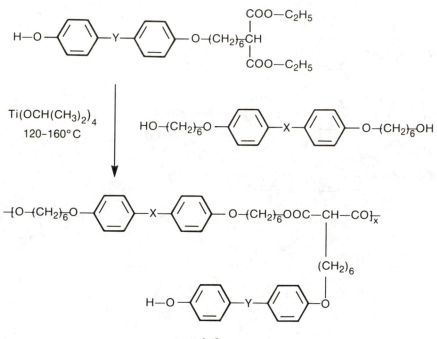

4-6

Scheme II

Table II: Phase transitions of the functionalized polymers
         4-6 (13) (see Scheme II)

No	X	Y	Phase transitions [a] in °C
4	–	–	c 141 i
5	–	-N(O)=N-	c 37 i
6	-N(O)=N-	–	g 24 lc 139 i

[a] see footnote to Tab. 1, c: crystalline, lc: liquid crystalline
phase, not further specified

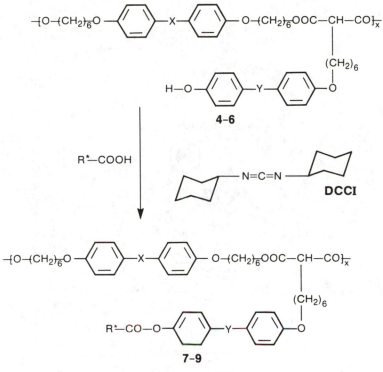

Scheme III

Table III: Phasetransitions of the chiral homopolymers $\underline{7}$-$\underline{9}$ (13-14) (see Scheme III)

No	R	X	Y	Phasetransitions[a] in °C
$\underline{7}$	C_2H_5- *$CH-$ *$CH-$ $\quad\quad CH_3 \quad Cl$	–	–	c 114 s_x 121 s_c^* 133 i
$\underline{8}$	C_2H_5- *$CH-$ *$CH-$ $\quad\quad CH_3 \quad Cl$	–	$-N(O)=N-$	s_y 84 s_A 112 i
$\underline{9}$	$\quad\quad O_2N$ $C_6H_{13}-$*$CHO-$ $\quad\quad CH_3$	$-N(O)=N-$	–	g 18 s_A 103 n* 125 i

[a] see footnote to Tab. 1; s_x or s_y: highly ordered smectic phase, not further identified, s_A : smectic A

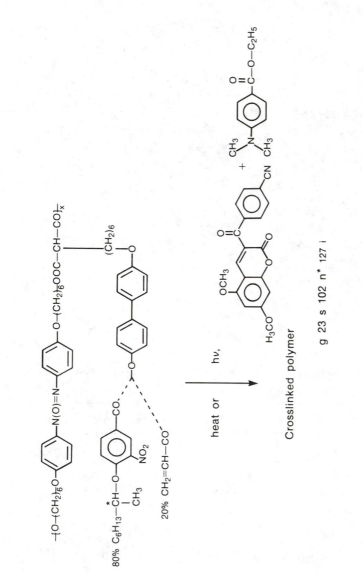

The lc-phases are retained during crosslinking

Figure 6. Photochemical crosslinking of an lc copolymer functionalized with acrylate groups.

In order to prepare photochemically crosslinkable lc polymers, copolymers were prepared, in which 70-80% of the mesogenic side groups were functionalized with chiral groups, while the remaining 20-30% were modified with acrylate groups (13) (see Figure 6). In order to prevent a thermally induced crosslinking, these polymers have to be stabilized with 1-2% of bis (3 tert. butyl-4-hydroxy-5-methylphenyl) sulfide. They can be photochemically crosslinked by adding a Kodak initiator system (15) and by illuminating the sample in the liquid crystalline phase (13) with a mercury lamp (see Figure 6). Further investigations of these photochemically crosslinked elastomers are in progress.

Literature Cited

1. Gleim, W.; Finkelmann, H.; In Side Chain Liquid Crystal Polymers, McArdle, C.B., Ed.; Blackie and Son, Glasgow, 1989, p.287
2. Zentel, R.; Adv. Mater. 1989, 321; Angew. Chem. Int. Ed. Engl., Adv. Mater. 1989, 28, 1407
3. Strzelecki,L.; Liebert, L.; Bull. Soc. Chim. Fr. 1973, 597
4. Bhadani, S.N.; Gray, D.G.; Mol. Cryst. Liq. Cryst. (Lett.) 1984, 120, 255
5. Zentel, R.; Liq. Cryst. 1988, 3, 531; Zentel, R.; Reckert, G.; Bualek, S.; Kapitza, H.; Makromol. Chem. 1989, 190, 2869
6. Beck, B.; Ringsdorf, H.; Makromol. Chem. Rapid Commun. 1985, 6, 291
7. Zentel, R.; Schmidt, G.F.; Meyer, J.; Benalia, M.; Liq. Cryst. 1987, 2, 651
8. Zentel, R.; Reckert, G.; Reck, B.; Liq. Cryst. 1987, 2, 83
9. Kapitza, H.; Zentel, R.; Makromol. Chem. 1988, 189, 1793
10. Finkelmann, H.; Kock, H.-J.; Rehage, G.; Makromol. Chem. Rapid Commun. 1981, 2, 317
11. Vallerien, S.U.; Zentel, R.; Kremer, F.; Kapitza, H.; Fischer, E.W.; Makromol. Chem. Rapid Commun. 1989, 10, 333
12. Terashima, K.; Ichihashi, M.; Kikuchi, M.; Furukawa, K.; Inukai, T.; Mol. Cryst. Liq. Cryst. 1986, 141, 237 (1986); Bahr, C.; Heppke, G.; Mol. Cryst. Liq. Cryst. Lett. 1986, 4, 31
13. Kapitza, H.; Ph.D. Thesis, Universität Mainz, FR-Germany, 1990
14. Vallerien, S.U.; Zentel, R.; Kremer, F.; Kapitza, H.; Fischer, E.W.; Proceedings of the Second Ferroelectric Liquid Crystal Conference at Göteborg, Sweden (June 1989), Ferroelectrics in press; Vallerien, S.U.; Kremer, F.; Kapitza, H.; Zentel, R.; Fischer, E.W.; Proceedings of the Seventh International Meeting on Ferroelectricity (IMF-7) at Saarbrücken, Germany (August 1989), Ferroelectrics in press.
15. Williams, J.R.; Specht, D.P.; Farid, S.; Polym. Eng. and Sci. 1983, 23, 1022

RECEIVED April 10, 1990

PHYSICS OF LIQUID-CRYSTALLINE POLYMERS: TEXTURE AND STRUCTURE

Chapter 17

Chemical Heterogeneity in Liquid-Crystalline Polyesters

C. K. Ober[1], S. McNamee[1], A. Delvin[1,3], and R. H. Colby[2]

[1]Materials Science and Engineering, Cornell University,
Ithaca, NY 14853–1502
[2]Corporate Research Laboratories, Eastman Kodak Company,
Rochester, NY 14650–2110

Random copolyesters based on bromoterephthalic acid, methyl hydroquinone, and hexane diol have been synthesized. Their mesophase properties were studied by differential scanning calorimetry, optical microscopy, real-time X-ray diffraction and melt rheology. At low molecular weight these copolymers exhibit triphasic behavior, where two mesomorphic phases coexist with an isotropic phase. Fractionation based on solubility in THF enables the identification of two components. Simple statistical arguments are employed to model the polymerization reaction and account for the observed phase behavior. When drawn into fibers from the melt, these polymers are observed to have two preferred orientations, with the director of one mesophase parallel to the flow direction and that of the other mesophase perpendicular to the flow direction. Also presented are the first linear viscoelastic data of polymers which form liquid crystalline glasses.

Aromatic liquid crystalline (LC) copolyesters of commercial importance are made by processes which produce compositionally heterogeneous chemical sequences along the polymer chain. Such chemical heterogeneity is responsible for a number of interesting properties of thermotropic LC polymers. These properties include a small amount of

[3]Current address: General Electric Plastics, Mount Vernon, IN 47620

0097–6156/90/0435–0220$06.25/0

crystallinity at low temperatures (1,2) and an extended clearing transition (3,4) at higher temperatures (the nematic/isotropic biphase).

The groups of Windle (1) and Blackwell (2) have demonstrated that particular sequences of monomers in thermotropic LC random copolyesters lead to very small crystalline regions in the solid state. Based on a statistical treatment of random copolymerization one concludes that the probability of sequences matching in two different chains, and hence the degree of crystallinity, increases as the chain length is lowered (1).

In the nematic/isotropic biphase both isotropic and nematic phases coexist at equilibrium over a wide range of temperature. Two origins of biphasic behavior in thermotropic main-chain LC polymers have been identified in the literature: chemical heterogeneity along the chain and polydispersity in chain length. Stupp and coworkers (3,4) have shown that a chemically ordered copolymer has a narrow biphasic region (of the order of 5 K) while the chemically disordered (random) copolymer of the same overall composition has a broad biphase spanning 120 K. Blumstein and coworkers (5) have separated the isotropic and mesomorphic phases in a thermotropic LC biphase and found that the mesomorphic phase contained higher molecular weight polymer than the isotropic phase. As predicted theoretically (6), the molecular weight dependence of the transition temperatures is observed to be quite strong at low molecular weights, but much weaker at high molecular weights (7).

We therefore expect low molecular weight LC polymers to be biphasic over a wide temperature range, as both the chemical heterogeneity and the molecular weight effects on the clearing temperature are large for short chains. In this paper we present an extreme example of the effect of chemical heterogeneity in LC polymers. In a low molecular weight thermotropic LC polyester we show evidence of three phases coexisting (an isotropic phase and two mesophases). One of the two mesophases is believed to be smectic, and the two mesophases were shown to have different responses to flow. During fiber drawing, the high temperature mesophase aligns parallel to the flow direction while the low transition phase aligns perpendicular to the flow direction. At higher molecular weight this "triphasic" behavior disappears. A Flory statistical treatment of the polymerization reaction is used to rationalize these observations.

Experimental

Synthesis of Model Compounds.

We have synthesized semi-rigid rod LC polyesters based on poly(methylphenylene bromoterephthalate-co-hexamethylene bromoterephthalate) (Fig. 1) by random copolymerization. The details of the synthesis of these polymers have been reported previously (8), and are shown schematically in Fig. 1. All chemicals were purchased from Aldrich; methyl hydroquinone, and 1,6-hexanediol were used as received. Bromoterephthalic acid (96%) was recrystallized with an 80/20 vol-% distilled water/ethanol solution. Pyridine (used a proton acceptor) was distilled and stored with desiccant. Solvents were used as received, and all glassware was oven dried.

In a typical reaction, bromoterephthalic acid, (0.0132 mol), was first converted to its acid chloride by refluxing with thionyl chloride and 4-5 drops of DMF as catalyst. After two hours, the unreacted thionyl chloride was distilled off. Each polymerization step was carried out in the same reaction flask as the remaining diacid chloride. After distillation, in a typical polymerization, 0.82 g (6.59×10^{-3} mol) of methyl hydroquinone was dissolved in a graduated funnel with 20 ml methylene chloride and 25 ml pyridine. The solution was slowly dripped into the reaction flask and stirred over night under N_2 atmosphere.

Finally, polymerization was completed by addition of 0.78 g (6.59×10^{-3} mol) hexamethylene diol that had been dissolved in pyridine and methylene chloride. This solution was dripped slowly into the reaction vessel over two hours; the reaction was then stirred for 24 hours under nitrogen atmosphere. The resulting polymer (BP6L) was washed in 1 N HCl, precipitated in methanol, extracted in a Soxhlet extractor with methanol and vacuum dried.

A higher molecular weight polymer (BP6H) was produced by precipitating the reaction product in dry hexane and heating it under vacuum at 150°C for 5 hours. The polymer was then dissolved in methylene chloride and extracted as above. The resulting polymers were fractionated by dissolving the soluble component in THF, centrifuging to isolate the insoluble component and precipitating both in methanol.

Physical Characterization.

^{1}H-NMR was carried out on a 200 MHz Varian XL 200 in hexafluoroisopropanol. Downfield shifts relative to TMS at δ = 4.4 ppm (CH_2-O), 1.5 ppm (CH_2) and 1.95 ppm (CH_2) were due to the hexyl spacer while the multiplet at δ = 2.3 ppm (Me) was due to the methyl group on the hydroquinone. The peak areas due to the hexyl methylenes and the hydroquinone methyl were evaluated to determine the relative amounts of methyl hydroquinone and hexane diol present in the polymers. These values could be used to calculate an average mesogen length, that is, the length of the average aromatic sequence found in the polymer. Differential scanning calorimetry with a Du Pont model 1090 DSC was done on samples annealed at 100°C for 2 hours, with a scan rate of 10°C/minute. Optical microscopy between crossed polarizers was used to visually determine the mesophases present and the clearing temperature.

Molecular characterization of the polymers consisted of intrinsic viscosity measurements and size exclusion chromatography. The intrinsic viscosities were measured in a solvent mixture of phenol and o-dichlorobenzene using Cannon-Ubbelhode viscometers. Polyethylene terephthalate equivalent molecular weights were determined using a Waters 244 GPC in a solvent mixture of methylene chloride and hexafluoroisopropanol. The results from NMR, DSC, and molecular characterization are summarized in Table I. Techniques for the dynamic x-ray diffraction studies are described elsewhere (9).

Table I. Properties of BP6 Fractions

Sample	BP6L	BP6Ls	BP6Li	BP6H
$[\eta]$ (dl/g)	0.20	--	--	0.33
M_n*	6200	7100	5700	14000
M_w/M_n	1.8	1.8	1.7	1.6
Solvent	CH_2Cl_2/ phenol	THF	CH_2Cl_2/ phenol	CH_2Cl_2/ phenol
T_g (°C)	42	39	--	40
Transition	120	120	160	--
Composition‡	Nominal	Nominal	2 MeHQ/diol	Nominal

* Molecular weights measured relative to PET in CH_2Cl_2/HFIP

‡ Composition by [1]H-nmr.

Viscoelastic measurements in oscillatory shear were performed using a Rheometrics System Four rheometer in the frequency range: 10^{-3} rad/sec $< \omega < 10^2$ rad/sec. 8 mm diameter parallel plates were used for temperatures below 80°C and 25 mm diameter parallel plates were used for T>80°C, with plate separations of 1.1±0.2 mm. Comparison of data from the two geometries (at 80°C for BP6L and at 90°C and 101°C for BP6H) indicate that the 8 mm data are a factor of 2.0 lower in modulus level than the 25 mm data. Geometry affects of this order have ben observed previously for LC polymers (10). A simple modulus scale shift of a factor 2.0 was required independent of frequency. No explanation of this shift is offered at this time, but such geometry effects are certainly worthy of further study.

Linear viscoelastic behavior was demonstrated at all temperatures by varying the strain amplitude by a factor of two (at least) in the same frequency range. This was easily accomplished at temperatures near Tg, where a wide range of strain amplitudes with measurable torque amplitudes delivered linear viscoelastic response. At higher temperatures (T>80°C) the empirical criterion that the torque amplitude be kept in the range 2-20 dyne-cm was found to give linear viscoelastic response. This corresponded to a maximum strain of 1% at 101°C and 5% at 150°C (the highest temperature used). Strongly nonlinear response was observed when the torque level was above 20 dyne-cm, which was apparently due to some structural change in the sample. Immediately after applying a large strain to the sample the modulus values at "safe" torques were too low, but eventually would build back up to their original values with time. The molecular origins of this reversible structure breaking/building is unknown, and only the linear viscoelastic response will be discussed here.

Results and Discussion

The synthetic scheme used for the preparation of the LC polymers reported here involves the random copolymerization of both aromatic and aliphatic components as shown in Fig. 1. Both the diol and hydroquinone components can react with nearly equal probability with the diacid chloride component. We therefore have three factors which can contribute to inhomogeneity in this polymer system: (i) the polydispersity of chain length expected in a condensation polymerization, (ii) the distribution of diol and hydroquinone components in the polymer chain, and (iii) the presence of methyl substituents and bromine substituents on the hydroquinone and terephthalate groups, respectively, means that many isomeric structures are also possible.

Several polymers were analyzed for this study and are listed in Table I. BP6L was synthesized as described in the experimental section. From this polymer both THF soluble (s) and THF-insoluble (i) fractions could be isolated and these samples are named BP6Ls and BP6Li respectively. Finally, a fourth sample which had been removed from the BP6L reaction mixture prior to workup by precipitation in dry hexane was further reacted to produce a higher molecular weight sample (BP6H) with the same nominal composition as BP6L.

Thermal Characterization. Preliminary thermal analysis (8) of BP6L and BP6H under repeated heating and cooling cycles revealed no detectable melting or clearing peak above T_g. BP6 therefore appeared to be a liquid crystalline glass showing an LC texture below T_g with no indication of crystallinity. Previously reported LC polymer glasses typically have been side chain LC polymers with siloxane backbones (11). The supposition that bulk BP6 is an LC glass is supported by the absence of sharp crystalline peaks in the x-ray diffraction patterns of these polymers at room temperature (below T_g).

Annealing of the bulk sample at 60°C did not lead to the presence of an endothermic peak. In the absence of an observable endotherm, transition temperatures of both BP6L and BP6H samples were determined optically. Above 127°C, flow was observed visually in these polymers and by 290°C clearing was essentially complete with only a small fraction of birefringent material remaining. The anisotropic BP6L exhibited the focal conic textures usually associated with smectic mesophases. No distinct texture was observed with BP6H. Birefringence returned to the isotropic phase upon cooling below 290°C, and the resulting texture was retained in the sample below T_g.

Between 127°C and 290°C biphasic behavior was observed and was marked by the presence of black isotropic regions of polymer melt under crossed polars. No change was observed in the depolarized light intensity of the isotropic domains during shearing of the sample between cover-slip and slide, and so homeotropic alignment was not the cause of the non-birefringent (black) regions.

Static and Real-time WAXD Studies. X-ray diffraction studies of BP6L are the subject of previous work (9), and the main results are summarized below. Observation of the static WAXD diffractogram of BP6L at room temperature revealed only a sharp peak at 12.6 Å and a very broad peak centered at 3.8 Å (9). Only a single diffuse peak was observable in the wide angle portion of the diffractogram. The 12.6 Å spacing observed in the BP6L WAXD pattern is associated with the distance between adjacent bromine groups in the polymer chain as shown in Fig. 2. This spacing has been observed previously in

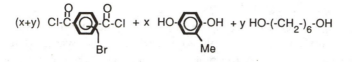

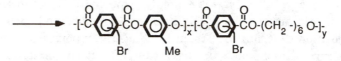

Figure 1.   Synthetic scheme for the preparation of BP6.

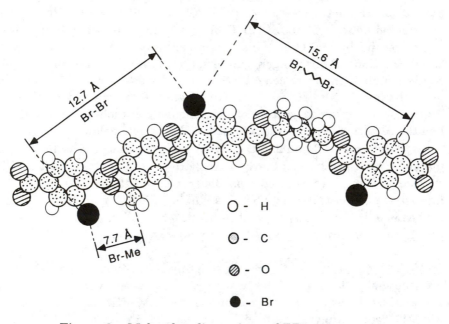

Figure 2.   Molecular dimensions of BP6 structure.

structurally similar substituted LC polyesters with no spacer (12). The observation of this sharp inner ring has not been established as being indicative of the presence or absence of a smectic mesophase. (See Fig. 2).

In our real-time studies, the 12.6 Å peak was monitored as a function of temperature for samples of BP6L fibers which showed two sets of perpendicular arcs. Fig. 3 shows a plot of intensity versus 2θ for the meridional arc as a function of temperature. As temperature was increased, the intensity of the 12.6 Å peak can be seen to decrease. This decrease, which occurred at 120°C for the meridional arc, was observed to take place at 160°C for the equatorial arc. Both transitions took place at temperatures where the sample clearly showed liquid crystallinity by optical microscopy, but no change in texture. Such observations suggested that different species were responsible for the two different transition temperatures; however, these transitions were not readily identified with any visually observed melting or clearing transition.

Additional syntheses were carried out to prepare a series of polymers with slightly different compositions (more or less spacer) from that of nominal BP6L. If we consider the mesogenic core length as the number of consecutive aromatic groups before separation by the hexane spacer, then those polymers with more spacer than methyl hydroquinone (MeHQ) would have a lower average mesogen length. These polymers also exhibited liquid crystalline behavior provided that the MeHQ (H) to diol (D) ratio was greater than 1:3; that is, the polymer had an average mesogen length greater than 2. Powder WAXD patterns were made for the entire compositional range of hexanediol to methylhydroquinone (MeHQ). A peak at 12.6 Å appeared in those polymers with H:D ratios of ranging from 2:3 to 1:0 or mesogen lengths of 2 and greater. Only those copolymers with little hydroquinone had no 12.6 Å peak. That the same d–spacing is present in polymers both with and without spacer is further support for the identification of the 12.6 Å spacing as being the bromine-to-bromine distance in the bromoterephthalate-MeHQ-bromoterephthalate sequence.

Linear Viscoelasticity of Unfractionated Samples. The BP6L and BP6H samples were found to give reproducible data at temperatures below 120°C if first exposed to 150°C for 5 minutes. After such a heat treatment measurements were made on these samples at T = 35, 41, 50, 60, 70, 80, 90, 101, and 120°C. The empirical time-temperature superposition principle (13) was found to be valid for BP6L between 60°C and 120°C and for BP6H between 40°C and 120°C, and was used to make master curves at a reference temperature of 101°C (Figs. 4 and 5). The modulus scale

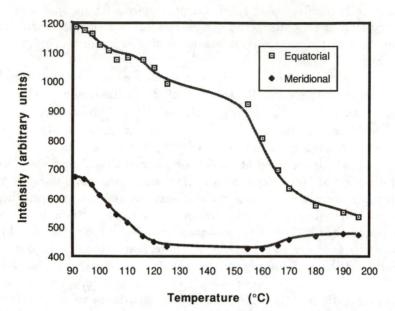

Figure 3. Temperature dependence of x-ray diffraction intensity for equatorial and meridional arcs (BP6L). Heating rate = 10 K/min.

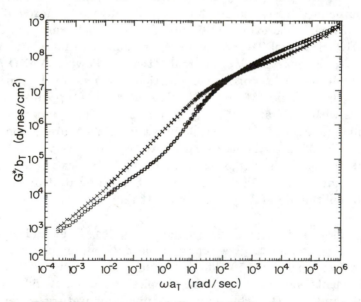

Figure 4. BP6L Master curve at T_{ref} = 101°C. Storage Modulus G' (O); Loss Modulus G"(×).

shifts were small and are not discussed here. The frequency scale shifts a_T were found to obey the WLF equation ([13]),

$$\log a_T = \frac{-c_1^g \, (T - T_g)}{c_2^g + T - T_g} \tag{1}$$

with c_1^g = 12. for BP6L, c_1^g = 15. for BP6H, and c_2^g = 80. K for both polymers. Data at lower temperatures did not superimpose, as is often observed for fully flexible polymers as well ([14]). The breakdown of time-temperature superposition was more obvious for the low molecular weight polymer (BP6L), which is again expected from results on flexible polymers ([14]). Time-temperature superposition also failed to work in the biphasic region (above 127°C) for the obvious reason that the phase structure changes with temperature.

There are three recognizable regimes of viscoelastic response in Figs. 4 and 5. BP6L and BP6H form liquid crystalline glasses on cooling. Focussing on Fig. 5, above $\omega \, a_T = 10^5$ rad/sec the sample exhibits a broad transition zone which ends at a glassy modulus at very high frequencies. The glassy modulus of BP6L is roughly 10^{10} dynes/cm^2. It is not shown in Figs. 4 and 5, because time-temperature superposition did not work at low temperatures. The complex moduli obey a power law in frequency in the frequency range 10^5 rad/sec $< \omega \, a_T < 10^7$ rad/sec,

$$G' \sim G'' \sim \omega^{0.45} \tag{2}$$

which is similar to the transition zone response of flexible polymers ([13]) except that the exponent is lower than typically observed (0.5-0.7). The broad transition zone, reflected in the surprisingly low value of the exponent, may be related to the chemically heterogeneous nature of this polymer.

At low frequencies (below $\omega \, a_T = 10^{-2}$ rad/sec for BP6L and below $\omega a_T = 1$ rad/sec for BP6H) there is a terminal response reminiscent of a branched polymer slightly beyond the gel point ([15]) with $G' \sim G'' \sim \omega^{0.7}$. This terminal response undoubtedly reflects the large scale structure of the sample, but unfortunately this structure is not well understood. Specifically, we do not know whether this is a characteristic of partly smectic LC polymers or if it is influenced by the trace amounts of crystallinity in the samples. (See Compositional Heterogeneity below).

At intermediate frequencies the two samples show differences which are presumably due to their different molecular weights. BP6H exhibits a rubbery plateau in the frequency range 10 rad/sec < $\omega\, a_T$ < 10^5 rad/sec which is not unlike the rubbery plateau of flexible polymers of similar molecular weight. BP6L exhibits a much shorter rubbery plateau (consistent with its lower molecular weight, 10^3 rad/sec < $\omega\, a_T$ < 10^5 rad/sec). BP6L also relaxes a great deal in the frequency range 10^{-2} rad/sec < $\omega\, a_T$ < 10^3 rad/sec before going into its terminal response. The result of this extra relaxation is a terminal response with the same power law as BP6H (as discussed above) but two orders of magnitude lower in modulus level.

Compositional Heterogeneity. We have observed that it is possible to extract two major constituents from the low molecular weight sample, based on solubility in THF. One fraction representing approximately 20 wt-% was insoluble in THF (BP6Li) while the remainder of the sample was THF soluble (BP6Ls). After annealing at 100°C for 2 hours, the THF insoluble fraction showed a thermal transition at 160°C while both the soluble fraction and the bulk sample possessed a thermal transition at 120°C. Thermal traces of annealed BP6L, and the annealed BP6Ls and BP6Li fractions as well as the high molecular weight BP6H are given in Fig. 6.

Powder diffraction patterns of the two fractions displayed characteristics nearly identical to those of BP6L, that is, a sharp peak at 12.6 Å and a broad peak centered at 3.8 Å. The powder diffraction patterns of BP6L, BP6Ls and BP6Li have been reported previously (9). Only the THF insoluble fraction showed any indication of significant crystallinity with additional peaks centered at 3.9 Å and 6 Å. The presence of the same 12.6 Å spacing in both patterns is due to the fact that all fractions have the same bromine-to-bromine distance in the mesogenic core.

Studies of these fractions by [1]H-nmr using HFIP (hexafluoro-isopropanol) solvent have shown that the soluble fraction has a composition with a 1:1 diol:MeHQ ratio whereas the insoluble component is enriched in the phenol monomer with a 1:2.8 diol:MeHQ ratio. These are equivalent to average mesogenic core lengths of approximately 3 and 6 for the soluble and insoluble fractions respectively, where the mesogenic core length is defined as the average number of directly-linked aromatic groups.

In the THF insoluble fraction (BP6Li), the observation of a focal conic texture as well as the presence of bâtonnets prior to clearing leads

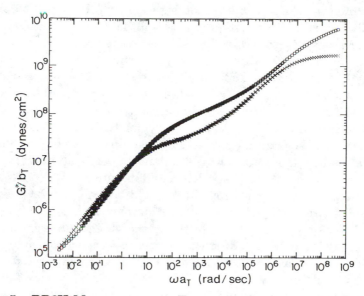

Figure 5. BP6H Master curve at $T_{ref} = 101°C$. Storage Modulus G' (○); Loss Modulus G"(✕).

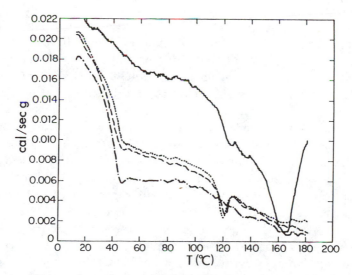

Figure 6. Thermal analysis of BP6 polymers after 2 hour anneal at 100°C. BP6L (·····); BPLs (- - -); BP6Li (–); BP6H (– · –); Heating rate = 10°C/min.

to the conclusion that this component of the polymer is smectic. A very interesting comparison can be made between the THF insoluble fraction and poly(methylhydroquinone bromoterephthalate), a liquid crystalline polyester recently reported by Jin and coworkers (16). This polymer has a structure identical to the BP6 polymer without the hexamethylene spacer. In their studies, a melting transition of 160°C was reported for the homopolymer. X-ray diffractograms of their sample also showed a sharp peak at ~12 Å as well as a broad peak centered at ~4 Å. Finally, optical microscopy showed that the polymer had a focal conic texture indicative of a smectic mesophase. These observations are virtually identical both to those made of BP6Li, and to poly(methylhydroquinone bromoterephthalate) synthesized in our laboratory. The THF-soluble fraction appears to be nematic which would be expected for mesophase behavior of a random copolymer. The difference in mesophase behavior would then be the driving force for the unexpected phase separation.

In dynamic x-ray diffraction studies reported earlier (17), perpendicular equatorial and meridional arcs with the same d-spacing were observed in drawn fibers. Both pairs of arcs showed different transition temperatures. DSC of annealed samples showed a small endotherm at 120°C which occurred at the same temperature observed for the transition in the BP6L meridional diffraction arc. The endothermic transition of the annealed THF insoluble fraction corresponds to the 160°C transition of the BP6L equatorial diffraction arc. Both fractions exhibit mesophase behavior above the observed thermal transitions, and only a subtle textural change is evident at that temperature under crossed polars, indicating that these thermal transitions are due to trace amounts of crystallization. The BP6Li fraction displays characteristics of the smectic mesophase while the texture and x-ray observations of BP6Ls do not allow conclusive identification of the mesophase (presumably nematic).

The presence of the two sets of perpendicular arcs at the same d-spacing can be attributed then to two chemically different fractions which are in different mesophases and respond differently to fiber drawing. Since the spacing of interest corresponds to the bromine-bromine distance of the mesogenic core, the orientation observed in the aligned sample is indicative of the chain direction. The transitions observed by dynamic x-ray diffraction, melt rheology and thermal analysis at 120°C and 160°C are most probably melting of a small fraction of crystalline material. The orientations observed in the drawn fiber are apparently due to crystalline material which was aligned in the triphase by the drawing operation.

Linear Viscoelasticity of BP6L Fractions. The BP6Ls sample (the fraction which is soluble in THF) has a nematic/isotropic biphase in the temperature range 127°C < T < 290°C. Below 127°C there is only one phase (the mesophase, which is presumably nematic). The BP6Li sample (the insoluble fraction) has a biphase above 240°C. Below 240°C BP6Li is in a purely smectic mesophase. As discussed in the previous section, annealing at 100°C alters slightly the phase behavior of each fraction. Figure 6 shows that a small amount of crystallization occurs on annealing in the BP6Ls fraction (with melting point 120°C) and a larger amount of crystallization in the BP6Li fraction (with melting range 120 - 160°C).

The linear viscoelastic response of the two fractions is very different. Fig. 7 shows the temperature dependence of the complex moduli at 1 rad/sec for both fractions. Each fraction shows sharp drops in modulus when their respective melting temperatures are reached (120°C for BP6Ls and 160°C for BP6Li, see Fig. 6).

Isothermal frequency sweeps at 104°C are compared in Figure 8. The viscoelastic response of the insoluble fraction is similar to that of a rubbery network with modulus $G' = 2 \times 10^6$ dynes/cm^2 independent of frequency at low frequency. The soluble fraction, on the other hand, relaxes significantly at 104°C, and its viscoelastic response is quite similar to the unfractionated BP6L. The similar viscoelastic responses of BP6L and BP6Ls are hardly surprising as BP6Ls is by far the major component of BP6L.

The rheological response of both BP6L and BP6Ls were observed to be time dependent at 104°C. This is demonstrated in Fig. 9 where the effect of a 27 hour anneal at 104°C is shown compared to an unannealed sample. Similar anneals in the DSC were made with only very subtle changes in the amount of crystalline material. The dramatic rheological change on annealing shown in Fig. 9 is completely reversible by heating to 150°C for five minutes. Apparently very small changes in the extent of crystallization can lead to significant changes in rheological properties.

Effect of Copolymerization on Monomer Sequence Distribution. As in any step growth copolymerization, a distribution of sequence lengths is generated. It is possible to calculate a distribution of mesogen lengths in the manner of Flory (18). In the two-stage synthesis employed, the oligomeric aromatic mesogenic groups were first built up. A similar approach has been used by Stupp to show the effect of a disordered sequence on the clearing transition (19). Based on the 2:1

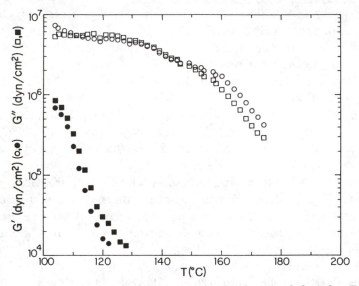

Figure 7. Temperature dependence of complex modulus for BP6Ls (●,■) and BP6Li (○,□) at a frequency of 1 rad/sec and 1% strain. Data were taken in 2K intervals (heating) with a 30 minute thermal equilibration at each temperature.

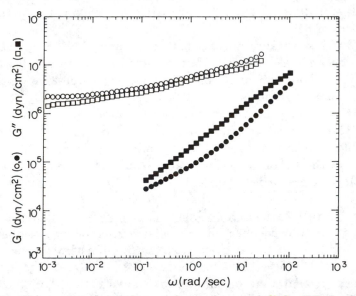

Figure 8. Frequency dependence of complex moduli for BP6Ls (●,■) and BP6Li (○,□) at 104°C and 1% strain.

terephthalate:MeHQ molar ratios used, a distribution of mesogen sizes was calculated using the relationship (18):

$$w_x = xr^{(x-1)/2}(1-r)^2/(1+r) \tag{3}$$

In this expression, r is the ratio of MeHQ to terephthalate molecules (= 0.5), w_x is the weight fraction of x-mers and x is the chain or mesogen length (here restricted to odd integral values). The results of the calculation are shown in Fig. 10 and represent the calculated distribution of lengths of the aromatic segment formed in the first stage of the polymerization. Clearly, the average value of mesogen length is approximately 3, but on a number basis there are many lone terephthalate groups and consequently on a weight basis there are many segments with a high mesogenic length.

Assuming that only mesogenic groups with acid chloride end groups will lead to chain extension in the second stage of the polymerization, then these odd length mesogenic groups can be added together based on relative population to form LC polymers. Using very simplistic simulations in which either 10 or 20 mesogenic groups were selected at random from the weighted population to form a polymer chain, the average mesogenic core length for all 5,000 resulting chains were calculated. These results are shown in Fig. 11 where average mesogen length is plotted for 10-mers and 20-mers as a function of relative population. As can be seen, with increased chain length the distribution of mesogenic lengths has narrowed.

These results are consistent with experimental results, since the simulation suggests that there is less of a high aromatic fraction in the longer chain LC polymer. We could not isolate more than 5 weight-% THF-insoluble fraction from BP6H whereas 20% of BP6L was insoluble in THF. The insoluble fraction of the 10-mers represented 20 weight-% of the sample, and as shown in Fig. 11 this is the component with mesogenic lengths (>~5) almost lacking in the simulated 20-mers. Bulk composition by [1]H-nmr was the same for both samples, but in the high molecular weight sample chemical heterogeneity was averaged out over the greater chain lengths. Clearly then, the phase separation behavior is more prominent for smaller chains.

Conclusion

In LC polymers possessing mesogenic cores linked by ester groups, it is possible for transesterification to occur at elevated temperatures. In

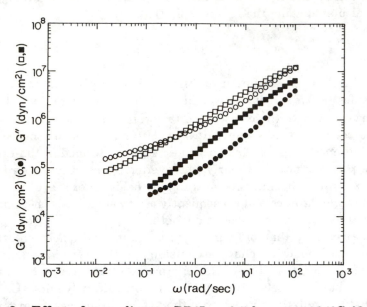

Figure 9. Effect of annealing on BP6Ls. 0.5 hours at 104°C (●,■);
27 Hours at 105°C (○,□).

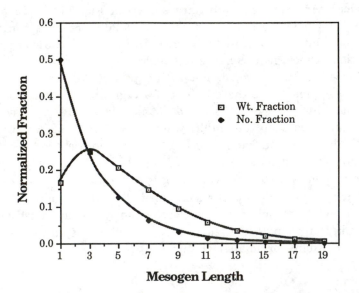

Figure 10. Weight and number distributions of mesogens. Mesogen
length is defined as the number of aromatic units per mesogen.

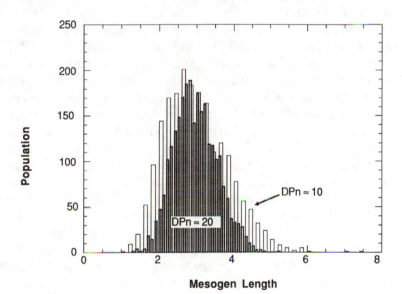

Figure 11. Population distribution for two 5,000 chain simulations corresponding to $DP_n = 10$ (open bars) and $DP_n = 20$ (shaded bars).

addition, LC polymers produced by random copolymerization will possess a distribution of composition introduced during polymer synthesis. As demonstrated by our Flory-style statistical treatment of the copolymerization, compositional heterogeneity is strongest in low molecular weight samples. Fractionation based on solubility in THF apparently separated the low molecular weight sample into compositionally distinct species which had roughly the same molecular weight and molecular weight distribution (see Table I). The mesophase behavior and rheological properties of these two fractions are very different. The phase behavior of these two fractions corresponds very well with that observed for the two components identified in BP6 using synchrotron diffraction. We conclude from this behavior that compositional heterogeneity has a very large effect on the mesophase behavior of LC polyesters, and is a significant source of their biphasic behavior, as suggested by Stupp (3,4). Presumably the distribution of molecular weights also contributes to the biphase (5), but this effect is secondary to that of compositional heterogeneity in this polymer system.

In preparing BP6, an addition sequence was used which created a distribution of mesogenic core lengths in the polymer. Low molecular weight polymer BP6L has an unusually complex phase behavior, with at least three phases existing in equilibrium (two mesomorphic phases and an isotropic phase). This complex phase behavior is apparently the cause of the observation of two preferred orientations in melt drawn fibers of BP6L. The higher molecular weight version of the same polymer (BP6H) also has a biphase over a broad temperature range, but does not exhibit the presence of multiple mesophases as does BP6L.

The addition of bromine groups to the mesogen imparted solubility to this LC polymer and gave it low transition temperatures. Unfortunately the bromine substituents dominate the x-ray diffraction characteristics of these polymers, making mesophase identification very complex and almost entirely dependent on texture identification. LC polymers with new structures having low clearing temperatures and solubility in common organic solvents are presently being synthesized which will not contain such large, electron-rich atoms. We expect that these polymers will be able to resolve some of the questions raised by these studies.

Acknowledgments

The authors would like to thank the Cornell Materials Science Center funded through NSF-DMR for financial support and use of the MSC facilities. Acknowledgement is made to the Donors of the

Petroleum Research Fund, administered by the American Chemical Society, for the partial support of this research. NSF grant DMR-8717815 is acknowledged for partial support of S. Mc. We would like to thank L. Williams (supported by an NSF/REU) for assistance during this project. Thanks also go to R. Miller (SEC), T. Mourey (molecular weight) and J. Gillmor (thermal measurements) of Kodak for their invaluable help. We would like to thank the Cornell High Energy Synchrotron Source (CHESS) funded through NSF DMR 84-12465 for use of the facility. Finally, we are indebted to Dr. Terry Bluhm (Xerox Corp.) for many helpful discussions.

Literature Cited

1. Hanna, S.; Windle, A.H. Polymer, 1988, 29, 207.
2. Gutierrez, G.A.; Chivers, R.A.; Blackwell, J.; Stamatoff J.B.; Yoon, H. Polymer, 1983, 24, 937.
3. Martin, P.G.; Stupp, S.I. Macromolecules, 1988, 21, 1222.
4. Stupp, S.I.; Moore, J.S.; Martin, P.G., Macromolecules, 1988, 21, 1228.
5. d'Allest, J.F.; Wu, P.P.; Blumstein, A.; Blumstein, R.B. Mol. Cryst. Liq. Cryst. Lett., 1986, 3, 103.
6. ten Bosch, A.; Maissa, P.; Sixou, P. J. Phys. Lett. (Paris), 1983, 44, L105.
7. d'Allest, J.F.; Sixou, P.; Blumstein, A.; Blumstein, R.B., Mol. Cryst. Liq. Cryst., 1988, 157, 229.
8. Delvin, A.; Ober, C.K. Polym. Bull., 1988, 26, 241.
9. Ober, C.K.; Delvin, A.; Bluhm, T.L. J. Polym. Sci., Polym. Phys. Ed., in press.
10. Wissbrun, K.F.; Kiss, G.; Cogswell, F. N. Chem. Eng. Comm., 1987, 53, 149.
11. Aguilera, C.; Bartulin, J.;Hisgen, B.; Ringsdorf, H. Makromol. Chem., 1983 , 184, 253.
12. Atkins, E. D. T.; Thomas, E.L.; Lenz, R.W.; Mol. Cryst. Liq. Cryst., 1988, 155, 263.
13. Ferry, J.D. Viscoelastic Properties of Polymers, Third Edition, Wiley: New York, 1980.
14. Plazek, D.J.; Rosner, M. J.; Plazek; D. L. J. Polym. Sci., Polym. Phys., 1988 26, 473.
15. Rubinstein, M.; Colby, R.H.; Gillmor, J.R. In Space-Time Organization in Macromolecular Fluids; pg. 66, Tanaka, F.; Ohta, T.; Doi, M. Ed.; Springer-Verlag: Berlin, 1989.

16.   Jin, J.-I.; Choi, E.-J.; Jo, B.W. <u>Macromolecules</u>, 1987, <u>20</u>, 934.

17.   Delvin, A.; Ober, C.K.; Bluhm, T.L. <u>Macromolecules</u>, 1989, <u>22</u>, 498.

18.   Flory, P. J. <u>Principles of Polymer Chemistry</u>, Cornell University Press: Ithaca, NY, 1953.

19.   Stupp, S.I. <u>Polym. Prep.</u>, 1989, <u>30</u>(2), 509.

RECEIVED March 15, 1990

Chapter 18

Light Microscopy of Liquid-Crystalline Polymers

C. Viney

Department of Materials Science and Engineering FB–10, and the Advanced
Materials Technology Program, University of Washington, Seattle, WA 98195

The light microscope can be used to identify liquid crystalline phases,
characterize molecular order, and quantify the distribution of defects. As
such, the technique is seldom exploited to maximum advantage,
particularly if studies are restricted to observations between crossed
polars in transmitted light. Also, there is much scope for misinterpreting
the observed contrast. This chapter offers a systematic description of the
options (and pitfalls) that are relevant to a research microscopist
studying the structure of liquid crystalline polymers.

Interpreting textures observed with a light microscope provided early clues to the
molecular order in low molecular weight liquid crystals (1). The microscopical
techniques used for such characterization were adapted from optical crystallography.
Formally, the term "texture" denotes the microstructure of a thin liquid crystalline
specimen viewed under bright field conditions and *between crossed polars*. Textures
usually are observed with the specimen confined between glass surfaces (e.g.
microscope slide and cover slip), though the upper glass surface may be absent in some
cases. Qualitative microscopical studies now are used routinely to identify liquid
crystalline phases: the observed texture is compared against an atlas of standards
obtained from materials for which the symmetry of molecular order is known already
(2,3).

Microscopists working with thermotropic low molecular weight liquid crystals are
rarely concerned about limits of resolution. The scale of microstructures, i.e. the
domain size or the separation of defects, typically is at least several microns and
therefore is resolvable. (In the case of lyotropic systems, where light microscopy may
be used to determine the critical concentration required for mesophase formation, it is
unlikely that the initial nucleation will be on a sufficiently coarse scale.) The
equilibrium mesophases of low molecular weight liquid crystals form at readily
accessible temperatures, which also facilitates microscopy. Furthermore, the molecules
have simple structures and correlations, so that the principal axes of optical properties
can be identified as the principal crystallographic axes. Light microscopy has the
advantage of being non destructive: the same sample can be used to study a number of
thermal histories, and only small (≤ 1 mg) sample sizes are required.

0097–6156/90/0435–0241$06.00/0

The technique also has limitations:
- We noted above that, in the case of *multi-component* systems, a liquid crystalline phase can only be detected microscopically if the volume fraction and degree of coarsening result in a *resolvable* microstructure.
- There are occasions when specimens exhibit a *paramorphotic* texture, i.e. one that reflects order inherited from the parent phase. At a molecular level, the transition between liquid crystalline phases typically occurs via a route that requires the minimum instantaneous rearrangement of molecules. Because textures are dictated by molecular order, the immediate post-transition texture may not easily be distinguishable from its pre-transition counterpart. Stable textures that are characteristic of the new phase may require a long time, sometimes months, to form. Transitions between highly ordered smectics are especially likely to favour paramorphoses.
- Some materials *crystallize* to give a microstructure that can be mistaken for a smectic mosaic texture. A simple test involves reheating the sample as soon as the "mosaic texture" has formed; if the transition is not reversed at close to the same temperature, we are dealing with crystallization (Neubert, M.E., Kent State University, personal communication, 1989). A reversible transition would be inconclusive, i.e. it would be consistent with the formation of either a crystalline or a liquid crystalline phase. However, reversibility is expected over a greater range of cooling rates in the latter case.

In all three cases just described, ambiguities can be clarified by using complementary characterization techniques. For example, x-ray diffraction gives a direct measure of order at the molecular level. However, such results represent an *average* over some bulk volume of sample that is greater than the typical domain size, and so do not give an indication of the *variation* in molecular orientation between domains. Electron microscopy may be preferred to x-ray diffraction in this respect, at least in the case of *polymeric* liquid crystals where the structure of the fluid mesophase often is quenchable. However, the interaction between the electron beam and the specimen can be destructive, leading to changes in the molecular weight distribution. Thermal analysis can be used to characterize the extent to which order changes during phase transitions.

Liquid Crystalline Polymers

Because the textures of liquid crystalline polymers (LCPs) are qualitatively similar to those of low molecular weight liquid crystals, they are interpreted in the same way. However, the microscopy of LCPs is less straightforward:
- Higher temperatures are needed to obtain equilibrium liquid crystalline phases.
- The microstructures are typically an order of magnitude finer in scale, and so are more difficult to resolve in the light microscope. (The reason for this difference in scale is still unknown.)
- The intrinsic birefringence typically is higher: the polarizability along polymer backbones can be larger than along the backbone of the corresponding monomer, if there is significant opportunity for electron delocalization in the polymer.
- Molecules in LCPs may develop correlated rotations about their chain axes, or correlated side chain rotations may develop, over optically resolvable distances. The optical properties of a correlated domain depend both on the configurational and conformational symmetry of the individual molecules and on the extent to which this symmetry is preserved on a larger scale by the correlations.

Together, these characteristics complicate the interpretation of contrast. The interactions between light and the microstructure must therefore be considered carefully, as in the next section of this chapter.

One mitigating feature of LCP polymer microstructures is that they often are quenchable, due to the relatively slow kinetics of crystallization. Thus one can examine microstructures at resolutions that are inaccessible while the specimen is enclosed in a heating stage. (Objectives with the highest numerical apertures have a short working distance, and all objectives should be protected from above-ambient temperatures.) However, the gain in resolving power may not be sufficient to offset the decrease in microstructural scale.

The Interpretation of Extinction

When liquid crystalline specimens are viewed between crossed polars, it is the positions of extinction bands, and how the positions change as the crossed polars are rotated, that is used to find the point-to-point variation in molecular orientation. It is appropriate to examine the principles on which this analysis is based:

We consider a monodomain sample viewed between crossed polars in monochromatic light (Figure 1). The formula for transmitted intensity is derived as follows:
  • Light transmitted by the polarizer is resolved into the vibration directions of the specimen.
  • Propagation through the specimen introduces a phase difference between the component rays, because the two vibration directions are associated with different refractive indices n_1 and n_2.
  • Both components are resolved into the transmission direction of the crossed analyzer, and are then superimposed.

For light of unit intensity and wavelength λ incident on the specimen (thickness d), the transmitted intensity is (4):

$$I = \sin^2\left[2\theta\right]\sin^2\left[\frac{\pi d}{\lambda}(n_2\text{-}n_1)\right]$$

(1)

Equation 1 predicts that a specimen *between crossed polars* will appear dark if:
 • The specimen is optically isotropic, i.e. the birefringence is zero;

$$n_2\text{-}n_1 = 0$$

In this case, the specimen will appear dark for all rotations of the crossed polars. It will also remain dark if it is tilted about a horizontal axis, i.e. one that lies in the plane of the microscope stage. These conditions hold for all wavelengths of monochromatic light as well as for white light.
 • The specimen is optically anisotropic and is being viewed along an optic axis, i.e. the birefringence again is zero. The specimen again will appear dark for all rotations of the crossed polars, but it will transmit light if tilted about an axis lying in the plane of the stage. These conditions hold for all wavelengths of monochromatic light as well as for white light.
 • The specimen is optically anisotropic and its vibration directions are parallel to the transmission directions of the polars;

$$\theta = \pm\frac{m\pi}{2} \; ; \quad \text{m is any integer}$$

The specimen should appear light if the crossed polars are rotated (or if the specimen is rotated in the plane of the microscope stage). These conditions hold for monochromatic light as well as for white light.
- The specimen is optically anisotropic and is being viewed in monochromatic light, and the combination of specimen thickness and birefringence represents a path difference of integral wavelength for the two refracted rays;

$$d\left(n_2 - n_1\right) = m\lambda$$

The specimen should appear light if the illuminating wavelength is changed.

A number of assumptions are often taken for granted by microscopists who use Equation 1:
- The optical properties of the specimen are constant at all points along a given normal.
- Light emerging from any one domain exhibits properties that characterize its passage through that domain only.
- The transmitted intensity depends only on interactions between light and the specimen, and between light and the polars.
- Contrast is due to specimen birefringence only.
- The vibration directions in the specimen can be identified with structural symmetry axes.

The first four assumptions are implicit to the derivation of Equation 1, while the fifth arises from misunderstanding. We will address the validity of each assumption individually in subsequent sections below:

Are the Optical Properties of the Specimen Constant at all Points Along a Given Normal? The assumption usually is valid if the specimen thickness clearly is smaller than the lateral extent of a typical domain.

If the specimen is thicker, domains are likely to overlap, and we must be concerned about whether extinction can ever occur when superimposed birefringent domains are viewed between crossed polars. By way of a simple example, we can consider the case of two domains with non-coincident vibration directions. The formula for transmitted intensity can be derived by analogy to the analysis leading to Equation 1:
- Light transmitted by the polarizer is resolved into the vibration directions of the first domain.
- Propagation through the domain introduces a phase difference between the component rays.
- Both components emerging from the first domain are resolved into the vibration directions of the second domain.
- A phase difference is introduced between the component rays in this domain too.
- Both components emerging from the specimen are resolved into the transmission direction of the crossed analyzer, and then superimposed.

Therefore, once light has entered the final layer in a multi-layer specimen, it will in general have been resolved into *two* vibration directions, *regardless of the vibration direction of the polarizer*.

So extinction by an analyzer is only possible if the components emerging from the specimen can specifically *interfere to give linearly polarized light*. This in turn depends on
- the orientation of the first layer relative to the polarizer, and the orientation of the two layers relative to each other;
- the phase differences introduced by the individual layers;
- the wavelength of light used.

We can conclude that *white light will not be extinguished* for any combination of polar vibration directions. If a specimen is viewed between crossed polars in white light, and a region fails to go into extinction as the specimen stage is rotated, the optical orientation in that region is not constant through the specimen thickness.

Algebraic calculation of transmitted intensity by the method outlined above is extremely cumbersome when applied to multi-layer specimens, though it does retain sight of the underlying physics. The effect of a sequence of birefringent layers on polarized light can be represented more efficiently by one of the following methods:

- The Poincaré sphere construction (5-7):
  This graphical technique is most useful for determining the qualitative effect of a given sequence of layers. (Quantitative calculations are possible too, but the necessary spherical trigonometry is no less cumbersome than the algebraic approach.)
- Matrix methods (6-9):
  Various matrix representations have been contrived to represent the orientation and optical properties of individual domains. The incident polarized light is resolved into "horizontal" and "vertical" components relative to the same reference axes, and is represented by a column matrix. The final polarization state is then calculated by matrix multiplication:

$$\begin{bmatrix} H \\ V \end{bmatrix} = \begin{bmatrix} a_{11} a_{12} \\ a_{21} a_{22} \end{bmatrix} \begin{bmatrix} b_{11} b_{12} \\ b_{21} b_{22} \end{bmatrix} \cdots \begin{bmatrix} h \\ v \end{bmatrix}$$

Several generic optical characteristics of multi-layer specimens can be derived with reference to the mathematical properties of matrices (10).

Either matrix theorems or the Poincaré sphere can be employed to prove that, **if** the individual layers are only weakly birefringent (e.g. due to being extremely thin), a multi-layer specimen is *optically equivalent to a single birefringent domain* in monochromatic light. However, the orientation that gives extinction will vary with wavelength, so that such a specimen can be distinguished from a true monodomain.

Does Light Emerging from Any One Domain Exhibit Properties that Characterize its Passage Through That Domain Only? Even if light is incident normally on the lower surface of the specimen, extraordinary rays can deviate from this direction and thus enter neighboring domains. The fraction of a domain's top surface that may transmit some light originally incident on a neighboring domain will increase as the specimen thickness increases. If this fraction is to be held below 50% for the smallest resolvable domain in a single-layer specimen, the thickness should not exceed (11)

$$\frac{\lambda \varepsilon \omega}{NA \left| \varepsilon^2 - \omega^2 \right|}$$

where NA = numerical aperture of objective;

λ = wavelength of light;

ω, ε = principal refractive indices of (optically uniaxial) specimen.

The maximum tolerable thickness therefore is predicted to decrease with increasing birefringence. The high degree of molecular alignment, and the conjugation that is often associated with rigid molecules, is responsible for the high birefringence of many LCPs. For example, Kevlar fibers have a birefringence of 0.45 (12), which is significantly higher than that of commonly encountered inorganic crystals or conventional polymer fibers.

We have now recognized two ways in which the intensity transmitted by a given domain between crossed polars is affected by adjacent material:

- If the light passes through material above and below the domain of interest, the final polarization state does not uniquely convey information about the optical properties of that domain.
- Extraordinary rays may enter the domain of interest from laterally adjacent material, even in a single-layer specimen.

Does the Intensity Depend Only on Interactions Between Light and the Specimen, and Between Light and the Polars? Equation 1 predicts that

$$I \propto \sin^2(2\theta)$$

i.e. the transmitted intensity is unaffected by 90° rotations of either the specimen stage or the crossed polars.

Exceptions to this prediction have been observed experimentally ([13]). They are due to one or more *interactions between light and components of the microscope* that are overlooked in the derivation of Equation 1:

- anisotropic transmission of light that is obliquely incident at the interface between the specimen and the immersion medium; this anisotropy is enhanced by optical pleochroism in the specimen, by oblique ray paths through the microscope (and therefore by fine microstructures), and by marked one-dimensional periodicity in the microstructure;
- anisotropic transmission at the beam splitter that divides light between the eyepieces and the camera;
- the amplitude transmitted by Polaroid sheet (i.e. the polarizer and analyzer) depends not only on the vibration direction of the incident light, but also on the azimuth and obliquity of the ray direction (Figure 2).

Is Contrast Due to Specimen Birefringence Only? Amplitude extinction between crossed polars, as described by Equation 1, may be only one source of extinction. Regions of the specimen may also appear dark for one or more of the following reasons:

- Amplitude contrast can arise from discontinuities in refractive index.
- Phase contrast can arise from discontinuities in refractive index.
- The specimen may be pleochroic as well as birefringent.

Because these sources of contrast do not require the specimen to be viewed between crossed polars, the contrast will persist if the analyzer is withdrawn.

Can the Vibration Directions in the Specimen be Identified with Structural Symmetry Axes? Simple observations of microstructure as a function of crossed polar rotation will reveal how the *in-plane* directions of maximum and minimum refractive index (i.e. the extinction directions) vary as a function of lateral position in a thin specimen. Microscopists are often tempted to identify one of these in-plane directions as coinciding with the local preferred direction of molecular alignment, commonly referred to as the *director*. However, at the very least one must allow for the director having a possible out-of-plane component. It is then appropriate to ask whether the *projection of the director onto the specimen plane* coincides with one of the in-plane vibration directions. This condition is met only if the corresponding point group includes a 3-, 4- or 6-fold rotation or inversion symmetry axis, i.e. if the *optical properties are uniaxial*. The existence of an out-of-plane component can then be inferred by using the light microscope alone, *if* the microstructure is coarse enough for an interference figure to be

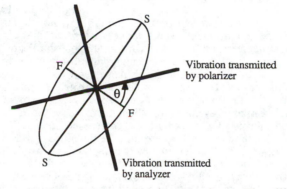

Figure 1.   The geometry relevant to deriving Equation 1. The elliptical outline represents the intersection between the optical indicatrix and the specimen plane. SS and FF respectively identify the slow and fast vibration directions in the specimen.

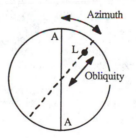

Figure 2.   Defining the azimuth and obliquity of a light ray (L) incident on Polaroid sheet that transmits vibrations parallel to AA. The transmitted intensity depends on both azimuth and obliquity, even if the incident light vibrates parallel to AA before arriving at the Polaroid sheet. Stereographic projection.

obtained. A variation in birefringence across the specimen *may* indicate an out-of-plane tilt, but does not suffice to demonstrate it (14): birefringence is representative of both the orientation *and shape* of the optical indicatrix, and therefore is affected by the direction *and perfection* of local alignment.

If the point group symmetry is equivalent to orthorhombic or lower, i.e. if the *optical properties are biaxial*, the local extinction directions are *not* simply related to the director as projected onto the specimen plane. It may be possible to use light microscopy to establish whether or not a region of the specimen is optically biaxial:

- The distinction is easy if the microstructure is sufficiently coarse to yield interference figures for the individual domains. However, sufficiently large domains are uncommon in the case of LCPs. Magnetic or electric fields are sometimes used to coarsen the microstructure (15,16), though this is only reliable if one can be certain that the field does not induce artificial biaxiality.
- If the microstructure is too fine, and the material is pleochroic, one can attempt to identify the azimuths for maximum and minimum absorption of linearly polarized light (17): in optically biaxial material, these need not coincide with the local extinction directions. The use of optical pleochroism as just described is one example of how studies limited to observations between crossed polars do not exhaust the information available from light microscopy.
- A specimen viewed in white light between crossed *circular* polars (Figure 3) exhibits local extinction only where the birefringence is zero. If a uniaxially drawn LCP specimen consists of optically uniaxial domains, their optic axes will be concentrated close to the macroscopic draw axis. If the domains are optically biaxial, their optic axes will be concentrated in a band of orientations distinct from the draw axis (Figure 4). The two possibilities can therefore be distinguished by sectioning the drawn specimen at various angles and examining the sections between crossed circular polars (18) - provided, of course, that the optic axial angle is significantly greater than the error in sectioning angle.
- In the case of optically biaxial materials, a magnetic field may alter the distribution of optical orientations at a temperature which is known to be too low to enable molecular reorientation (19).

The possibility of optical biaxiality should be considered whenever molecules (or side chains) have a lath-like or "sanidic" conformation that might support correlated rotations about their chain axes. If the suggested tests involving optical microscopy are impractical or inconclusive, one must resort to characterization by complementary techniques. For example, optical biaxiality can be inferred if there is no preferred orientation of optical extinction directions in a specimen that consists of globally aligned molecules as demonstrated by x-ray or electron diffraction (18).

Checklist When Interpreting Extinction

The preceding sections of this chapter have described various origins of extinction as observed between crossed polars in the light microscope. The following list summarizes the additional questions that should always be addressed in order to minimize possible ambiguities of interpretation:

- Does extinction persist if the crossed polars are rotated? Areas that remain in extinction under these conditions will also uniquely exhibit extinction in white light between crossed circular polars.
- Does extinction persist if the specimen is tilted about an axis that lies in the plane of the microscope stage?
- Does extinction depend on the illumination having a particular wavelength, or does it also occur if white light is used?
- Are the extinction orientations, as determined between crossed polars and in monochromatic light, sensitive to wavelength?

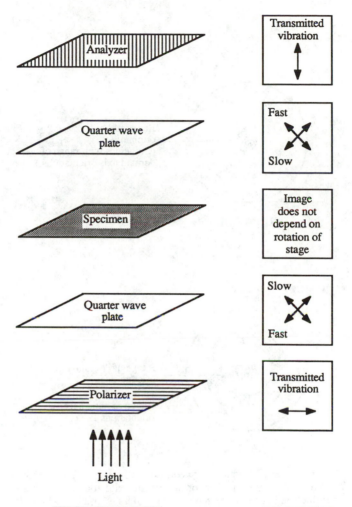

PERSPECTIVE VIEW
OF OPTICAL TRAIN

TOP VIEW OF
INDIVIDUAL ELEMENTS

Figure 3. The sequence of optical hardware needed to view a specimen between crossed circular polars. A quarter wave plate typically introduces a path difference in the range 0.15μm to 0.16μm (approximately 1/4 of the wavelength of Na_D light) between the slow and fast rays. The use of crossed circular polars to identify regions of zero birefringence unambiguously is most sensitive if the illumination has a wavelength of exactly four times the path difference introduced by the quarter wave plates.

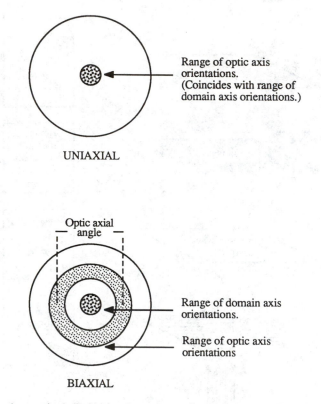

Figure 4.   A practical distinction between optically uniaxial and optically biaxial drawn (or extruded) material. For optically uniaxial material, the area fraction exhibiting extinction between crossed circular polars is greatest when the normal to the plane of the thin section is parallel to the draw axis. For optically biaxial material, the greatest area fraction is observed in a section cut so that the angle between its normal and the draw axis is equal to half the optic axial angle of a monodomain.

- Is the observed contrast affected by 90° rotations of the crossed polars or the specimen?
- What contrast is obtained when the analyzer is withdrawn and the specimen stage is rotated?
- Is the specimen optically uniaxial or biaxial?

Contrast That Does Not Depend on Optical Anisotropy

Up to this point, we have been concerned with interpreting optical textures, i.e. with exploiting the optical anisotropy of the specimen to study molecular order correctly. The light microscope can also be used to quantify the distribution of *defects* in the order, in which case details of the point-to-point variation across the specimen are superfluous. The optical discontinuity associated with line disclinations and inversion walls is sufficiently abrupt to scatter light, so that these features appear as dark lines on a light background if specimens are observed under simple bright field illumination conditions. (This scattering accounts for the turbidity of liquid crystalline fluids.) However, it is difficult to discern fine detail in this contrast mode. While there are several methods of improving contrast in the light microscope, so that fine detail is highlighted, they have been greatly under-utilized in the context of liquid crystalline materials.

In the following summary of contrast enhancement techniques, it is assumed that specimens are being observed in transmission, that they are not self-luminous, and that the light source is not imaged onto the specimen by the microscope condenser. All these assumptions describe typical conditions for LCP microscopy. Figures 5 and 6 show ray diagrams for a normally incident and obliquely incident beam of parallel rays, respectively. In both cases, the objective back focal plane contains the Fraunhofer diffraction pattern of the specimen.

According to the Abbe criterion for maximum attainable resolution, at least two diffraction maxima must be collected by the objective in order to form an image; information about structure in the object must be present in the objective back focal plane if it is to be reproduced in the image. For this reason, the resolution obtainable in obliquely incident light is finer than that which can be achieved by using normally incident light. With reference to Figures 5 and 6, we note that there are two approaches to enhancing the contrast of fine detail:

- Because the finest resolved detail concentrates light into diffracted maxima near the perimeter of the objective back focal plane, contrast can be enhanced by artificially and selectively operating on the characterizable properties (amplitude, phase, frequency, polarization state) of either the undiffracted or the diffracted intensity. This involves placing apertures and stops, with appropriate optical properties, in the objective back focal plane and in the plane of the condenser iris. (With the microscope set for Köhler illumination, these planes are conjugate, i.e. the image of one occurs at the plane of the other.) Contrast enhancement is most efficient if the illuminating beam is incident normally on the specimen, because the perimeter of the objective back focal plane then contains *only* diffracted light, i.e. the undiffracted and diffracted components are separated in that plane. However, this approach does not give maximum resolution.
- Resolution and image brightness are maximized if a cone of illuminating rays is incident on the specimen. The contrast of fine detail can be enhanced by artificially and selectively operating on the characterizable properties of the oblique illumination, relative to the normally incident component.

Based on these broad principles, a large number of practical contrast-enhancing accessories have been developed for the light microscope. Specific techniques include:

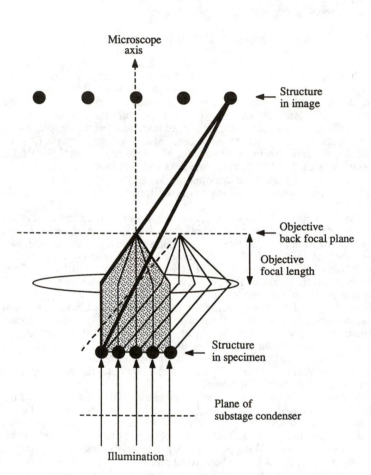

Figure 5.   The process of image formation by the objective in a transmitted light microscope when light is normally incident on the specimen. A diffraction pattern of the specimen arises at the objective back focal plane, as illustrated by ray paths drawn as thin lines. The "zero order", consisting of undiffracted rays, has been identified by shading. A finer structure in the specimen leads to a greater divergence of diffracted orders, eventually placing them beyond the aperture of the objective. Principal ray paths drawn as thick lines illustrate the formation of detail in the image.

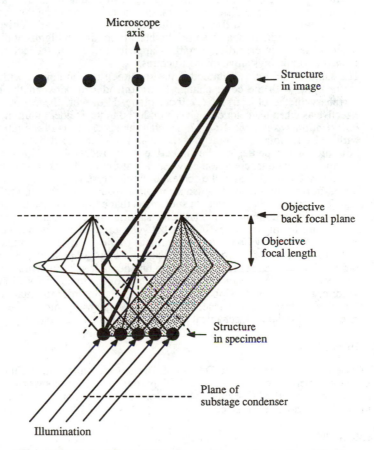

Figure 6. When light is incident obliquely on the specimen, the objective can accommodate a greater divergence between the zero and first diffracted orders than when the light is incident normally. This results in finer resolution.

- *Dark field microscopy* (20), in which the relative *amplitude* of the diffracted orders is changed relative to that of undiffracted light. Usually this involves complete elimination of the undiffracted light, so that the image is featureless apart from the fine detail.
- *Phase contrast microscopy* (20,21), in which the relative *phases* of diffracted and undiffracted light are changed. The best contrast is obtained if, additionally, the amplitude of the zero order is reduced in relation to that of the others.
- *Rheinberg differential color contrast* (22,23), in which the normal and oblique illuminating rays have different colors. Fine detail in the image of the specimen appears with a color different to that of the coarse detail. This technique maximizes illumination, and is useful when attempting to highlight disclinations without loss of intermediate detail. It appears to be a novel technique in the context of liquid crystalline microstructures.
  The same resolution is obtained if the specimen is illuminated with a cone of *white* light, with the color change being introduced between the center and periphery of the objective back focal plane. However, the contrast is not as effective as when the color change is introduced in the incident illumination.
- *Polarization-based contrast* (21) in which the specimen is illuminated normally with linearly polarized light and the analyzer is replaced by a small disc of Polaroid in the center of the objective back focal plane. This allows any intermediate contrast condition between bright field and dark field to be selected, and appears to be untried in the context of liquid crystals.

These different contrast mechanisms can all be used to reveal the scale of liquid crystalline polymer microstructures. In specimens that exhibit a mosaic texture, and in those that contain predominantly planar defects, domain size is easily defined in terms of areas that uniformly show extinction between crossed polars. However, if the defects are predominantly linear, as in specimens that exhibit schlieren textures, such simple characterization of microstructural scale is no longer possible. Here it is more convenient to look at the length of disclination line per unit volume, which is equivalent to the number of lines intersecting unit area, and analogous to the dislocation density as defined for crystalline solids. Good contrast is essential in order to obtain an accurate count. Technologically, microstructural scale is of growing interest because of its relevance to processability, mechanical properties and optical transparency.

Acknowledgments

Acknowledgment is made to the Donors of The Petroleum Research Fund, administered by the American Chemical Society, for partial support of this work. Support was also provided by the IBM Corporation as part of a block grant on the microdesigning of ceramics and polymer / ceramic composites.

Literature Cited

1. Kelker, H. Mol. Cryst. Liq. Cryst. 1973, 21, 1-48.
2. Demus, D.; Richter, L. Textures of Liquid Crystals; Verlag Chemie: Weinheim, 1978.
3. Gray, G.W.; Goodby, J.W. Smectic Liquid Crystals; Leonard Hill: Glasgow, 1984.
4. McKie, D.; McKie, C.H. Crystalline Solids; Nelson: London, 1974; p 403.
5. Ramachandran, G.N.; Ramaseshan, S. In Encyclopaedia of Physics; Flügge, S., Ed.; Springer: Berlin, 1961; Vol. 25/1.
6. Shurcliff, W.A. Polarized Light; Harvard University Press: Cambridge (Mass.), 1962.
7. Shurcliff, W.A.; Ballard, S.S. Polarized Light; van Nostrand: Princeton, 1964.
8. Yeh, P. Surface Science 1980, 96, 41-53.

9. Yeh, P. J. Opt. Soc. Am. 1982, 72, 507-513.
10. Hurwitz, H.; Jones, R.C. J. Opt. Soc. Am. 1941, 31, 493-499.
11. Viney, C. Microscope 1988, 36, 35-47.
12. Roche, E.J.; Wolfe, M.S.; Suna, A.; Avakian, P. J. Macromol. Sci. (Phys.) 1985, B24, 141.
13. Viney, C.; Windle, A.H. Phil. Mag. A 1987, 55, 463-480.
14. Viney, C. In Handbook of Polymer Science and Technology; Cheremisinoff, N.P., Ed.; Marcel Dekker: New York, 1989, Vol.1.
15. Hessel, F.; Finkelmann, H. Polymer Bull. 1986, 15, 349-352.
16. Hessel, F.; Herr, R.-P.; Finkelmann, H. Die Makro. Chem. 1987, 188, 1597-1611.
17. Donald, A.M.; Viney, C.; Windle, A.H. Phil. Mag. B 1985, 52, 925-941.
18. Windle, A.H.; Viney, C.; Golombok, R.; Donald, A.M.; Mitchell, G.R. Farad. Disc. Chem. Soc. 1985, 79, 55-72.
19. Viney, C.; Marcher, B.; Chapoy, L.L. Mol. Cryst. Liq. Cryst. 1988, 162B, 283-299.
20. Haynes, R. Optical Microscopy of Materials; Blackie: Glasgow, 1984; Chapter 6.
21. Pluta, M. In Advances in Optical and Electron Microscopy; Barer, R.; Cosslett, V.E., Eds; Academic Press: London, 1975.
22. Abramowitz, M. Am. Lab. 1983, 15, 38-41.
23. Delly, J.G. Photography Through the Microscope; Eastman Kodak Company: Rochester N.Y., 1988.

RECEIVED March 15, 1990

Chapter 19

Solid-State Structures of a Homologous Series of Mesogenic Aromatic–Aliphatic Azomethine Ether Polymers

A. Biswas[1,3], K. H. Gardner[1], and P. W. Wojtkowski[2]

[1]Central Research and Development Department, Experimental Station, E. I. du Pont de Nemours and Company, Wilmington, DE 19880–0356
[2]Agricultural Products Department, Experimental Station, E. I. du Pont de Nemours and Company, Wilmington, DE 19880–0356

The solid state structure of a homologous series of mesogenic aromatic-aliphatic azomethine ether polymers (AZMEP-n), with n (=1-16) methylene spacers in the main chain, has been studied by X-ray fiber diffraction techniques. Unit cell parameters determined from the fiber patterns of heat-treated single filaments have been used to classify the polymers into distinct groups having different chain conformations and crystal systems. The even members of the series, where $n = 2$ & 4, have C-centered monoclinic unit cells while the unit cells for polymers with $n = 6$ & 8 are I-centered monoclinic. When n is even and ≥ 10 the unit cells are (primitive) triclinic. The fiber repeat, in the triclinic cells correspond to one chemical moiety while in the monoclinic cells the fiber repeat corresponds to two chemical residues. In addition, the fiber repeat distances are consistent with the methylene spacers adopting an all-trans conformation for all even AZMEP-n polymers.
    When n is odd and ≤ 5, the unit cells are orthorhombic and contain either two ($n = 1$ & 5) of four (n = 3) chains. The fiber repeat consists of two chemical moieties and the methylene spacers are extended. When $n = 7$ & 9, the unit cells are I-centered monoclinic and similar to the those observed for $n = 6$ & 8. In the series where n is large and odd (AZMEP-n, $n = 11$, 13 & 15), the unit cells are triclinic and closely resemble the triclinic cells found for the high even polymers (n = ≥ 10). However, the observed fiber repeats in polymers where n is odd and ≥ 7 necessitate the presence of non-trans conformers in the aliphatic portions of the chains. The thermal and mechanical properties of the homologous series of polymers are discussed in light of the observed solid state structures.

Thermotropic liquid crystalline polymers with alternating rigid and flexible units along the backbone are a class of polymers which have been studied extensively over the years. The rigid units are usually aromatic moieties connected to each other by a variety of functional groups while the flexible units are most commonly methylene or

[3]Current address: Max Plank Institute für Polymerforschung, Postfach 3148, D–6500 Mainz, Federal Republic of Germany

oxyethylene oligomers. The synthesis and characterization of these polymers are well documented in numerous reviews (1-3). A common characteristic of these mesogenic polymers is the 'odd-even' alternation in thermodynamic properties such as transition temperatures and changes in enthalpy and entropy as a function of the number of flexible spacers (4-6). The focus of most of the literature has been directed towards the liquid crystalline phase and the transition to the isotropic phase. The solid state and the solid-liquid crystal transition have received relatively less attention. There have been studies on the orientation of samples quenched from the melt (with or without the influence of magnetic or shear fields) using techniques such as X-ray diffraction (7, 8). Despite the achievement of significant molecular alignment, the quality of the data has not been adequate to construct detailed molecular models of the structure in the solid state. More importantly, a detailed study of the influence of the rigid and flexible units on the solid state structure for a homologous series of polymers has not been reported. To this end, we have used X-ray diffraction techniques to characterize the heat treated oriented fibers from a homologous series of mesogenic azomethine ether polymers described below.

The polymers under study have a chemical repeat comprised of three phenyl rings linked by azomethine groups forming the rigid unit and a flexible polymethylene spacer (1 to 16 methylene units) linked to the rigid units by ether oxygens.

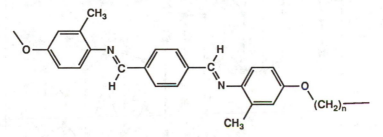

These polymers have been prepared from terephthaldehyde and 1,n-bis(4-amino-3-methyl phenoxy)alkanes, where n = 1-16; details of the synthesis and initial characterization of some members of the series have been reported previously (9). Calorimetric studies on the as-precipitated polymers indicated a series of endotherms including a crystal to nematic transition (K-N) and nematic to isotropic transition (N-I) (AZMEP-2 & 4 decompose before the N-I transition). While some of the endotherms correspond to crystal-crystal phase transitions, other endotherms may correspond to solid state polymerization and endgroup effects. Because of the difficulty in determining the origin of some of the endotherms, we have concentrated on the structure of the highest melting phase. By annealing fibers of polymers near the K-N transition temperature it was possible to isolate the highest melting phase for each polymer. (The only exception was AZMEP-4 which could not be isolated in a single crystalline phase.)

Figure 1 shows the K-N and N-I transition temperatures as a function of the number of methylene units in the chemical repeat. Data was derived from DSC scans (20°/min) of heat-treated filaments (Gardner, K. H., Biswas, A. & Wojtkowski, P. W., in preparation.). While the transition temperatures decrease as n increases and a general 'odd-even' alternation in transition temperatures is observed, there are serious perturbations on the trend. Besides a molecular weight dependance, the mechanical properties and the K-N transition temperature are influenced profoundly by the conformation and packing of the polymer chain in the solid state. We hope to gain a better insight into the correlation between the thermal and mechanical properties and the observed solid state structures for the homologous series of liquid crystalline polymers described above.

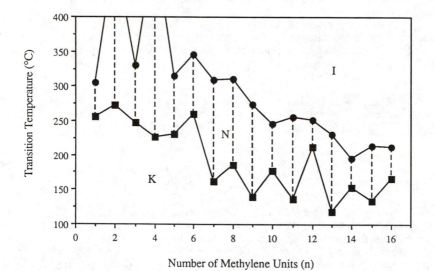

Figure 1. Plot of K-N and N-I transition temperatures versus the number of methylene units in the aliphatic segment.

Experimental

Oriented fibers of each polymer were obtained by pulling filaments from a pool of the anisotropic melt with tweezers. The fibers were subsequently heated treated close to the melting temperature under vacuum. X-ray fiber diffraction patterns were obtained from single filaments using a 57.3mm radius vacuum cylindrical camera and Ni-filtered CuKα radiation. Diffraction patterns were digitized with a Photomation P1700 rotating drum film scanner (Optronics International) and processed on a Jupiter 7+ raster graphics terminal using the Fiber Diffraction Processing Program (FDPP) (Gardner, K. H. & Hilmer, R. M., unpublished work). The unit cell parameters were subsequently determined from the d-spacings of the observed reflections.

Molecular modeling of chain conformation and packing was done using the software packages LALS (10) and MOLEDITOR (11). Bond lengths and angles derived from the crystal structures of suitable model compounds (12-15) were used in the models.

Results

AZMEP-*n, n*=even. Fiber diffraction patterns from single filaments of the AZMEP-*n* polymers with an even number of methylene units ($n = 2k$, k =1, 8) indicate the presence of very high crystallinity and very good molecular orientation. Unit cell parameters obtained from the diffraction data are listed in Table I. The even members of the AZMEP-*n* series can be divided into three distinct groups on the basis of the number of chains in the unit cell, the crystal system and centering type.

AZMEP-*n*, *n* = 2 & 4	two chain <u>C-centered monoclinic</u> (mC) unit cell
AZMEP-*n*, *n* = 6 & 8	two chain <u>I-centered monoclinic</u> (mI) unit cell
AZMEP-*n*, *n* = 10, 12, 14 & 16	one chain <u>Primitive triclinic</u> (tP) unit cell

A discussion of representative members of each group follows.

AZMEP-2. The fiber diffraction pattern for AZMEP-2 is shown in Figure 2. The observed reflections were indexed by a C-centered monoclinic unit cell (mC, b axis unique). The unit cell contains two chains (located at 0,0,0 and 1/2,1/2,0) and has a crystallographic repeat along the fiber (c) axis corresponding to two chemical repeats. Molecular modeling indicates that the fiber repeat distance is consistent with an all-trans conformation of the ethylene segment. This fully extended conformation results in a small tilt of the rigid groups with respect to the c axis. A schematic drawing of the ac projection of the AZMEP-2 structure is shown in Figure 3. The residues related by the centering operation have been shaded. The rigid groups are aligned parallel to each other and the monoclinic angle β=31.2° leads to an axial displacement of 6.85Å between adjacent chains at 0,0,0 and 1/2,1/2,0 It is also evident from the ac projection that this results is a partial overlap between the rigid and flexible segments of the two chains in the unit cell.

AZMEP-4. The fiber pattern of AZMEP-4 exhibited a large number of sharp diffraction maxima on well defined layer lines, reminiscent of a well oriented and highly crystalline polymer. However, all of the reflections could not be indexed by a single unit cell, indicating the presence of multiple crystal forms. This fact is supported by by the DSC data described earlier. Nevertheless, the majority phase has diffraction features similar to those observed in AZMEP-2. The presence of all the reflections on the layer lines enables the measurement of the c-spacing and suggests

TABLE I. Unit Cell Parameters and Characteristics for AZMEP-n Polymers where n is even

n	a (Å)	b (Å)	c (Å)	α (°)	β (°)	γ (°)	Z	Crystal system	# Chains
2	14.81	6.12	40.24	90.0	31.2	90.0	4	mC	2
4			46.0					mC	2
6	10.68	9.27	51.93	90.0	27.8	90.0	4	mI	2
8	11.14	9.36	57.11	90.0	26.7	90.0	4	mI	2
10	14.2	5.4	31.0	92.5	25.0	74.0	4	tP	1
12	14.73	5.54	33.12	101.1	24.2	82.5	1	tP	1
14	15.3	5.4	35.6	90.0	23.0	73.0	1	tP	1
16	14.60	5.34	36.63	93.2	23.8	75.6	1	tP	1

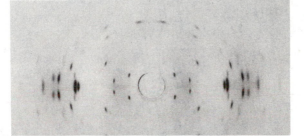

AZMEP-2

AZMEP-8

AZMEP-10

Figure 2. Fiber diffraction patterns of AZMEP-2, AZMEP-8 and AZMEP-10 (fiber axes vertical).

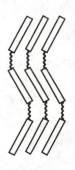

n = 1, 3, 5
oP
extended flexible
spacer

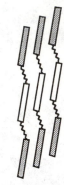

n = 2, 4
mC
extended flexible
spacer

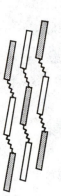

n = 7, 9
mI
shortened flexible
spacer

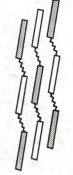

n = 6, 8
mI
extended flexible
spacer

n = 11, 13, 15
tP
shortened flexible
spacer

n = 10, 12, 14, 16
tP
extended flexible
spacer

Figure 3. Schematic drawings of the ac-projections of the different packing modes found in AZMEP-*n* polymers. (See text for explanation of shading.)

that the polymorphism derives from differences in chain packing and not conformational differences.

AZMEP-8. The fiber diffraction pattern of AZMEP-8 is shown in Figure 2. The pattern is significantly different from that of AZMEP-2 in that there are are a series of lower angle reflections and a prominent 'horizontal' four-point pattern. The reflections were indexed by a I-centered monoclinic unit cell (mI, b axis unique). Like in the case of AZMEP-2, the unit cell contains two chains (now at 0,0,0 and 1/2,1/2,1/2), has two residues in the fiber repeat and has a fully extended flexible spacer. A schematic drawing of the ac projection of the AZMEP-8 structure is shown in Figure 3. Residues related by the I centering are shaded.

AZMEP-10. The fiber pattern for AZMEP-10, shown in Figure 2, is representative of the patterns for the higher even series. The characteristic features of the fiber pattern are three strong equatorial reflections and a four-point reflection pattern at low angle centered on the direct beam. The polymer packs in a one chain triclinic unit cell with a fiber repeat corresponding to the length of a single chemical repeat. As is the case for the lower even series, the flexible aliphatic spacers have an extended conformation and the rigid groups are aligned parallel to each other. There is an axial stagger of 12.87Å between adjacent chains in the ac plane as dictated by the unit cell parameters. As a consequence, the rigid unit of the chain repeat overlaps completely with the flexible spacers and partly with the rigid unit of its neighbors. A schematic of the packed structure in the ac plane is shown in Figure 3. As the number of even methylene spacers increases, the overlap between adjacent rigid units decreases. The resulting interdigitated structure is unique for polymers having rigid and flexible segments along the backbone. Recently, such a structure has been reported as a metastable crystal phase in a class of main chain liquid crystalline polyethers (16).

AZMEP-n, n=odd. Fiber diffraction patterns from single filaments of the AZMEP-n polymers with an odd number of methylene units in the backbone ($n = 2k-1$, $k = 1-8$) indicate that the polymers can have high crystallinity with good to moderate orientation. Unit cell parameters derived from the diffraction data are listed in Table II. The odd AZMEP-n's can be divided into three distinct groups on the basis of the crystal lattice type:

AZMEP-n, $n = 1, 3$ & 5 two (or four) chain Primitive orthorhombic (oP) unit cell

AZMEP-n, $n = 7$ & 9 two chain I-centered monoclinic (mI) unit cell

AZMEP-n, $n = 11, 13$ & 15 one chain Primitive triclinic (tP) unit cell

AZMEP-1. Figure 4 shows the fiber pattern for AZMEP-1. The observed reflections can be indexed by a two chain orthorhombic unit cell with two chemical moieties in the fiber repeat. The rigid mesogenic units in a given chain are tilted in opposite directions with respect to the fiber axis giving the chain a zig-zag conformation. In this structure the methylene spacers are in an extended conformation. A schematic of the chain packing for AZMEP-1 in the ac plane is shown in Figure 3. As the number of methylene spacers increases to 3 and 5, the crystal type remains orthorhombic but the ratio a/b changes. A decrease in the orientational order, as evidenced by the arcing of the reflections, is also observed. However, when $n = 7$, there is a sudden change in the solid state structure.

AZMEP-7. The fiber diffraction pattern for AZMEP-7 (Figure 4), is significantly different from those of the lower odd series. (Indeed, it bears a strong resemblance to the fiber pattern of AZMEP-8 (see Figure 2) except for its lower orientation.) The unit cell parameters indicate that AZMEP-7, like AZMEP-8, has a I-centered monoclinic (b axis unique) lattice. The c repeat length is significantly larger

TABLE II. Unit Cell Parameters and Characteristics for AZMEP-n Polymers where n is odd

n	a(Å)	b(Å)	c(Å)	α (°)	β (°)	γ (°)	Z	Crystal system	# Chains
1	11.63	4.36	36.00	90.0	90.0	90.0	4	oP	2
3	15.36	6.73	39.59	90.0	90.0	90.0	8	oP	4
5			43.95						2
7	10.72	9.34	53.67	90.0	28.3	90.0	4	mI	2
9			58.0				4	mI	2
11	14.17	5.84	31.04	101.2	26.12	80.7	1	tP	1
13	13.54	6.29	32.56	115.4	26.7	94.0	1	tP	1
15								tP	1

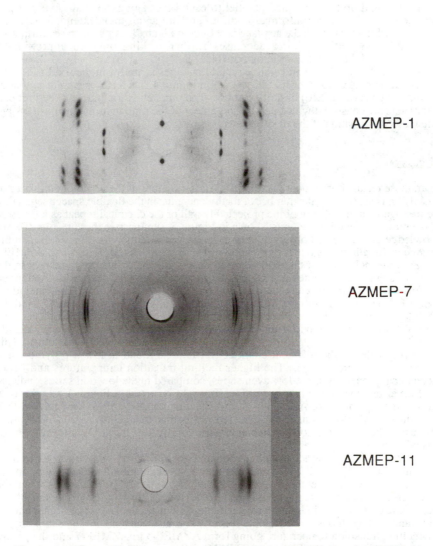

Figure 4. Fiber diffraction pattern of AZMEP-1, AZMEP-7 and AZMEP-11 (fiber axes vertical).

than what would be expected from the trend of the lower odd AZMEP-n's. From molecular packing models and on the basis of the packed structures for the low even series describe above, the most probable structure for AZMEP-7 corresponds to one where the rigid units are aligned parallel to each other along the chain axis but the flexible spacers are in a conformation different from an all-trans extended state. A schematic of the packing in the ac plane is shown in Figure 3. As n increases further to AZMEP-9, orientational order decreases further making only an approximate measurement of the layer line spacing possible.

   AZMEP-11. The solid state structure changes again for higher odd AZMEP-n's. The fiber pattern for AZMEP-11, shown in Figure 4, has features similar to that of the high even series ($n \geq 10$) described above. The unit cell is triclinic and the unit cell parameters are consistent with a parallel alignment of the rigid units and a non-trans but extended conformation of the flexible units.

Discussion

From these results it is evident that the solid state structure of the homologous series of AZMEP-n is sensitive to the number of methylene units in the flexible spacer separating the mesogenic units. A plot of the projected length of the chemical repeat as a function of the number of methylene units, shown in Figure 5, summarizes these results. To our knowledge this is the first observation of an 'odd-even' alternation of a crystallographic parameter for a homologous series of polymers. For the low odd series, i.e., AZMEP-1, 3 & 5, the inclination of the mesogenic cores in a zig-zag manner with respect to the fiber axis (in the ac plane) results in low projected lengths despite the extended conformations of the flexible spacer. For the low even series up to and including AZMEP-8, in the ac projection, all rigid mesogenic units are aligned in parallel and the flexible spacers are fully extended. The mechanical properties of AZMEP-n fibers have been shown to have an 'odd-even' alternation (with n even polymers having tensile moduli and tensile strengths much higher than the polymers where n is odd). This effect originates in the different chain conformations found in low n polymers and dampens as n increases.(9). The higher melting transition temperatures and better mechanical properties of the low even series, compared to the low odd series, indicate the former to be structurally more stable. It might be noted in Figure 1 that the K-N transition temperature for AZMEP-4 is lower that that of AZMEP-6. One might recall that while AZMEP-2 & 4 have a C-centered monoclinic unit cell, AZMEP-6 & 8 are I-centered. Perhaps, the former becomes structurally less stable as the number of methylene spacers increases.

   The solid-state structure of both odd and even AZMEP-n polymers, when $n \geq 6$, is dominated by the parallel alignment of the mesogenic units in the ac plane. In the AZMEP-7 polymer, the trend established by the low odd series of packing the mesogenic groups in a zig-zag manner is reversed. The parallel alignment of the mesogenic cores results in longer projected lengths but at the expense of higher energy non-trans conformations in the flexible spacer. This also accounts for the sharp drop in the melting transition temperature going form AZMEP-5 to AZMEP-7 and the poorer orientation observed in the fibers of AZMEP-7 & 9.

   All the AZMEP-n polymers with $n \geq 10$ have a one chain triclinic unit cell in which the extended flexible spacers and the mesogenic cores of adjacent chains overlap to form a stable structure irrespective of whether or not there are odd or even number of methylene spacers. The odd series, not unexpectedly, still exhibits lower melting points although the differences with the higher evens are less pronounced. AZMEP-12 however exhibits a higher melting transition in comparison to its near neighbors. A possible explanation for this perturbation in the general trend could be that the

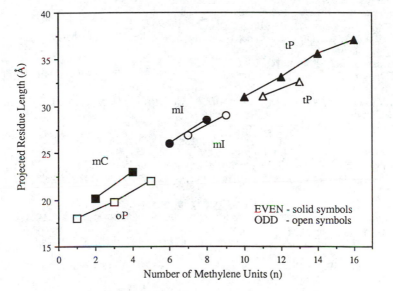

Figure 5. Plot of the projected length of the chemical repeat (along the c axis) versus the number of methylene units in the flexible spacer.

molecular lengths of the rigid and flexible segments in AZMEP-12 are optimal for the interdigitated structure exhibited by higher n series of polymers.

Further studies on the detailed structure of the polymers in this series are in progress and will be reported shortly. Synthesis of higher homologues of this series and low molecular weight model compounds is also being pursued in an effort to determine the role of the rigid and flexible units in influencing the formation of liquid crystalline phases.

Literature Cited

1. Blumstein, A.(ed.), Liquid Crystalline Order in Polymers; Academic Press, N.Y., 1978.
2. Ciferri, A.; Krigbaum, W. R.; Meyer, R. B. (eds.), Polymer Liquid Crystals; Academic Press, N.Y., 1982.
3. Ober, C. K.; Bluhm, T. L. in Current Topics in Polymer Science; Vol. 1, Ottenbrite, Ultracki, Inoue (eds.), 1987, p. 249.
4. Roviello, A.; Sirigu, A. Makromol. Chem. 1982, 183, 895.
5. Griffin, A. C.; Thomas, R. B.; Hung, R. S. L.; Steele, M. L. Mol. Cryst. Liq. Cryst. 1984, 105, 305.
6. Blumstein, A. Polymer J. 1985, 17, 277;
7. Capasso, R.; Roviello, A.; Sirigu, A.; Iannelli, P. J. Polym. Sci., Part B. Polym. Phys. 1987, 25, 2431.
8. Chin, H. H.; Azaroff, L. V.; Griffin, A. C. Mol. Cryst. Liq. Cryst. 1988, 157, 443.
9. Wojtkowski, P. W. Macromolecules 1987, 20, 740.
10. Smith, P. J. C.; Arnott, S. Acta Crystallogr. 1978, A34, 3.
11. Hilmer, R. M. J. Mol. Graphics 1989, 7, 212.
12. Burgi, H. B.; Dunitz, J. D. Helveta Chemica Acta 1970, 53, 1747.
13. Doucet, P. J.; Mornon, J. P.; Chevalier, R.; Lifchitz, A. Acta Crystallogr. 1977, B33, 1701.
14. Vani, G. V.; Vijayan, K. Mol. Cryst. Liq. Cryst. 1977, 42, 249.
15. Gardner, K. H.; Biswas, A.; Wojtkowski, P. W.; Calabrese, J. C. Acta Crystallogr., Section C. 1990, submitted.
16. Ungar, G.; Keller, A. Mol. Cryst. Liq. Cryst. 1987, 155, 313.

RECEIVED April 10, 1990

Chapter 20

Copolymerlike Structures of Nematic Nitroimine Dimers

Leonid V. Azároff[1], Jack M. Gromek[1], Agya R. Saini[1], and Anselm C. Griffin[2]

[1]Institute of Materials Science, University of Connecticut,
Storrs, CT 06269–3136
[2]Departments of Chemistry and Polymer Science, University of Southern
Mississippi, Hattiesburg, MS 39406

X-ray diffraction diagrams of a homologous series
of nitroimine dimers were recorded at their nematic
temperatures in a magnetic field of 2,000 G. Two
kinds of well aligned monodomains could be distinguished. When the central aliphatic chain contained an even number of methylene groups, the
degree of alignment was higher than when n was odd.
The aperiodic spacing of up to eight meridional
maxima suggested that the dimers form aperiodic
linear arrays containing two recurring repeat
distances whose lengths were determined by point-
model calculations to approximate extended coupled
dimers having singly and doubly overlapping phenyl
rings, respectively. Line-model calculations based
on complete intensity profiles showed that the
double overlap was favored when n is even and
increases with increasing length of the central
spacer chain.

In recent years considerable effort has been directed toward the
study of the structures of liquid-crystalline polymers and copolymers. Using x-ray diffraction techniques developed for the
examination of fibers (1,2), detailed structural information can be
obtained provided the polymers are aligned in a strong magnetic field
or by mechanical means. As reported more recently (3), quite similar
x-ray intensity distributions have been obtained from nematic monodomains of dimer molecules aligned in relatively weak magnetic
fields.

The tendency of rod-like (small) molecules to form linear arrays
in the nematic state has been recognized for some time (4). Using
steric arguments, it can be shown that such packing models can
account for the one or two disk-shaped intensity distributions
recorded along the meridian in reciprocal space. More recently,
nematic alignments have been reported (5) in which neighboring
Siamese-twin molecules tended to form linearly extended pairs through

0097–6156/90/0435–0269$06.00/0

dipole-dipole interactions between adjacent dimers with the probability of such pairing increasing as the length of the terminal aliphatic chains increased.

The present investigation was undertaken to verify that such basically liquid systems could organize themselves into structures closely resembling the linear arrays more typical of polymeric fibers. A previously described (3) homologous series of nitroimine dimers consists of two aromatic cores joined by a variable-length aliphatic spacer and has the general structure

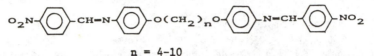

$$n = 4\text{-}10$$

Assuming that neighboring pairs of extended dimer molecules can "couple" by means of dipole-dipole interactions as proposed by Chin et al. (5), two coupling modes are most likely in the present case as illustrated in Figure 1. If such couplings join a succession of dimers into an extended array, an aperiodic chain-like structure results as was confirmed by point-model (6) calculations. This previously reported (3) finding is reexamined in the present paper by comparing the continuous intensity distribution function along the meridion in reciprocal space with that calculated from molecular transforms of the proposed chain (line) model. The relative proportion of singly (Figure 1a) and doubly (Figure 1b) overlapped couplings is also established thereby.

Experimental

The synthesis and characterization of the nitroimine dimers has been reported elsewhere (3). Flat-film transmission photographs were recorded for each member of the homologous series for n = 4 to 10. The samples were heated to about 10°C above their respective melting points in a magnetic field of 2000 G and exposed to a crystal-monochromated beam of Cu Kα radiation (7). The resulting diffraction patterns (Figure 2), typical of aligned nematic monodomains, were converted to digital optical densities using an Optronics micro-densitometer operating at 100 micron resolution. The resulting digitized intensity distributions were subsequently manipulated and analyzed using software developed at the Institute of Materials Science for the analysis of x-ray scattering by liquid-crystalline arrays.

In this way, separate meridional intensity profiles were generated for each dimer, corrected for polarization, and plotted in the range of $0 < s (=2\sin\theta/\lambda) \le 0.3$ in constant s-width steps. These profiles then were available for comparison with calculated intensity distributions. Similarly, azimuthal intensity distributions passing through the equatorial reflections (measured as a function of azimuthal angle from the meridian) were used to determine the Hermans order parameters (8).

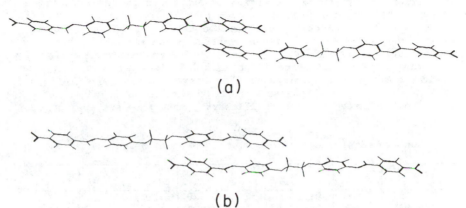

(a)

(b)

Figure 1.

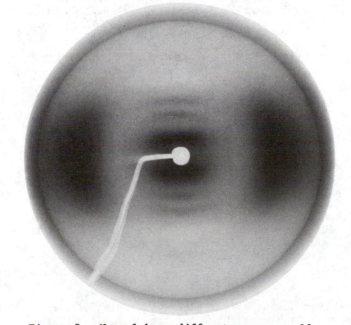

Figure 2. Normal-beam diffractogram, n = 10.

Results and Discussion

Probably the most distinguishing property of the homologous dimer series studied is the unusually pronounced alternation in the N-I transition entropies and enthalpies as evidenced by the much higher clearing temperatures when n is even (3). In the present case, this suggests that the degree of ordering in the nematic liquid is considerably greater for the even-numbered spacers. A qualitative indication of this can be judged from the x-ray photographs directly by noting a relative sharpening of all diffraction maxima when n is even. The reciprocal spacings of the observed maxima are listed in Table I by order of appearance along the meridian in diffraction (reciprocal) space in order to illustrate that they are aperiodic.

TABLE I
Reciprocal spacings of meridional reflections (A^{-1})

n	1	2	3	4	5	6	7	8
2	0.051	0.087	0.135	0.189	0.226	0.260	–	–
4	0.048	0.082	0.123	0.163	0.211	0.244	0.278	–
5	0.047	–	0.121	0.162	0.209	0.238	–	–
6	0.045	0.078	0.112	0.149	0.187	0.228	0.258	–
7	0.044	–	0.110	0.149	–	0.226	–	–
8	0.042	–	0.103	0.139	–	0.216	0.241	0.271
9	0.041	–	0.101	0.136	–	0.211	0.235	–
10	0.038	–	0.094	0.129	0.159	0.200	0.227	0.254

In the case of well-ordered crystals, it is possible to deduce their atomic structures by appropriate manipulation of diffraction intensities. In the case of x-ray scattering by liquids, direct use of measured intensities yields, at best, very limited structural information (radial distribution functions). For ordered liquids, however, it is possible to posit structural models and to calculate what their scattering intensities would be so that it is more productive to conduct the comparisons in diffraction space. To this end, it is possible to devise a point model to represent the spatial repetition of the constituent units in the ordered array and to compare its scattering maxima to the observed ones (6,9). More sophisticated analyses (10-12) make use of the complete electron densities (or projections onto the chain axis z), usually by calculating their Patterson functions P(z) since the scattering intensity function is its Fourier transform.

$$I(s) = T\{P(z)\} \qquad (1)$$

If the selfconvolution of the electron density of a molecule (projected onto z) is denoted M(z) and the point-distribution function prescribing how the molecules are repeated along z by D(z), the desired Patterson function is their convolution

$$P(z) = M(z) * D(z) \tag{2}$$

Making use of the properties of Fourier transforms

$$\begin{aligned} I(s) &= T\{M(z) * D(z)\} \\ &= T\{M(z)\} \ X \ T\{D(z)\} \end{aligned} \tag{3}$$

As is well known, if $D(z)$ is a periodic function with period d, then its Fourier transform is a periodic set of delta functions of period 1/d and the intensity distribution in reciprocal space consists of discrete maxima also spaced 1/d apart. If the distribution function is not periodic but consists of two randomly arrayed periods, then it is necessary to consider all possible combinations and permutations, suitably weighted by their probability of occurrence.

Consider first the distribution function $D(z)$ for a set of point molecules, i.e., their electron densities are projected onto points so that $M(z)$ is a delta function. For the dimer with n = 4, the most probable repeat distance for point molecules joined by a single overlap (Figure 1a) is d_1 = 23.9 A and for double overlap (Figure 1b) d_2 = 17.5 A. Assuming that both are equally likely to occur in a chain consisting of ten coupled dimers, the point Patterson has the form shown in Figure 3a. If one assumes further that the two repeat distances may themselves vary by a few per cent due to thermally induced or other displacements, then the Patterson will be modified as shown in Figures 3b to 3d for (Gaussian) width variations of 0.2 to 1.0 A, respectively. The dimunition of the Patterson peaks with increasing distance from the origin is a charactereistic of all aperiodic copolymer systems; the appropriate distribution width being a function of the rigidity and conformational regularity of the constituent monomers.

In the present case of nitroimine dimers, the liquid nature of the system and the relative flexibility of the aliphatic spacers favor a fairly broad distribution width. Thus the liquid crystalline monodomain exhibits discernable (high-probability) translational repetitions only for a fairly small number of aligned molecules so that increasing the actual length of the chain assumed in calculating the distribution function beyond, say, ten molecules has little effect on the resulting distribution function. Three other variables have to be considered, however: the relative fractions of the two kinds of coupled dimers present, their relative orientations, and the resulting relative repeating lengths. A Monte Carlo calculation was used to generate many random chains from which an "averaged" normalized nitroimine chain was determined.

Calculation of a predicted intensity distribution (Equation 3) requires a knowledge of the individual molecular self transforms. Assuming that the dimers favor a fully extended conformation, the self transforms for the even-numbered spacers are shown in Figure 4. Any deviations from the assumed structures introduce obvious sources of error in these calculations, however, as also noted by others (11), small changes in molecular configurations or orientations have

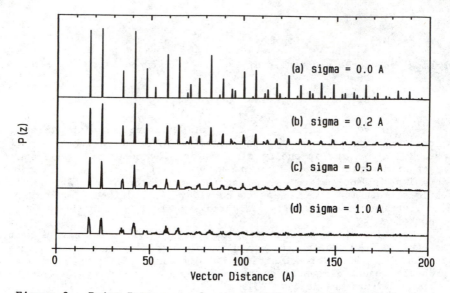

Figure 3.   Point Pattersons for ten overlapping dimers with n = 4.

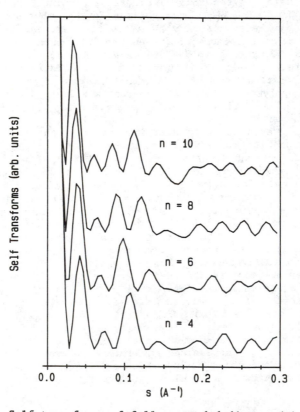

Figure 4.   Self transforms of fully extended dimers with n even.

negligible effect on the calculated self transforms probably because they are less discernible in their projections on the chain axis.

In the previously reported analysis (3), based on a point-model calculation of the locations of the observed diffraction maxima, a set of optimal repeating lengths was determined for each member of the homologous series. These were used as starting values in the present calculations. Next, the relative ratios of the coupled dimers were varied in steps of 10%, deemed to be the limit of sensitivity of this model. Utilizing the self transform shown in Figure 4, the resulting intensity distributions along the meridion in reciprocal space for n = 6 are compared in Figure 5 to the observed distribution (bottom of figure) for singly-to-doubly overlapped ratios of 50:50, 30:70, and 10:90, respectively. Since instrumental factors were not incorporated in the calculated intensities, the peaks are too sharp but the peak positions should be unaffected. Similarly, it is not possible to determine a reliable scale factor for relating observed to calculated intensities for a liquid. To facilitate comparisons, therefore, the total integrated intensity for each dimer was considered to be a constant and used to scale the curves in Figure 5. By comparing the positions and the integrated intensities of individual diffraction maxima, it is then possible to select the curve for 30:70 as representing the best agreement with the observed intensity distribution. Table II lists these ratios for all the dimers examined.

TABLE II

Single and double overlap in coupled nitroimine dimers

n	% single	% double
4	40	60
5	70	30
6	30	70
7	60	40
8	20	80
9	50	50
10	10	90

Despite the relatively low reliability of the percentages in Table II (at best, plus or minus 10%), several deductions can be safely made:

1. Double overlap is clearly favored when n is even.
2. Single overlap is favored when n is odd.
3. In all cases, double overlap increases progressively as n increases.

The above results can be understood from steric considerations. For n even, the meso-group vector is parallel to the extended chain vector. Thus for n = 4, a fully extended aliphatic spacer is just long enough to permit double overlaps at both ends of a dimer without steric interferences of the nitro groups. As n (even) increases, such

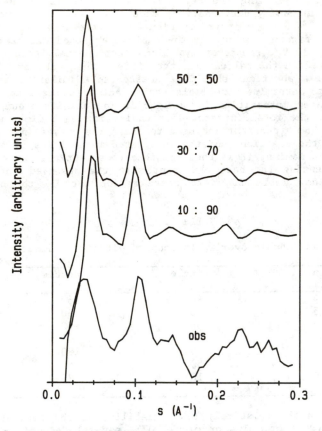

Figure 5.   Intensity distributions along the meridian for n = 6
            calculated using self transform in Fig. 4.

possible steric hindrances rapidly diminish further. For n odd, the meso-group vectors are not parallel to the extended chain vector so that single overlap is geometrically and entropically easier. As n increases, however, it becomes possible for the aliphatic spacer to adopt conformations which can better accommodate double overlap.

Although the above analysis is based on diffraction data from nematic monodomains aligned in a magnetic field, it is highly probable that the observed tendencies of the dimers to couple persist in the nematic state of unaligned nitroimine dimers as well. It is thus possible to explain the unusually large differences between the clearing temperatures of dimers having odd and even numbered spacers by noting that the stronger interactions formed by double overlap require much higher temperatures to separate them. Similarly, the extreme alternations in the transition entropies are doubtlessly related to the relative proportions of doubly overlapped coupled dimers as is the progressive rise in the transition entropy with increasing n.

The intensity distributions in the equatorial diffraction maxima were used to determine the inter-chain separations by noting the d (= $\lambda/2\sin\theta$) values at the opposite half-maximum intensities measured along the equator and the Hermans orientation factor (8) from the azimuthal extent of these maxima. Both sets of values are presented in Table III. The Hermans orientation factor does not appear to change significantly as n increases although the slight drop in the orientation factor as n goes from 8 to 10 may account for the comparable drop in the respective transition entropies from 8.86 to 8.10 J/mole.Kelvin. As expected, the degree of orientation (parallel alignment) is greater for the dimers having even-numbered spacers.

TABLE III

Alignment of dimer chains

n	Inter-chain d (A)	Orientation factor
4	$3.8_3 - 5.5_9$	0.66
5	$3.7_2 - 5.7_5$	0.54
6	-- --	0.67
7	$3.8_8 - 5.6_5$	0.54
8	$3.9_7 - 5.4_7$	0.69
9	$3.8_3 - 5.6_8$	0.54
10	$3.8_5 - 5.4_7$	0.65

Acknowledgments

The authors appreciate very much the help that Professor John Blackwell proffered in setting up the original point-model cal-

culations and in his helpful subsequent discussions of this study.
L.V.A., J.M.G., and A.R.S are grateful to the Defense Advanced
Projects Agency for a URI Grant establishing the LCP Research Center
at the University of Connecticut while A.C.G. acknowledges the
support of his efforts by the National Science Foundation under Grant
DMR 8417834.

Literature Cited

1. Vainshtein, B. K. Diffraction of X-Rays by Chain Molecules;
    Elsevier: Amsterdam, 1966.
2. Alexander, L. E. X-ray Diffraction Methods in Polymer Science;
    Wiley-Interscience: New York, 1969.
3. Azároff, L. V., Saini, A. R., Hari, U., Britt, T. R., and
    Griffin, A. C. Polymer Prepr. 1989, 30, 515-16.
4. Azároff, L. V. Mol. Cryst. Liq. Cryst. 1980, 60, 73-96.
5. Chin, H. H., Azároff, L. V., and Griffin, A.C. ibid 1988 157,
    443-53.
6. Blackwell, J., Cageao, R. A., and Biswas, A. Macromolecules,
    1987, 20, 667-71.
7. Azároff, L. V. and Schuman, C. A. Mol. Cryst. Liq. Cryst. 1985,
    122, 309-19.
8. Hermans, J. J., Hermans, P. H., Vermaas, D., and Weidinger, A.
    Rec. Trav. Chim. Pays-Bas 1946, 65, 2165-85.
9. Davies, G. R. and Jakeways, R. Polymer Comm. 1985, 26, 9-10.
10. Blackwell, J., Gutierrez, G.A., and Chivers, R. A., J. Polym.
    Sci.; Phys. Ed. 1984, 22, 1343-46.
11. Biswas, A. and Blackwell, J. Macromolecules, 1987, 20,
    2997-3001.
12. Mitchell, G. R. and Windle, A. H. Coll. & Polym. Sci. 1985, 263,
    230-44.

RECEIVED March 15, 1990

Chapter 21

Simulation and Evaluation of Polymorphism in the Solid State of a Rigid Rod Aramid

G. C. Rutledge[1], C. D. Papaspyrides[1,2], and U. W. Suter[1]

**[1]Department of Chemical Engineering, Massachusetts Institute
of Technology, Cambridge, MA 02139
and
Institut für Polymere, Eidgenössische Technische Hochschule,
CH–8092, Zürich, Switzerland**

The application of atomistic modelling methods to the study of microstructure in the solid state of the rigid rod aramid poly(p-phenylene
terephthalamide) provides insight into the molecular source and
specificity of structure formation. The simulation results, based on the
criterion of minimum potential energy, suggest that such rodlike
polymers may realize multiple packing geometries of polymer chains due
to the considerable intermolecular forces operating between chains
combined with the anisotropy of intermolecular specificity. However,
each allomorph possesses the common features of chain extension and
hydrogen-bonded sheet formation, with polymorphism hinging on the
variation of packing between sheets. Further, simulation results are
compared at the phenomenological level to experimental data from wide
angle x-ray scattering of fibers produced in-house which are consistent
with fibers reported in the literature.

Rigid backbone macromolecules of the aromatic polyamide and aromatic polyester types
have attracted considerable interest for their near-theoretical ultimate tensile properties
when processed in such a way as to obtain largely parallel alignment of molecular axes
in load bearing geometries. However, the same structural features which are responsible
for the molecular rigidity and desirable thermal and mechanical properties of the
processed polymer are also accountable for its relatively troublesome processing and
intractability. Poly(p-phenylene terephthalamide) (PPTA) is one such polymer that, in
commercial form as DuPont's Kevlar, has attracted considerable attention to the nature
of microstructure in the bulk solid of such materials. To date, two crystal forms for
this polymer have been reported in the literature, resulting from different processing
environments.(1-3) Atomic models based on first principles enable one to anticipate the
source of such polymorphic behavior on a thermodynamic basis. Furthermore, from the
results of such modelling efforts, one can predict a priori complex macroscopically
observable phenomena for direct comparison to and analysis of experimental data. In
this paper, we report results from a static atomistic analysis constructed specifically to
elucidate the source of molecular scale order in PPTA and provide insight into the
nature of polymorphism in such rigid rod systems, and consider them in relation to
experimental x-ray data on fibers.

[2]On leave from the Laboratory for Special Chemical Technology, Department of Chemical
Engineering, National Technical University of Athens, Athens 106 82 Greece

Simulation Procedure

The method of simulation follows the methodology of molecular mechanics. As such, it is a static model which calculates atomic force interactions, with the assumption that those geometries having the lowest potential energy are most stable in the solid state. The critical features of the model, (4) which are central to the accurate representation of local structure in such rigid rod systems, are the generalized representation of helical, and degenerate helical (ie. zig-zag and rodlike), structure and the simultaneous minimization of the potential energy of a multichain ensemble with respect to intramolecular and intermolecular degrees of freedom. The close packing and unusually high intermolecular energies realized in the relatively dense solid state of rigid rod polyamides creates a condition for considerable influence of intermolecular parameters on local chain conformation, which in turn must readjust to accommodate nearest neighbor interactions. As a result, one must anticipate nonideal or near-crystalline structures whose preferred organizations cannot realize perfect translational periodicity over long distances, but may exhibit conformational helix discommensurations of the type suggested by Saruyama, (5,6) with energies comparable to the large intermolecular energies operative in the chain packing. Our calculations allow for the smooth variation of chain conformation as a function of internal chain parameters through all commensurate and incommensurate helices during the process of energy minimization. The individual chains are modelled with fixed bond lengths and planar hexagonal phenylene rings; only those bond angles and torsion angles which describe the helical structure of the chain, plus those torsions required to describe phenylene ring rotation, are allowed to vary. This view of the chain structure is illustrated in Figure 1. ϕ_1 and ϕ_3 represent torsions about the virtual bonds spanning the rigid phenylene moieties, while ϕ_2 and ϕ_4 represent torsions about the C-N bonds of the amide moieties; these four torsions influence the helical structure of the chain directly. ϕ_5 and ϕ_6 represent rotations of the ring planes about the C1-C4 axes, independent of the larger helical structure of the chain.

We have modelled PPTA using nine chain and twenty-five chain ensembles, within which local conformations and packing structures are represented explicitly in terms of the finite set of bond angle, torsion angle, and chain packing degrees of freedom. We have found that the nearest-neighbor interactions realized within this explicit ensemble representation are the predominant influence determining the *geometric* features of conformation and packing, in accord with the observations of Tripathy et al. (7) The calculation of interactions with the longer range environment is expedited by the assumption of true crystalline periodicity of atomic positions outside of the explicitly represented ensemble; this allows for the use of rapid lattice summations of interactions, typically up to 40 Å distant, which provide the tail correction terms necessary to accurately estimate the *total potential energy* of the system.

The force field consists of empirical bond angle and bond torsion potentials, deduced from infra-red measurements and crystal data of low molecular weight analogues of the aromatic amide repeat unit, (8,9) and two-body van der Waals and atom-centered electrostatic interactions, parameterized by means of available crystallographic data (10) and programs such as AM1 (11) for semi-empirical quantum mechanics analysis, coupled with two-body interactions between the explicit ensemble and the idealized long range crystalline environment. A comprehensive search and minimization procedure entailed first the identification of low energy regions of the potential energy hypersurface by means of initial grid scans of parameter space, and was then followed by progression along an energy minimization trajectory with respect to geometry within each region to identify the set of most significant local energy minima. From this set, we discard as irrelevant those local minima whose statistical significance, according to a Boltzmann-weighted probability of occurrence on a per repeat unit basis, falls below our cut-off criterion of 0.2, or 20% probability of occurrence. The final list provides the complete set of significant candidate structures to be expected in the stable bulk ordered solid state, based on a priori considerations.

Experimental Procedure

Fiber Spinning. Monofilaments of PPTA were spun in our lab from anisotropic solutions (10% by weight) of Kevlar 29 (η_{inh}=6.1 dl/g in 96% H_2SO_4 at 25°C), supplied by the Dupont Company, redissolved in 100% sulfuric acid by slow stirring under inert atmosphere at 70°C. Fibers of 20 μm dry diameter were spun via the conventional dry-jet wet-spinning technology (12,13) using a spinnerette of diameter 60 μm, extrusion rates between 4 and 40 m/min, and nominal draw-down ratios, defined as the ratio of fiber collection velocity to fiber extrusion velocity, up to 3.2; an air gap of 8 to 10 mm and an aqueous coagulant bath at 24°C were used in all cases. The residual acid in the fiber was thoroughly neutralized, and the fibers washed and dried. Subsequent annealing of the fibers was performed by passing the filaments under nominal tension through a tubular furnace at 500°C under inert atmosphere for 1 to 2 seconds.

X-Ray Analysis. The resulting as-spun and annealed fibers were analyzed via Wide Angle X-Ray Scattering (WAXS) using a Siemens D500 diffractometer with a four circle goniometer and a scintillation counter. Circular-collimated CuKα radiation and angular scans in Bragg angle 2θ from 10° to 55° and azimuthal angle β (defined relative to the normal to the fiber axis) from 0° to 90° were used to generate the first quadrant of the fiber diagram, from which the remaining three quadrants were determined by mirror symmetry; signal error was less than 5%.

Cohesive Energy and Polymorph Geometry

For PPTA, the simulation suggests up to eight possible structural geometries having comparable cohesive energies (E_{coh}), defined as the increase in internal energy per mole of a substance upon removal of all intermolecular forces (Equation 1), on the order of 39±1 kcal/mole. Predicted densities (ρ) range between 1.42 and 1.57 g/cm³. The Hildebrand solubility parameter (δ) may then be calculated as the square root of the cohesive energy density, as given in Equation 2. We obtain solubility parameters between 15 and 16 [cal/cm³]$^{1/2}$, or 31×10³ to 33×10³ [J/m³]$^{1/2}$. Experimental values for solubility parameters of aromatic polyamides are not currently available for comparison. However, these predictions are in accord with an expected slight increase in δ over values for the corresponding aliphatic polyamides, poly(8-amino-caprylic acid) and poly(hexamethylene adipamide), at 12.7 and 13.6 [cal/cm³]$^{1/2}$, respectively. (14)

$$E_{coh} \quad = \quad E_{total} \ - \ E_{isolated\ chain} \tag{1}$$

$$\delta \quad = \quad (E_{coh} \ / \ V)^{1/2} \quad = \quad (E_{coh} \ \rho \ / \ M_w) \tag{2}$$

Table I presents the detailed packing geometries for ensembles consisting of up to two independently orientable chains of like conformation. Although analysis by simulation suggests several distinguishable candidates of comparable energy for chain packing in the oriented PPTA solid state, the predominant trends in structure formation may be deduced from the common elements across the set of packed geometries. The

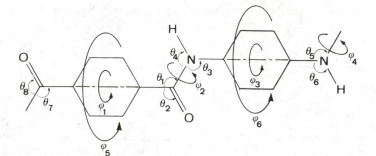

Figure 1. A segment of poly(p-phenylene terephthalamide) with all torsion angles in their zero positions.

Table I. Multichain energy minimization results:
Structural parameters for eight most probable unit cells

Structure ID →	1	2	3	4	5	6	7	8
Structural Parameters ↓								
a (Å)	6.8	7.1	8.3	8.4	8.4	8.5	8.4	4.8
b (Å)	6.2	5.9	5.0	4.9	4.9	4.6	4.9	5.0
c (Å)	13.2	13.1	13.1	13.1	13.1	13.1	13.1	13.1
α (degrees)	90	93	90	86	80	79	85	55
β (degrees)	90	92	90	78	89	94	93	95
γ (degrees)	88	92	92	89	90	93	90	96
Chain Locations (ab projection)	[0,0] [½,½]	[0,0] [½,½]	[0,0] [½,½]	[0,0] [½,½]	[0,0] [½,0]	[0,0] [½,0]	[0,0] [½,0]	[0,0] -
Chain Setting Angles (degrees), relative to the bc facet	6 -174	2 2	18 -161	12 13	-2 -1	-9 -9	2 1	2 -
Inter-sheet Translation (Å)	6.2	6.6	5.9	6.6	2.4	3.1	6.6	2.7
Helix Twist Θ (degrees)	0	0	4	6	3	0	1	4
Monomer Phenylene Ring Rotation (degrees)								
diacid ring ϕ_5:	-16	-30	-26	-35	-43	39	43	34
diamide ring ϕ_6:	15	21	43	36	40	38	33	29
Cohesive Energy (kcal/mol)	40.5	38.5	38.3	38.2	38.1	39.3	39.7	39.6
Density (g/cm³)	1.42	1.45	1.45	1.49	1.50	1.57	1.47	1.56

first of these is the consistent retention of low but, significantly, in some instances nonzero helical twist elements within the chain conformations, despite the possibility for the chain to assume a less extended conformation. This is a direct consequence of the rigid trans-planar conformation of the amide moiety and the planar para-linked phenylene ring structure. Other conformations such as the zig-zag or helix conformations, while viable for the isolated chain, are eliminated due to poor packing energies, relative to the extended chain conformation. Deviations from perfect chain extension reflect the impact of intermolecular forces on the individual chain moieties. Nevertheless, such deviations remain small, indicative of the extreme rigidity of the aromatic polyamide chemical structure; this rigidity ensures that chain alignment plays a dominant role in packed structure formation.

The second priority in ordering the packing of chains is of intermolecular origin and arises as a consequence of the extended single chain conformation; as such, it also serves to stabilize the extended conformation in a packed chain ensemble. This feature is the propensity for neighboring chains to associate into sheets wherein the chains form multiple intermolecular hydrogen bonds between successive amide moieties, made possible by the alternating direction of amide dipoles all lying roughly in a plane containing the chain axis in the extended chain conformation. The register between chains within such sheets is quite specific. Hydrogen bonding energies may be deduced to be of the order 3.5 to 7.5 kcal/mole of chain repeat units, or 1.8 to 3.8 kcal/mole for the hydrogen bond itself. This single specific interaction accounts for roughly 45% to 75% of the total cohesive energy of 8 to 10 kcal/mole binding chains together within a hydrogen-bonded sheet formation. Thus each favorable packing geometry exhibits chain setting angles which orient the amide planes close to one of the crystal facets and which lead to packed sheet structures.

The third and final aspect of ordered chain packing is that defining the relation between hydrogen-bonded sheets. Herein lies the greatest disparity between the various distinct packing geometries. Perhaps surprisingly, the total cohesive energy between chains in neighboring sheets is again quite high, on the order of 6 to 9 kcal/mole. However, in contrast to the interaction energy between chains within a sheet, the sheet-to-sheet interactions are evenly distributed over the components of the chain; the total contribution of inductive and dispersive interaction forces accounts for 45% to 65% of the cohesive energy between sheets, while the electrostatic interactions between amide moieties contributes less than 20%, or 0.8 kcal/mole for each amide-amide interaction, in all cases. This lack of moiety-to-moiety interaction specificity is largely responsible for the variability of sheet packing behavior and the multiplicity of crystal forms witnessed in the model analysis.

WAXS Fiber Diffraction Patterns

Predictions by Simulation. Theoretical (WAXS) reflection intensities for each simulated structure were calculated by assuming the interior unit cell to be representative of the bulk and then applying the conventional summation of atomic scattering over lattice indices, h, k, and l, to arrive at the set of structure factors $F(hkl)$, which were then corrected for polarization, Lorentz scattering, and isotropic thermal motion to obtain observable intensities $I_{obs}(hkl)$: (15)

$$I_{obs}(hkl) = \quad M(\theta)\ P(\theta)\ L(hkl)\ |\ F(hkl)\ |^2 \tag{3}$$

$$F(hkl) \quad = \quad \sum_{j=1}^{n} f_j(\sin\theta/\lambda) \left[\cos 2\pi(hx_j + ky_j + lz_j) + i\sin 2\pi(hx_j + ky_j + lz_j)\right] \quad (4)$$

n is the number of atoms in the unit cell and h, k, and l are the indices defining the crystallographic planes. f_j is the atomic scattering factor, expressed as a function of the Bragg angle θ and the wavelength of radiation, for atom j located at the fractional coordinate position (x_j, y_j, z_j). The correction factor employed for polarization effects was:

$$P(\theta) \quad = \quad (1+\cos 2\theta)/2 \tag{5}$$

For the Lorentz factor one takes (from de Wolff (16)):

$$L(hkl) \quad = \quad (\sin^2\theta \, \cos\theta \, \sin\phi)^{-1} \quad \text{for general reflections} \tag{6}$$

$$L(00l) \quad = \quad (t \, \sin^2\theta \, \cos\theta)^{-1} \quad \text{for meridional reflections} \tag{7}$$

Here, ϕ is the angle between the normal to the reflecting planes and the fiber axis; t is given to a reasonable approximation in terms of the azimuthal angle $\beta_{1/2}$, in radians, at which the intensity of the reflection falls to one half of its peak value (at $\beta=0°$), by

$$t \quad = \quad 0.815\beta_{1/2} \tag{8}$$

The correction for thermal motion, finally, is:

$$M(\theta) \quad = \quad \exp[\, -8\pi^2\mu_s^2(\sin^2\theta/\lambda^2)\,] \tag{9}$$

where μ_s^2 is the isotropic mean square displacement of atoms due to thermal motions, assumed to be the same for all atoms.

For purposes of display, the intensity of each calculated reflection was assumed to be distributed about its mean position according to Gaussian distributions in 2θ and β. The breadths of these distributions represent the expected peak broadening due to finite crystallite sizes, the crystal mosaic, and paracrystal distortions, which are not available a priori from the model calculations. Additionally, we require an estimate for $\beta_{1/2}$ for the Lorentz correction of meridional reflection intensities. The peak broadening distributions were selected of the form

$$y \quad = \quad A \exp[-\ln 2 \, (2(x-x_0)^2/\omega_x^2)\,] \tag{10}$$

where A is the amplitude of the distribution at x_0 and ω_x is the breadth of the distribution at $y=A/2$. Then the intensity at any point of the diagram may be calculated by summation of the contributions from all reflections at that point; the integrated volume of the distribution in 2θ and β equals the calculated intensity I_{obs} of each reflection i located at $(2\theta_i, \beta_i)$:

$$I_{obs}(hkl) \quad = \quad A_{hkl}\omega_{2\theta}\omega_{\beta}\pi/4\ln2 \qquad (11)$$

$$I_{obs}(x_1,x_2) \quad = \quad \sum_{i=1}^{np} A_i\,(\exp\{\,-\ln2[2(x_1-2\theta_i)/\omega_{2\theta}]^2\})$$
$$\times (\exp\{\,-\ln2[2(x_2-\beta_i)/\omega_{\beta}]^2\}) \qquad (12)$$

We require only three parameters not available through the model, the root mean square thermal displacement amplitude and estimates of $\omega_{2\theta}$ and ω_{β}, in order to recreate the entire fiber diffractogram. For our universal isotropic thermal displacement factor we employed the B value used successfully by Northolt (1) for the experimental analysis of scattering from PPTA:

$$B \quad \equiv \quad 8\pi^2\mu_s^2 \quad = \quad 5\times10^{-16}\ cm^2 \qquad (13)$$

corresponding to a root mean square displacement of 0.25 Å. For $\omega_{2\theta}$ and ω_{β} values of 1.5 degrees and 15 degrees, respectively, were found to be appropriate, corresponding to $\beta_{1/2}$ equal to 7.5 degrees. At this level of prediction, it was not deemed justified to incorporate anisotropy of the thermal motion correction into the pattern simulations.

Experimental Data Analysis. The data grids for x-ray diffraction from actual fibers were analyzed quantitatively by selecting a finite number of Gaussian peaks to fit to the two-dimensional experimental pattern, of the same form as were used to display the simulated structure patterns. The fitting procedure was performed through minimization of the relative deviation R_D of the complete pattern intensity distribution, as a function of the peak heights A_i, positions $(2\theta_i,\beta_i)$ and widths at half intensity $(\omega_{2\theta},\omega_{\beta})$ for all peaks $i=1,n_p$.

$$R_D \quad = \quad \frac{\displaystyle\sum_{i=1}^{n2\theta}\sum_{j=1}^{n\beta}(I_{exp,i,j}-I_{calc,i,j})^2}{\displaystyle\sum_{i=1}^{n2\theta}\sum_{j=1}^{n\beta}(I_{exp,i,j})^2} \qquad (14)$$

Allowing all parameters to vary independently describes a set of $5n_p$ independent variables; this number was initially limited by requiring all peaks to have the same set of peak widths, reducing the independent parameter set to $3n_p + 2$. The fitting procedure was initiated with different numbers of peaks and different initial peak assignments, until a best fit (ie. minimum relative deviation R_D) was achieved; the variance was generally less than 10%. Subsequent reoptimization allowing each distribution to vary independently in width did not produce significant improvements in fitting. For the final set of Gaussian peaks describing the best fit to the experimental pattern, the total intensity attributable to each reflection was calculated using Equation 11.

Comparison of Diffraction. Figures 2a and 2b show three-dimensional contour plots of one quadrant of the WAXS fiber diffraction patterns, calculated from the simulation results for structures #3 and #4, respectively; these two structures differ primarily in the orientation of amide dipoles in successive hydrogen-bonded sheets. For comparison, Figure 3a demonstrates the corresponding quadrant of the WAXS fiber diffraction pattern for the experimental measurements from our annealed fiber, spun as described above at a draw-down ratio of 3.2. It is important to stress here that the calculated structures referred to hereafter are structures of *minimum potential energy*, identified through the simultaneous minimization of intramolecular and intermolecular energies, using interatomic force fields based on first principles, with *no adjustable parameters*. This is in contrast to the crystal structures previously reported in the literature, which were derived by fitting the x-ray scattering patterns from atomic models *which are consistent with, but by no means unique to*, the observed scattering of x-rays at wide angles; these latter are not necessarily minimum energy configurations.

With the exception of the overlap of the intense [200] and [110] reflections in both of the simulated patterns in Figure 2, the agreement in position and intensity of the remaining significant reflections is encouraging. Structure #3 most closely approximates the experimental pattern; one notes that in the simulated pattern for Structure #4, the [00l] reflections lie off the meridian, due to the deviation in the crystallographic angle β from 90°, indicative of a slight shift between successive sheets of hydrogen-bonded chains as discussed previously. The annealed fiber appears to exhibit characteristics of both polymorphic forms; the major reflections are represented in each calculated structure, while the minor reflections could derive from the presence of either or both possible contributors. Table II contains the intensities calculated for the two simulated structures #3 and #4. Due to the nature of the simulations, the predicted structures are inherently triclinic, but the reflections may be grouped into "families" of overlapping reflections. Also shown in Table II for comparison are the integrated intensities for the annealed fiber deduced from the two-dimensional Gaussian deconvolution procedure. The weaker meridional reflections predicted by simulation may be traced to the distortion, dictated by local packing constraints, of the 2/1 screw symmetry of the chain conformation. Unit cells similar to Structure #3 but possessing a higher apparent crystal symmetry were deduced by Northolt (1) and Tashiro et al. (2) directly from WAXS data on well-annealed fibers. Comparable diffraction patterns that may be deduced from the remaining simulation predictions are distinguished by the shift in location of major reflections due to a significant alteration of the lattice dimensions (ie. structures #1 and #2), the presence of odd meridional reflections, in cases where the register between successive hydrogen-bonded sheets is shifted by 2.5 to 3.0 Å (ie. structures #5 and #6), or unusually intense off-meridian second layer reflections (ie. [112] or [212] families of reflections) in those cases where both phenylene rings in the polymer chain repeat unit are similarly rotated with respect to the amide bond planes (ie. structures #6, #7, and #8).

The experimental WAXS fiber intensity contour for the as-spun fiber is shown in Figure 3b. This is reproduced here for the purpose of illustrating the appearance of equatorial reflections near 18° and 29° in 2θ, in accord with the crystal allomorph initially reported by Haraguchi et al.; (3) the latter was also deduced from structure factor fitting to limited WAXS data by employing an analogy to the previously reported crystal structure. In agreement with the simulation results, the meridional reflections, determined primarily by chain conformation, remain essentially unchanged, but the new equatorial reflections are indicative of changes in the lateral alignment of sheet packing. In addition to imperfections in the crystalline order, the increased peak breadth could be attributed to the overlap of reflections from multiple packing structures, as has been suggested by others. (1,17) In cases where peak resolution is of sufficient quality to justify more in-depth analysis, the complex experimental pattern may be reconstructed by appropriate combination of contributions from energetically stable allomorphs.

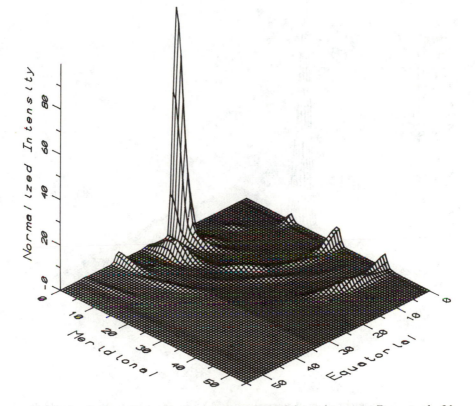

Figure 2a. Surface contours of normalized WAXS intensity versus Bragg angle 2θ and azimuthal angle β in polar coordinate form. Only the upper right hand quadrant is shown. Calculated pattern from Structure #3.

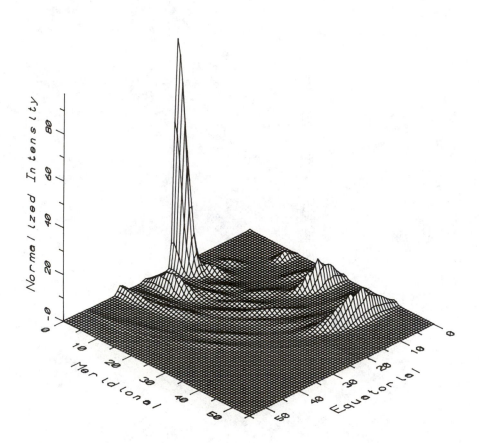

Figure 2b. Surface contours of normalized WAXS intensity versus Bragg angle 2θ and azimuthal angle β in polar coordinate form. Only the upper right hand quadrant is shown. Calculated pattern from Structure #4.

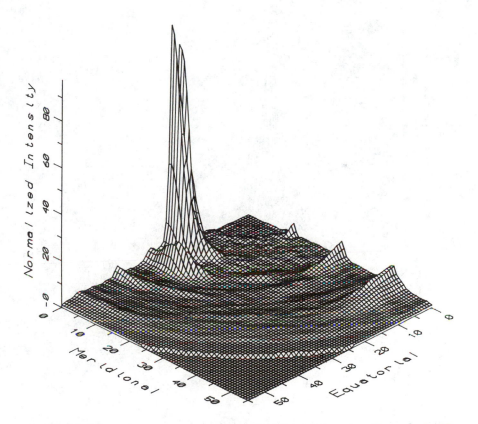

Figure 3a. Surface contours of normalized WAXS intensity versus Bragg angle 2θ and azimuthal angle β in polar coordinate form. Only the upper right hand quadrant is shown. Expermental data on annealed fiber.

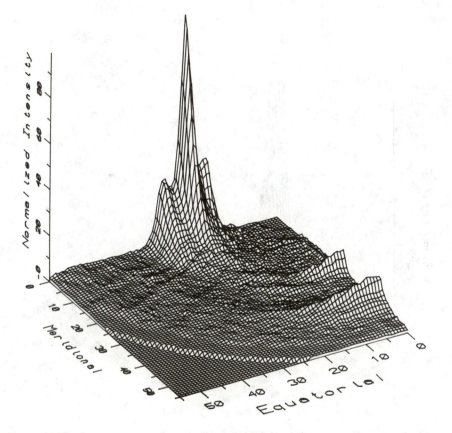

Figure 3b. Surface contours of normalized WAXS intensity versus Bragg angle 2θ and azimuthal angle β in polar coordinate form. Only the upper right hand quadrant is shown. Experimental data on as-spun fiber.

Table II

Summary of x-ray reflections in molecular simulation and experiment

hkl	Structure #3			Structure #4			Annealed Fiber		
	2θ	β	I	2θ	β	I	2θ	β	I
110	21.0	0	74⌉	21.0	0	50⌉	20.4	0	81
1̄10	20.5	0	23⌋	21.2	0	21⌋			
200	21.4	0	100	21.5	0	100	22.5	0	100
310	37.7	0	7⌉	37.3	0	4⌉	38.4	0	10
3̄10	36.7	0	1⌋	37.7	0	1⌋			
							49.9	0	8
1̄01				13.8	29	3			
011				19.0	21	2⌉			
01̄1				19.9	20	4⌋			
111	22.1	18	1⌉						
1̄11	22.1	18	2⌋						
201	22.5	18	2⌉						
2̄01	22.5	18	2⌋						
211	29.2	13	4⌉	27.6	14	2⌉			
2̄11	28.4	14	5⎸	30.0	13	6⎸	29.1	13	23
21̄1	28.5	14	5⎸	28.6	14	4⎸			
2̄1̄1	29.2	13	4⌋	30.3	13	7⌋			
121				38.0	10	1			
3̄01				34.7	11	2			
002	13.5	90	5	13.8	78	5	13.7	84	3
1̄12				24.9	33	2			
2̄02				27.9	29	2			
312	40.2	20	1⌉	37.0	22	1	37.3	23	6
3̄12	40.2	20	1⌋						
103				21.4	72	2⌉			
1̄03				25.3	54	1⌋			
013				26.7	50	2			
203	29.8	43	2⌉						
2̄03	29.7	44	1⌋						
213				31.6	40	2			
303				35.1	35.8	2			
004	27.2	90	13	27.8	78	17	28.0	86	12
1̄14				34.5	52	3	33.4	57	5
105				34.5	82	1	29.8	74	5
006	41.3	90	9	42.3	78	17	43.4	85	26
106	42.8	75	4⌉				45.6	80	15
1̄06	42.8	75	4⌋						
116				44.3	69	3			
206				43.7	72	2			

Conclusion

The packing geometry of the rigid rod polymer PPTA in the solid state appears to be polymorphic and process-dependent. Simulation of such structures by atomistic methods suggests that this polymorphism is influenced by the magnitude and directional specificity of intermolecular forces, which are sufficient to distort the local chain symmetry and fix the hydrogen-bonded sheet geometry, but which allow for considerable variation in inter-sheet correlation and packing. From the minimum energy geometries predicted by simulation, one can generate the x-ray scattering phenomena deterministically. These calculated patterns may serve as viable starting points for the construction of models for further refinement of pattern fitting to experimental data and structure determination. Alternatively, an a priori knowledge of the competitive polymorphic forms based on minimization of potential energy and consideration of their combined contributions to scattering behavior could provide a key to deciphering complex WAXS patterns. This is especially important to the analysis of geometrically constrained polymeric chains such as aramids displaying high intermolecular interaction energies.

Acknowledgments

We gratefully acknowledge the financial support of a Graduate Fellowship to GCR from the National Science Foundation and grants from the Office of Naval Research and the Materials Research Laboratory at the Massachusetts Institute of Technology. A Fulbright Senior Research Scholarship to CDP is also greatly appreciated.

Literature Cited

1.   Northolt, M.G. Europ. Polym. J. 1974, 10, 799.
2.   Tashiro, K.; Kobayashi, M.; Tadokoro, H. Macromol. 1977, 4, 731.
3.   Haraguchi, K.; Kajiyama, T.; Takayanagi, M. J. Appl. Polym. Sci. 1979, 23, 915.
4.   Rutledge, G.C.; Suter, U.W. Polymer Preprints 1989, 30, 2, 71.
5.   Saruyama, Y. J. Chem. Phys. 1985, 83, 413.
6.   Saruyama, Y.; Miyaji, H. J. Polym. Sci.: Polym. Phys. 1985, 23, 1637.
7.   Tripathy, S.K.; Hopfinger, A.J.; Taylor, P.L. J. Phys. Chem. 1981, 85, 1371.
8.   Jorgensen, W.L.; Swenson, C.J. J. Amer. Chem. Soc. 1985, 107, 569.
9.   Tashiro, K.; Kobayashi, M.; Tadokoro, H. Macromol. 1977, 10, 413.
10.   Hagler, A.T.; Huler, E.; Lifson, S. J. Amer. Chem. Soc. 1974, 96, 5319.
11.   Austin Model 1; Ampac version available through the Department of Chemistry, Indiana University, Bloomington, Indiana, 47405, QCPE Program #506.
12.   Kwolek, S. US Patent 3 671 542, 1972.
13.   Blades, H. US Patent 3 767 756, 1973.
14.   van Krevelen, D.W. Properties of Polymers; Elsevier: New York, 1972, p88.
15.   Alexander, L.E. X-Ray Diffraction Methods in Polymer Science; Wiley-Interscience: New York, 1969, Chapter 1.
16.   de Wolff, P.M. J. Polymer Sci. 1962, 60, S34.
17.   Dobb, M.G.; Johnson, D.J.; Saville, B.P. J. Polym. Sci.: Polym. Phys. 1977, 15, 2201.

RECEIVED March 20, 1990

PHYSICS OF LIQUID-CRYSTALLINE POLYMERS: TRANSITIONS AND PROPERTIES

Chapter 22

Calorimetric Investigation of the Glass Transition in Main-Chain Nematic Polymers

Molecular Mass Dependence of the Heat Capacity Increment at the Glass Transition

R. B. Blumstein, D. Y. Kim, and C. B. McGowan[1]

Polymer Science Program, Department of Chemistry, University of Lowell, Lowell, MA 01854

We present a DSC investigation of the glass transition in nematic poly[oxy(3-methyl-1,4-phenylene)azoxy(2-methyl-1,4-phenylene)oxy(α,ω-dioxo-α,ω-alkanediyls)] with spacer length $(CH_2)_7$ and $(CH_2)_{10}$. Influence of spacer parity and molecular mass on T_g, the shape of the heat capacity curve $C_p(T)$ and heat capacity increment at T_g ($\Delta C_p = C_{p,liquid} - C_{p,solid}$) is discussed. Within the range of chain lengths studied (fractions with $M_n = 2,000-20,000$) ΔC_p decreases with increasing M_n. Temperature dependence of $C_{p,solid}$ is the same for all fractions, within experimental error; $C_{p,liquid}$ decreases with increasing chain length while $d(C_{p,liquid})/dT$ increases to approach the value characteristic of the solid below T_g. These results reflect a continuous increase in the level of molecular ordering within the nematic phase, at least over the range of chain lengths investigated.

This paper is part of a series devoted to an investigation of the glass transition in main chain thermotropic liquid-crystalline polymers (LCPs). We present here results from a calorimetric study of the glass transition in nematic glasses of LCPs formed by regularly alternating mesogens and aliphatic spacer groups. Although such LCPs are extensively studied in many laboratories, investigations of their glass transition by calorimetry (1-3) or dynamic methods (4) are few. The data presented herein are part of an attempt to provide a systematic correlation between macroscopic properties of these ordered glasses and their molecular structure. Specifically, we discuss two poly[oxy(3-methyl-1,4-phenylene)azoxy(2-methyl-1,4-phenylene)oxy(α,ω-dioxo-ι,ω-alkanediyl)]s with spacer length $(CH_2)_7$ and $(CH_2)_{10}$.

These polymers are designated AZA9 and DDA9, respectively, and belong to a homologous series which has been extensively studied in our laboratory (for a review see (5) and references therein). The chains are inherently flexible, e.g. they behave as random coils in the isotropic melt or solution. The isotropic/nematic (I/N) transition is coupled with extensive orientational and conformational ordering. Spacer extension and concomitant chain rigidity increase as the temperature is lowered through the nematic

NOTE: This chapter is part 1 in a series.

[1]On leave from Department of Chemistry, Wellesley College, Wellesley, MA 02181

0097–6156/90/0435–0294$06.00/0
© 1990 American Chemical Society

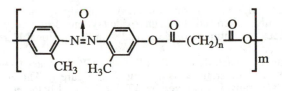

AZA9   n=7
DDA9   n=10

phase. As T_g is approached, spacer conformational order order approaches the value observed in the crystal. The resulting highly ordered nematic glasses (below T_g) and highly ordered "rubbery" phases (above T_g) have interesting material properties: high values of tensile modulus (6) which decrease by a mere 30-50% as T_g is crossed on heating (Blumstein, A.; Jegal, J. G.; Lin, C. H. unpublished results.), in contrast to the usual drop of several decades which is a characteristic of conventional glasses. Macroscopic alignment is easily achieved by application of external fields and does not relax in the supercooled mesophase above T_g within the usual experimental time-scale. Such systems provide models for preparation of highly ordered and easily processed materials (from melt or solution).

We have previously investigated twin model compounds (9DDA9 and 9AZA9) as well as sharp fractions of AZA9 and DDA9, in order to elucidate influence of molecular weight on phase behavior, orientational and conformational order, and viscoelastic properties. Fractions of the same materials will be used here together with some polydisperse samples, the twin models, and polystyrene (PS) standards as a reference conventional glass.

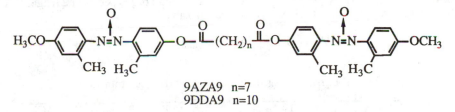

9AZA9   n=7
9DDA9   n=10

Influence of chain length and spacer length on T_g, on the heat capacity increment at Tg (ΔC_p), and the shape of the $C_p(T)$ curve will be presented in the first paper (preliminary results can be found in (3)). Subsequently, we will address influence of thermal history in the isotropic phase and N+I biphase, physical aging below T_g and T_g in blends of LCPs. Finally, an interpretation of the macroscopic data in terms of molecular organization in these and other nematic LCP glasses will be attempted.

Materials and Methods

Sample preparation and characterization has been described previously (see references cited in (5)). The DDA9 fractions used here are the same as in Reference (7), where representative GPC chromatograms are illustrated. Polydispersity of fractions is $M_w/M_n < 1.1$; unfractionated samples and polystyrene standards are characterized by polydispersity indices 1.6-1.8 and < 1.05, respectively. The molecular masses of AZA9 and DDA9 are listed in Table IV and are M_n values as obtained by GPC.

Heat capacity measurements were carried out on a Perkin-Elmer DSC2C with a TADS data station. Absolute heat capacity (Cp) was determined with the Specific Heat

Measurement Kit supplied by Perkin-Elmer. Output signal and analysis procedure were calibrated with indium and aluminum oxide (synthetic sapphire). The measured absolute C_p values of aluminum oxide agreed with the National Bureau of Standard values to better than 1%.

Following the approach described in Reference 8 the glass transition region is characterized by five temperatures, as illustrated on Figure 1. The width of the major portion of the glass transition is given by ΔT, the interval between the extrapolated beginning (T_1) and the end (T_2). The first perceptible beginning, T_b, represents the first deviation in the heat capacity curve from the solid line. The end, T_e, is judged by the return to the liquid line following the transition. The glass transition temperature, T_g, corresponds to the half-devitrification point where the heat capacity curve intersects with the mid-line of the extrapolated solid and liquid lines. The heat capacity increment is measured at T_g as shown on Figure 1. Finally, a sixth temperature (T_{max}) records the enthalpy recovery peak maximum when present.

All the LCP samples have been given a uniform thermal history in the isotropic melt, namely annealing for 15 minutes at 450 K. DDA9 samples were then quenched in liquid nitrogen and placed into the calorimeter at 210 or 230 K. The remaining samples were cooled in the calorimeter to 230 K at 320 K/min (nominal) or as otherwise indicated. Heating rates were 20 K/min.

Results

Twin models. Figure 2 illustrates the temperature dependence of heat capacity for the two twin models and Table I gives the corresponding numerical data. Figure 2 typifies the $C_p(T)$ curve of conventional glasses with a well defined enthalpy relaxation peak and smooth solid and liquid lines. From the extrapolated solid and liquid lines we can measure the heat capacity jump at T_g, by equation 1. Within our experimental range, the data fit a straight line with slopes (B) as listed in Table II.

$$\Delta C_p(T_g) = C_{p,solid} - C_{p,liquid} \tag{1}$$

Table I. Thermal analysis of twin models at the glass transition

	MW	T_b (K)	T_1 (K)	T_g (K)	T_2 (K)	T_e (K)	T_{max} (K)	ΔT (K)	ΔC_p (cal/g K)
9AZA9	696	254.8	264.4	266.6	270	274.8	268.8	4.4	0.108
9DDA9	738	250.8	258.4	260	263.6	268	262	3.6	0.107

The slopes of the solid and liquid lines are typical of values observed in conventional glasses. The values of $\Delta C_p(T_g)$ are 0.108 and 0.107 cal/g K for 9AZA9 and 9DDA9, respectively (314.5 and 330.4 J/mol K).

Polystyrene fractions. Numerical data for the PS reference fractions are summarized in Table III; $C_p(T)$ curves are not shown, as they resemble the conventional glass behavior illustrated on Figure 2.

Molecular weight dependence of T_g follows the Fox-Flory equation (9) (Equation 2), with $T_{g,\infty} = 378.5$ K and $K' = 8.55 \times 10^4$, as expected. The width $\Delta T = T_2 - T_1$ is

$$T_g = T_{g,\infty} - K'/M_n \tag{2}$$

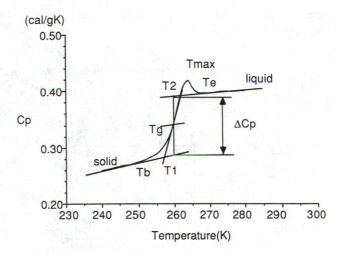

Figure 1. Schematic representation of the glass transition region

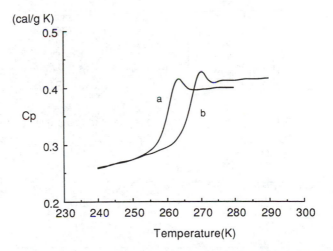

Figure 2. Glass transition region of models: a)9DDA9; b)9AZA9

5.2 ± 0.72 K.   The slope of the line from $C_p(T_1)$ to $C_p(T_2)$ (Figure 1) is 0.0183 ± 0.0003 cal/g K^2. The enthalpy relaxation endotherm does not change with chain length and $\Delta C_p(T_g)$ is constant at 0.070 ± 0.001 cal/g K, in agreement with literature (10). This is typical of the behavior of conventional polymeric glasses, where $\Delta C_p(T_g)$ varies by less than 5% over the entire range of molecular weights (Wunderlich, B., University of Tennessee, personal communication, 1989.). The $C_p(T)$ curves can be superimposed by a shift along the temperature axis. Interestingly, the values of $T_{max} = 384.6 - 8.88 \times 10^4/M_n$.

Table II.   Solid and liquid heat capacity of twin models and PLCs.

$$Cp = A + B (T - 240)$$

(Solid)	A	B x 10^3
	(cal/g K)	(cal/g K^2)
9AZA9	0.2582	1.6018
9DDA9	0.2582	1.4643
AZA9 polymer	0.2613	1.2540
DDA9 polymer	0.2781	1.5900
(Liquid)		
9AZA9	0.3339	0.0361
9DDA9	0.3738	0.0305
AZA9M2300	0.3828	0.8823
AZA9M4200	0.3693	0.9677
AZA9M7200	0.3635	0.9523
AZA9M20000	0.3472	1.1330
DDA9M2550	0.4083	0.4545
DDA9M5500	0.3830	0.7760
DDA9M6400	0.3757	1.0870
DDA9M8400	0.3625	1.0652
DDA9M11300	0.3560	1.2500
DDA9M18900	0.3607	1.1667

Table III. Thermal analysis of polystyrene standard at the glass transition

	Mn	T_b (K)	T_1 (K)	T_g (K)	T_2 (K)	T_e (K)	T_{max} (K)	ΔT (K)	ΔC_p (cal/g K)
Poly-	2550	328	342.2	345.2	348.8	356	350	6.0	0.069
Styrene	7600	354	363.6	366.4	369.2	381	372	5.6	0.070
	19600	359	372.4	363.8	374.8	388	381.5	4.8	0.070
	47000	362	374.4	377.2	380.0	391	382	5.6	0.072
	115000	370	375	378	379	392	384	4.0	0.071

DDA9 and AZA9. Figures 3 and 4 illustrate the $C_p(T)$ curves for some representative AZA9 and DDA9 fractions; Table IV summarizes numerical data for representative fractions as well as for some polydisperse DDA9 samples. Let us now comment on the similarities and differences between conventional glasses and our LCPs on the one hand, and AZA9 and DDA9 on the other.

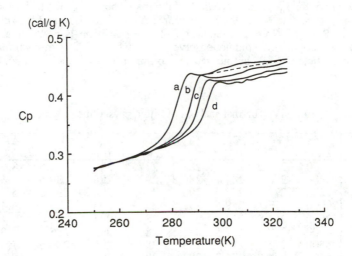
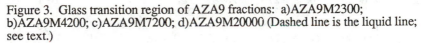

Figure 3. Glass transition region of AZA9 fractions: a)AZA9M2300; b)AZA9M4200; c)AZA9M7200; d)AZA9M20000 (Dashed line is the liquid line; see text.)

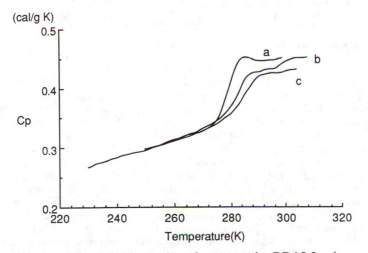

Figure 4. Glass transition region of representative DDA9 fractions: a)DDA9M2550; b)DDA9M6400; c)DDA9M18900

Both sets of fractions follow the Flory-Fox equation albeit with a small value of K', especially in the case of DDA9. The values of $T_{g,\infty}$ are 294.2 and 285.7 K for AZA9 and DDA9, respectively, while $K' = 3.013 \times 10^4$ and 1.25×10^4.

The total width, T_e-T_b, and the width of the major portion of the transition $\Delta T = T_2$ - T_1 are about the same as in the PS fractions; ΔT is 6.40 ± 1.36 and 8.85 ± 1.24 K for AZA9 and DDA9. Note that the width of the transition is not increased in the polydisperse samples. We did not observe the very broad T_e-T_b interval reported for another nematic LCP (2), but no general comment can be made in the absence of further studies.

Table IV. Thermal analysis of liquid crystalline polymers

	Mn	T_b (K)	T_1 (K)	T_g (K)	T_2 (K)	T_e (K)	T_{max} (K)	ΔT (K)	ΔC_p (cal/g K)
AZA9	2300	265.2	277	281.2	287.4	294	285.6	8.6	0.112
	4200	268	284	286.8	292	298	289.6	5.6	0.092
	7200	272	287.6	290	294	304	292.6	5	0.082
	20000	277	289.6	292.8	298	306	296	6.4	0.074
DDA9	2550	264	274.6	278.8	284	294	282	9.4	0.114
	4000	267.5	277.5	281.5	286	295	285	7.5	0.104
	5500	271	279	283	288	295	286.5	7.5	0.072
	6400	266.8	276.7	282	288	296	286.8	10.1	0.066
	7900	272.5	278	284	288	293	288.5	10.5	0.064
	8400	270	278	282.4	290	296.8	287.6	9.6	0.058
	11300	277.5	281	285.5	289.5	292	289	7	0.063
	18900	272	279.2	284	294	299.6	288.4	9.2	0.056
DDA9[a]	3200	270	278.5	282.5	286	294.5	286.3	7.5	0.103
	6500	270	280	285.1	289	294	290	9.0	0.075

[a] unfractionated samples.

The values of $\Delta C_p(T_g)$ are listed in the last column of Table IV and plotted on Figure 5. They are seen to decrease significantly with increasing chain length, at least in the range of molecular weights studied. Polydisperse samples listed in Table IV (as well as others not shown here) follow the same trend as fractions. This large drop in $\Delta C_p(T_g)$ is unexpected and cannot be attributed to molecular mass dependence of the crystalline fraction w_c, as all samples were judged to be amorphous by WAX diffraction. Figure 6 shows that AZA9 does not crystallize significantly on cooling from the melt: the enthalpy relaxation endotherm increases with decreasing cooling rate, as expected, but $\Delta C_p(T_g)$ does not change. A very small, broad endotherm, barely discernible from the baseline is present in all samples, even those quenched in liquid nitrogen (see Figure 7). This endotherm moves to higher temperatures with increasing chain length (Figure 8), and its value is in the range 0.12-0.16 cal/g. It is tentatively attributed to the presence of imperfect crystallites, too small or to few to be seen by X-ray diffraction. LCP morphology is associated with low surface energy at the crystal-mesophase boundary and microcrystallites might be stable and easily nucleated (11).

In contrast to AZA9, DDA9 crystallizes readily on cooling from the melt. The decrease in $\Delta C_p(T_g)$ with decreasing cooling rate at fixed chain length is illustrated on Figure 9. At fixed chain length such curves can be used to calculate an approximate

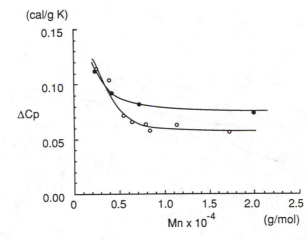

Figure 5. C_p at T_g versus molecular mass: ● AZA9; ○ DDA9

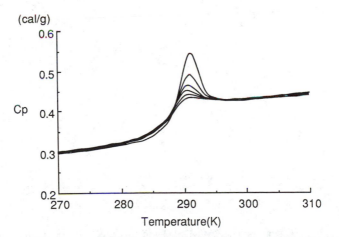

Figure 6. Influence of cooling rate on enthalpy recovery peak in AZA9M4200. From top to bottom cooling rates are: .62; 2.5; 5; 10; 20 K/min; quenched.

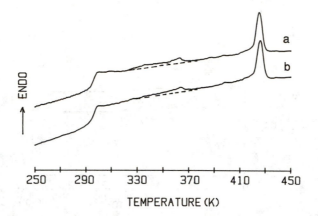

Figure 7. DSC scan of AZA9M4200 a)quenched in liquid nitrogen; b)cooled at 320K/min (nominal). T_g is followed by a very small melting endotherm(above dashed line) and the N/I peak.

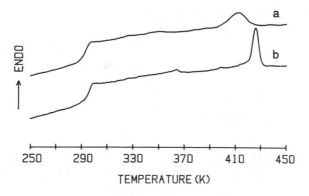

Figure 8. DSC scan of a)AZA9M2300 and b)AZA9M20000. Cooling rate 320K/min.

value of the crystalline fraction, w_c from the familiar Equation 3, where

$$(\Delta C_p)_{experimental} = (\Delta C_p)_{amorphous} (1-w_c) \tag{3}$$

$(\Delta C_p)_{amorphous}$ is as listed in Table IV. Quenching of DDA9 in liquid nitrogen produces a LC glasss, based on X-ray evidence (12). This is corroborated by calorimetry, as illustrated below: Figure 10 shows normalized DSC scans of two DDA9 fractions heated at 20 K/min after quenching in liquid nitrogen. The corresponding thermal data are listed in Table V. Transition temperatures, cold crystallization (T_{CC}), melting (T_{KN}), isotropization (T_{NI}) and isotropization enthalpy ΔH_{NI} increase with increasing mass, as expected from a previous study of influence of molecular weight on phase behavior (13). DDA9M2550 is typical of short chain lengths which display premelting followed by recrystallization (its heat of recrystallization is -1.98 cal/g); DDA9M8400 is typical of higher masses in that it shows two melting peaks. For DDA9M2500 $\Delta H_{KN} = \Delta H_{CC} + \Delta H_{recrystallization}$; for DDA9M8400 $\Delta H_{KN} \approx \Delta H_{CC}$, within experimental error. Similar results are obtained for other samples (see Table V); in no case can the observed decrease in $\Delta C_p(T_g)$ be explained by the presence of crystallinity. In fact, as illustrated in Table V, cold crystallization kinetics is slowed by increasing chain length. The same trend applies to crystallization from the melt (Kim, D. Y., Ph.D. Thesis, University of Lowell, in preparation.).

Table V.   Thermal data for selected DDA9 fractions.

	T_{CC} (K)	ΔH_{CC} (cal/g)	T_{KN} (K)	ΔH_{KN} (cal/g)	T_{NI} (K)	ΔH_{NI} (cal/g)
DDA9M2550	321.1	6.42	373.0	8.45[a]	402.8	2.34
DDA9M6400	326.8	6.71	395.0	7.32[b]	432.5	3.20
DDA9M8400	327.5	6.58	395.7	6.81[b]	435.1	3.39
DDA9M18900	336.9	4.99	396.0	4.28[b]	442.5	3.15

[a] $\Delta H_{KN} = \Delta H_{CC} + \Delta H_{recrystallization}$
[b]   Sum of two melting peaks in ratio 0.3/0.7

The $C_p(T)$ traces of AZA9 resemble that of conventional glasses and the enthalpy recovery peak area appears relatively independent of chain length. Within our experimental temperature range the absolute value of $C_{p,s}$ fits the values illustrated in Table II. The slope is typical of that reported for conventional glasses below T_g (14), as already mentioned in the case of the twin models. The slope of the line from $C_p(T_1)$ to $C_p(T_2)$ is independent of chain length, as in the case of PS fractions, and equals 0.0147 ± 0.0012 cal/gK^2. The observed decrease in $\Delta C_p(T_g)$ is due to a decrease of the liquid heat capacity $C_{p,l}$ with increasing chain length, as illustrated on Figure 3. The slope of the liquid lines increases with increasing chain length (Table II). Table VI shows representative values of $C_{p,l}$, $C_{p,s}$ and $\Delta C_p(T_g)$ calculated for AZA9. While the calculated values of $C_{p,s}$ increase with increasing T_g, as expected, the values of $C_{p,l}$ at T_g do not.

   Contrary to all the samples discussed so far, the shape of $C_p(T)$ curves of DDA9 is strongly dependent on chain length, as illustrated on Figure 4. The curves can no longer be matched by a shift along the temperature axis. As the molecular mass is increased, the $C_p(T)$ curve in the glass transition region of DDA9 develops a composite

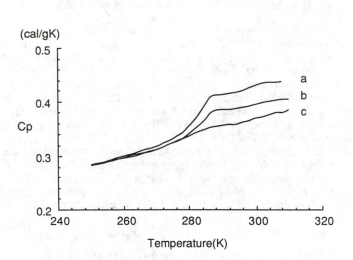

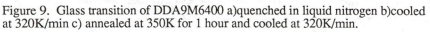

Figure 9. Glass transition of DDA9M6400 a)quenched in liquid nitrogen b)cooled at 320K/min c) annealed at 350K for 1 hour and cooled at 320K/min.

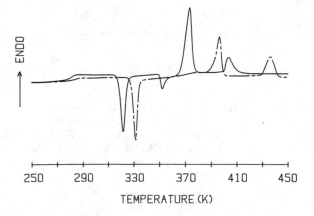

Figure 10. Normalized DSC scans: full line DDA9M2250; dashed line DDA9M8400. See text and Table VI for details.

Table VI.   Solid and liquid heat capacity of AZA9 fractions at Tg.

	T_g (K)	$C_{p,l}$ (cal/g K)	$C_{p,s}$ (cal/g K)	ΔC_p (cal/g K)
AZA9M2300	281.2	0.4199	0.3081	0.112
AZA9M4200	286.8	0.4146	0.3197	0.095
AZA9M7200	290.0	0.4111	0.3280	0.083
AZA9M20000	292.8	0.4070	0.3310	0.076

shape with the appearance of a broad peak whose position shifts to the right with increasing chain length (compare Figure 4a, 4b and 4c). Such enthalpy relaxation curves are sometimes found in thermally quenched samples (15). The evolution illustrated on Figure 4 is reminiscent of what one might observe for a given sample in a conventional glass upon increasing physical aging time below T_g. Concommittantly with a change in ΔC_p, we expect a change in the relaxation behavior of the glass and this study will be adressed in a forthcoming publication. As in the case of AZA9, the observed decrease in $\Delta C_p(T_g)$ is due to a decreasing heat capacity of the liquid mesophase. The slope of the liquid line (Table II) increases with increasing chain length, approaching the values found in the solid phase.

In summary, the observed decrease in $\Delta C_p(T_g)$ is quite unexpected, as the dynamics of devitrification involve a very local motion. The only other report of $\Delta C_p(T_g)$ decrease with increasing chain length, to our knowledge, is for poly(propylene glycols) up to $M_n = 4,000$ (16). The authors explain their results on the basis of a dielectric relaxation study which suggests an increasingly large amplitude of the ß-process at temperatures below T_g. Our data, on the other hand, suggest a systematic decrease in the liquid heat capacity within the range of chain lengths investigated (up to Mn = 20,000). In contrast to this decrease one might note that the specific heat of n-paraffins at 25°C (n-nonane through n-hexadecane) is constant at $C_{p,l}$=0.5285±0.0005 cal/g (17), exactly the value reported for $C_{p,l}$ of polyethylene at the same temperature (18).

Discussion

Both AZA9 and DDA9 are characterized by a strong decrease in $\Delta C_p(T_g)$ with increased chain length in the ordered glassy structure, the effect being significantly more pronounced in the case of the latter polymer. Influence of spacer length must undoubtedly be related to the well documented odd-even oscillation in the physical properties of this homologous series (5). Orientational and conformational order are higher for n=even (DDA9 for example) then n=odd (AZA9). For n=even the mesophase is cybotactic nematic, as opposed to a "normal" level of nematic order for n=odd. Phase transition temperatures (with the as yet unexplained exception of T_g, which follows an inverse even-odd oscillation), specific volume change at T_{NI}, induced magnetic birefringence, anisotropy of magnetic susceptibility and other properties follow the same oscillation.

Since the twins provide excellent models for investigation of LCP behavior, let us first consider 9AZA9 and 9DDA9. If we use an empirical group additivity approach to the calculation of the heat capacity increase at Tg, by dividing the molecule into mobile structural moieties ("beads") with a contribution of 11.3 J/Kmol for a small "bead" (-CH_2-, -CO_2-, etc) and approximately twice that amount for "large" beads (-Ph-for example), (19-20) the "expected" values are observed for a number of conventional low molecular mass nematic glasses such as 4-butyl-4'-methoxyazoxybenzene, a close relative of our mesogenic core (21). In the case of polymers the same holds true for nematic poly(azomethines) with flexible $(CH_2-CH_2-O)_n$ spacers and relatively low level of nematic order (2). One must point out, however, that molecular weight dependence of $\Delta C_p(T_g)$ in these poly(azomethines) has not been reported.

The experimental values of ΔC_p are 330.5 J/K mol and 314.6 J/K mol for 9AZA9 and 9DDA9, respectively. Two observations can be made:

i) The difference $\Delta C_{p,9DDA9} - \Delta C_{p,9AZA9} = 15.9$ J/K mol is less than half the value expected for full devitrification of three $-CH_2-$ units.

ii) A major contribution to the heat capacity increment appears to originate with the mesogen. Assuming full participation of the methylene sequence in the devitrification process of 9AZA9 and a value of 11.3 J/K mol each per mole of $-CH_3$, $-O-$, and $-CO_2-$ groups, the experimental value of $\Delta C_{p,9AZA9}$ yields 83.9 J/K mol for the 2,2'-dimethyl-azoxybenzene core. With 83.9 J/K mol of core, an "upperbound" value for full devitrification of a repeating unit of polymer would be 185.6 and 219.5 J/K mol for AZA9 and DDA9, respectively. Table VII lists the deviation (in %) of the experimental $\Delta C_p(T_g)$ with respect to this "upperbound" value. As chain length increases the mesophase structure appears to progressively change to a "strong" liquid which exhibits a small value of ΔC_p due to "restrictions on the configurational states which the system can adopt" (22).

Table VII.   Comparision of experimental heat capacity with values estimated from an empirical additivity approach.

	Mn	ΔC_p(J/mol K)	%deviation
AZA9	2300	193.1	-
	4200	158.6	14.5
	7200	141.4	23.9
	20000	127.6	31.2
DDA9	2550	227.0	-
	4000	207.1	5.6
	5500	143.4	34.7
	6400	131.4	40.1
	7900	127.5	41.9
	8400	115.5	47.3
	11300	125.5	42.8
	18900	111.5	49.2

%deviation = $(1 - \Delta C_{pexp}/\Delta C_{pcal})$ x 100

Influence of chain length on the level of molecular ordering in AZA9 and DDA9 has been investigated by a combination of methods, including NMR and SANS, as summarized in Reference 5. The data suggest that level of ordering continues to increase beyond the commonly recognized "plateau" value at DP10. In addition, an investigation of biaxility of molecular orientational order as a function of chain length has recently been carried out via a combination of NMR (DMR and PMR) and magnetic susceptibility measurements (23). The data indicate that increase of order is due to a progressive reduction of the orientational fluctuations of the aromatic core in one plane only; fluctuations in the perpendicular plane are not affected by an increase of chain length. The observed decrease in ΔC_p, which reflects a drop in $C_{p,l}$ alone, is thus a natural consequence of the rigidification of the liquid phase brought about by decrease of orientational fluctuations coupled with increased conformational order.

Acknowledgments

This work was supported by NSF grant DMR-8823084. One of us (C.B. McGowan) is indebted to the National Science Foundation for partial support under NSF grant DMR-8908762. The authors wish to thank Dr. F. Volino for stimulating discussions.

Literature Cited

1. Grebowicz, J.; Wunderlich, B. J. Polymer Sci., Polym. Phys. Ed. 1983, 21, 141.
2. Cheng, S. Z. D.; Janimak, J. J.; Sridhar, K.; Harris, F. W. Polymer 1989, 30, 494.
3. Kim, D. Y.; Blumstein, R. B. Polym. Prep. 1989, 30 (2), 472.
4. Müller, M.; Meier, D.; Kothe, G. Prog. N.M.R. Spect. 1985, 17, 211.
5. Blumstein, R. B.; Blumstein, A. Mol. Cryst. Liq. Cryst. 1988, 165, 361.
6. Blumstein, A.; Lin, C. H.; Mithal, A. K.; Tayebi, A. J. Appl. Polym. Sci. (in print).
7. Klein, T.; Hong, X. J.; Esnault, P.; Blumstein, A.; Volino, F. Macromolecules 1989, 22, 3731,.
8. Cheng, S. Z. D.; Cao, M. Y.; Wunderlich, B. Macromolecules 1986, 19, 1898.
9. Fox, T. G.; Flory, P. J. J. Appl. Phys. 1950, 21, 581.
10. Karasz, F. E.; Bair, H. E.; O'Reilly, J. M. J. Phys. Chem. 1965, 69, 2657.
11. Blundell, D. J. Polymer 1965, 23, 3591.
12. Maret, G.; Blumstein, A. Mol. Cryst. Liq. Cryst. 1982, 88, 295.
13. Blumstein, R. B.; Stickles, E. M.; Gauthier, M. M.; Blumstein, A.; Volino, F. Macromolecules 1984, 17, 2, 177.
14. Mathot, V. B. F. Polymer 1984, 25, 579.
15. Berens, A. R.; Hodge, I. M. Macromolecules 1982, 15, 756.
16. Johari, G. P.; Hallbucker, A.; Mayer, E. J. Polym. Sci., Part B, Polym. Phys. Ed. 1988, 26, 1923.
17. Shaw, R. Chem. Eng. Data. 1969, 14, 461.
18. Loufakis, K.; Wunderlich, B. J. Phys. Chem. 1988, 92, 4205.
19. Wunderlich, B. J. Phys. Chem. 1960, 64, 1052.
20. Gaur, U.; Wunderlich, B. Polym. Prep. 1979, 20(2), 429.
21. Grebowicz, J.; Wunderlich, B. Mol. Crys. Liq. Cryst. 1981, 76, 287.
22. Angell, C. A. Relaxation in Complex Systems; Ngai, K. L.; Wright, G.B., Eds.; U.S. Government Printing Office, Washington, D.C., 1985. (Available from National Technical Information Service, P.O. Box 5285, Royal Rd., Springfield, MA 22161.)
23. Esnault, P.; Galland, D.; Volino, F.; Blumstein, R. B. Macromolecules 1989, 22, 3137.

RECEIVED March 22, 1990

Chapter 23

Liquid-Crystalline Polymers
A Unifying Thermodynamics-Based Scheme

A. Keller[1], G. Ungar[1,3], and V. Percec[2]

[1]H. H. Wills Physics Laboratory, University of Bristol, Tyndall Avenue,
Bristol BS8 1TL, United Kingdom
[2]Department of Macromolecular Science, Case Western Reserve University,
Cleveland, OH 44106

It is shown that conditions for realizing mesomorphic
states in polymers are readily expressed in terms of
the relative thermodynamic stabilities of the crystalline, liquid crystalline and liquid phases. The simple
scheme here presented has the merit of providing: i) a
unifying perspective over a wide range of mesophase
behavior which otherwise may be considered in isolation
as disconnected occurrences, ii) a signpost for purposeful attainment (or enhancement) of the mesomorphic
state also in the case of systems which are not
normally considered liquid crystal forming, and iii) an
explanation for the dependence of the mesomorphic
temperature range as a function of the degree of polymerization. Examples are quoted from a range of materials most comprehensively from polyethylene, bringing
together diverse past and some rather striking new observations illustrating the validity and usefulness of
the scheme.

In recent years the mesomorphic state has come increasingly to the
forefront of polymer science, particularly since the purposeful
synthesis and study of polymeric liquid crystals. Most frequently
it is associated with the presence of mesogenic groups, (such as on
their own would form small molecular liquid crystals) built in, or
attached to, the macromolecular chain where the mesomorphic state
is usually attributed to the stiffness imparted by these groups. In
other instances of mesophase forming stiff molecules the chains are
too irregular to crystallize in which case the suppression of
crystallization is considered as the factor which promotes the
mesophase. However, there are chains, including quite regular
ones, which can give rise to the liquid crystalline state without

[3]Current address: Division of Ceramics, Glasses, and Polymers, University of Sheffield,
Sheffield S10 2TZ, United Kingdom

any constituent capable of forming liquid crystals as separate small molecules. Such are e.g. the polymers based on the flexibly jointed diphenyl compound diphenyl ethane (1), the totally flexible main chain polysiloxanes with side groups beyond the length of methyl (2), main chain polyphosphazenes (3) and even the fully flexible and chemically simplest long chain compound polyethylene under circumstances to be discussed in specific detail below. Many examples of mesophases in polymers of the latter type have been listed in the review by Wunderlich et al. (4). The purpose of the present publication is to lay out a simple thermodynamic scheme which embraces all the above categories. It will be purely diagrammatic and qualitative, along familiar lines, yet we hope that in the context presented it will provide a coherent thread linking together the various manifestations of mesophases, thus contributing towards the understanding of their interrelationship and providing guidance for their purposeful design. Illustrative examples will be quoted from available experimental material such as are not readily found all together in the literature at least in the present context. Such will embrace cases with chemical constitution and/or physical parameters as variables. They will include some examples of chains with varying ratios of rigid and flexible constituents but mostly polyethylene in its diverse forms and circumstances. Finally, the dependence of phase transition temperatures on molecular weight will be discussed based on the same principles.

Equilibrium States

For the considerations to follow it will suffice to consider the basic thermodynamic relationship

$$dG = Vdp - SdT \qquad (1)$$

where G is the free energy, S the entropy, V the volume, p the pressure and T the temperature. For the scheme in question take first the melting of a true crystal at constant pressure (dp=o). As seen from Figure 1a, and as follows from Equation 1, the free energies of both crystal (G_k) and isotropic liquid (G_i) decrease with increasing temperature where the decrease in G_i is the steeper, due to $S_i > S_k$. Where G_i crosses G_k the crystal melts, which of course is at $T_{k-i}^i = T_m$, the melting point of the crystal. In this case, as drawn in Figure 1a, the free energy of any hypothetical mesophase (if such can exist at all), G_{1c}, cannot fall below both G_k and G_i, hence correspond to a stable state at any temperature. While G_{1c} decreases faster with T than G_k it will only cross at G_k at a point which is above G_i, i.e. where the isotropic liquid is already the stablest phase. Thus the mesophase is virtual and remains unrealizable as a stable phase.

In order to create a stable mesophase a section of the G_{1c} versus T curve will need to be brought beneath both G_k and G_i (thus to a state of greatest stability). This can be achieved either a) by raising G_i (Figure 1b), or b) by raising G_k (Figure 1c), or by a combination of both a) and b). As seen in Figures 1b and 1c the mesophase will be "uncovered" in a temperature range bounded by T_{k-1c} and $T_{1c-i} = T_i$ corresponding to temperatures of crystal melting and isotropization respectively. (The lowering of G_{1c} would have

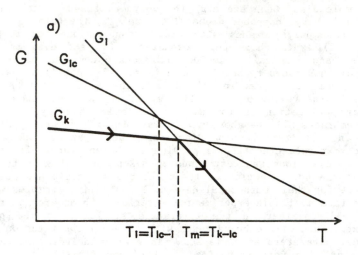

Figure 1a. Schematic plot of free energies vs. temperature for a scheme that does not show a mesophase. G_k, G_{lc} and G_i are, respectively, the free energies of the crystalline, mesomorphic (virtual) and isotropic liquid states. $T_{k-i} = T_m$ is the crystalline melting point. Here, as in subsequent Figures 1b and 1c, the heaviest lines correspond to the stablest state at a given temperature.

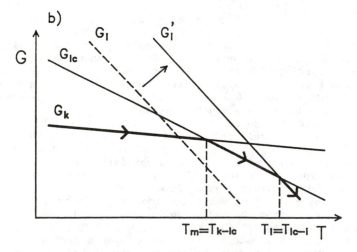

Figure 1b. Schematic plot of free energies vs. temperature for the system in Figure 1 but with G_i raised (to G_i') so as to "uncover" the mesophase. T_{k-lc} and T_{lc-i} are the crystal-mesophase transition and the isotropization transition temperatures.

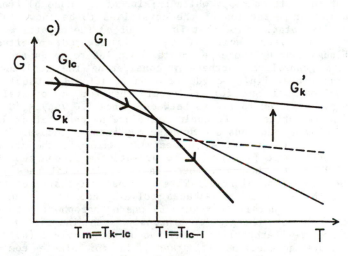

Figure 1c. Schematic plot of free energies vs. temperature for the system in Figure 1a but with G_k raised (to G_k') so as to "uncover" the mesophase.

the same effect; but changes in G_{lc} are expected to be small compared to those in G_i and G_k and will be disregarded in what follows.)   In general, raising of G_i (case a) arises from the lowering of the melt entropy, while raising of G_k (case b) from the reduction in the perfection of the crystal as to be shown in specific examples later. Note that in case of a) the crystal melting point, T_m, is raised, while in b) it is lowered, a situation clearly brought out by the experimental examples to be quoted.
combination of both a) and b). As seen in Figures 1b and 1c the mesophase will be "uncovered" in a temperature range bounded by T_{k-lc} and $T_{lc-i}=T_i$ corresponding to temperatures of crystal melting and isotropization respectively. (The lowering of G_{lc} would have the same effect; but changes in G_{lc} are expected to be small compared to those in G_i and G_k and will be disregarded in what follows.)   In general, raising of G_i (case a) arises from the lowering of the melt entropy, while raising of G_k (case b) from the reduction in the perfection of the crystal as to be shown in specific examples later. Note that in case of a) the crystal melting point, T_m, is raised, while in b) it is lowered, a situation clearly brought out by the experimental examples to be quoted.

We can generalize further by considering the influence of change in pressure, i.e. the Vdp term in Equation 1. Usually the specific volume is larger for the liquid than for the crystal with the mesophase expected to lie in between; hence $(\delta G_k/\delta p)_T < (\delta G_{lc}/\delta p) < (\delta G_i/\delta p)_T$. In principle it could therefore happen that at some p G_{lc} falls below G_k and thus a mesophase may become "uncovered". It has been found that in most experimental systems (5) the effect of increased hydrostatic pressure is to promote the mesophase ("barophyllic" behavior). However, there is no fundamental reason that would make this a general rule. This is clearly illustrated by the example of the sequence of alkanes→polyethylene (as it will be described later in detail in Figure 6) where "barophobic" behavior of short n-alkanes changes continuously with increasing chain length toward the "barophyllic" behavior of polyethylene (6).

The above scheme, as it stands, is for single component systems leading to thermotropic mesophases. It can readily be extended, with appropriate elaborations, not to be pursued here, to two-component systems which would then also embrace the lyotropic state of matter.

Metastable States

Figures 1a-c refer to states of thermodynamic equilibrium. However, systems may not respond immediately when passing from one stable regime to another within the phase diagram, hence the metastable phases can often arise.

In what follows we shall digress into the possibility of metastable states, as they are frequent in polymeric liquid crystals, and in particular, as they generate an ambiguous nomenclature which we are anxious to clarify. The most commonly encountered metastability is that arising on crystallization. As familiar, crystallization only sets in at certain supercooling. In polymers in particular, crystallization temperature $T_{i \to k}$ can be appreciably below the equilibrium melting point $T_{k \to i}=T_m$. On the other hand, the formation of a mesophase generally requires less supercooling. Now, if the temperature of the transition from the isotropic liquid to a normally unstable mesophase lies somewhere between $T_{k \to i}$ and $T_{i \to k}$, such a "virtual" mesophase may materialize on cooling. As the temperature is lowered still further, crystallization will occur. At this state two extreme situations may be envisaged, as depicted in Figures 2 and 3. On crystallization the free energy either drops from G'_k to G_k, i.e. the value for the perfect crystal (Figure 2, where the dotted line indicates a possible pathway), or else there is no discontinuous change in G, i.e. a highly imperfect crystal is formed with its free energy remaining at G'_k (Figure 3). The realistic path would be somewhere in between these two extremes, i.e. some decrease in G is expected, which may not quite reach the level of G_k.

We shall first consider the extreme situations of Figures 2 and 3. When the perfect crystal of Figure 2 is reheated, it melts directly into the isotropic liquid at $T_{k \to i}$: thus such a system displays the mesophase only on cooling, and is called "monotropic". On the other hand, the imperfect crystal of Figure 3 first changes back into the mesophase at $T_{k' \to lc}$ and then into the isotropic liquid at $T_{lc \to i}$ on reheating; thus the mesophase occurs both on cooling and heating and is called "enantiotropic". The latter case clearly illustrates that an enantiotropic mesophase does not necessarily mean stability of the mesophase, as sometimes implied, although a stable mesophase, naturally, must be enantiotropic. As mentioned before, real systems are in between those described by Figures 2 and 3. Some decrease below G'_k will occur upon crystallization, the magnitude of the drop depending, among others, on crystallization kinetics. Accordingly, neglecting possible perfectioning on subsequent heating, mono- and enantiotropic behavior are distinguished by the magnitude of the drop in G on crystallization; if G stays above a critical value (G^c_k) the system is enantiotropic, if it falls below it is monotropic, the definition of G^c_k being apparent from Figure 2. It is easily seen how crystal perfectioning on annealing can lead to a "conversion" of an enantiotropic into a monotropic mesophase, an effect frequently observed in both polymeric and small molecule liquid crystals (1).

While it was assumed above that only G_k is affected by thermal history, in the case of main chain polymeric liquid crystals pronounced time dependent variability in G_{lc} has recently also been observed (7,8). It was shown that the lack of equilibrium perfection in the nematic phase can lead to substantial depression of the isotropization temperature $T_{lc \to i} = T_i$. Thus non-equilibrium mesomorphic states can also, in principle, affect the phase sequence (enantiotropic, monotropic) in the case of polymeric liquid crystals.

It is worth noting further that, under certain conditions, polymers will also display superheating effects, in which case the mesophase may only appear on heating; this can be regarded as "monotropic" behavior in the reversed sense. An example of this in connection with polyethylene will be quoted further below.

As seen from the above, kinetics will always play part in any LCP phase diagram as determinable in practice. Its influence on our present considerations will depend on its magnitude. If it is small enough, so as not to alter the sequence of appearance (or disappearance) of the different phases with changing temperature, the equilibrium situations in Figures la-c will be taken to apply

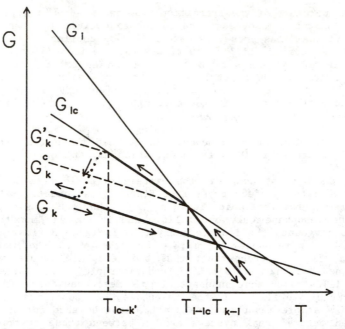

Figure 2. Schematic plot of the free energy diagram illustrating the origin of a monotropic liquid crystal (see text). Here, as in Figure 1, the arrows represent cooling and heating pathways.

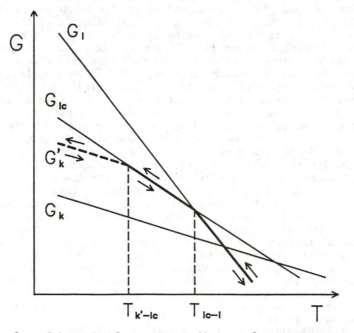

Figure 3. Schematic free energy diagram for an enantiotropic liquid crystal, where the mesophase is metastable (see text).

with the kinetics as an overlay, affecting only the exact numerical values of the actual divides. If, however, the phase sequence itself is affected, and in fact new phases are created due to metastability, then the whole phase behavior will become kinetically determined and consideration in Figures 2, 3 will pertain.

Finally a further kinetic factor, the glass transition (T_g), needs invoking, particularly pertinent to LCP's. On cooling the system becomes immobilized at T_g (more precisely also dependent on the rate of cooling), hence phase transformations will be arrested or altogether prevented. The inverse will apply when a previously immobilized system is heated above T_g when the system will again be able to follow its course towards the equilibrium state. In practice, this will lead to inaccessibility of certain portions of the phase diagram, or conversely, lead to the freezing in of the liquid or of the liquid crystal state enabling their attainments at temperatures where by thermodynamic criteria they would be unobtainable otherwise (isotropic or LC glass respectively (4)). Even if T_g is not a thermodynamic quantity the indication of its location in the phase diagram can therefore serve a useful purpose.

Having placed metastability, glass transition and related kinetic issues on the map we shall resume our argument as referring to thermodynamic equilibrium alone.

EXAMPLES - GENERAL

Mesogenic Polymers

The scheme embodied by Figures 1a-c encompasses most known situations. However, we have to be aware that this treatment considers that we are above the molecular weight below which phase transitions are still molecular weight dependent. The dependence of phase transitions on molecular weight will be discussed in the last part of this paper. Thus high (or raised) G_i (Figure 1b) corresponds to low melt entropy which is consistent with the familiar tendency of stiff chains (worm-like or rigid rods) to give rise to mesophases. From this in turn follows the mesophase promoting role of mesogenic groups, the most often quoted single factor responsible for polymeric liquid crystals (LCP) and principal subject area of present day LCP research. In fact, the whole presently active subject area of chemical stiffening of chains, so as to give rise to liquid crystals, all relies in the first instance on appropriate shifting of G_i in the context of Figure 1b.

The extreme limit of stiffness is the fully rigid rod such as results from the polymerization of all phenyl or other condensed ring containing monomers of which poly(phenylene benzobisthiazole) (PBZT) of currently increasing topicality is a salient example (9). In such cases of fully rigid rods G_i (or even G_{1c}) becomes so high that T_i (or even T_m) may lie beyond the chemical stability limit of the molecule, in other words the material becomes infuseable (and also practically insoluble). As well known this fact has given rise to the widespread activity of chemically loosening the liquid crystal forming rigid rods so as to make them more readily fuseable and/or soluble (i.e. processable) without, however, losing the mesophase "window" in the phase diagram. For processing purposes fluidity alone is required. This is achieved once T_m is reached.

Hence, in practice the attainment of T_i for the technologically important rigid rod type molecules (e.g. Kevlar, PBZT, Vectra - see below) is not usually pursued.

In present practice the loosening of rigid rod polymers is being pursued by copolymerizing rigid phenyl containing mesogenic groups with flexible alkane or ethylene oxide spacers in varying predesigned ratios (10,11). Here the number of CH_2 (or CH_2CH_2O) groups within the spacer in the different preparations determines the overall flexibility/rigidity, longer spacers leading to more flexible molecules and vice versa. The corresponding T_i will be higher with shorter spacers (7,8) due to decreasing entropy of the isotropic liquid.

The above type of molecular tailoring of mesophases may well appear self evident from the chemical point of view. Even so, its incorporation in an underlying thermodynamic scheme should be helpful for a purposeful planning of chemical design and perceiving wider interconnections. For this purpose phase diagrams with chemical (e.g. spacer) composition as variable should be particularly appropriate, especially when the limiting conditions for the existence of mesophases are approached. Explicit reference to such phase diagrams will be made in the next section (and beyond in ref. 16), in that case in connection with the influencing of crystal perfection with which this issue becomes combined.

Returning to Figures 1a-c the position of G_i, hence the existence or creation of mesophases, can also be influenced by physical means. Thus orienting chains, which would otherwise be randomly arranged in the isotropic phase, and also stretching them (in the case of chain flexibility) will reduce the configurational entropy, hence raise G_i, with consequent facilitation of liquid crystal formation. This LC promoting effect of orientation is familiar among rigid or partially stiff liquid crystal polymers, as manifest e.g. by the drop of viscosity when passing through orifices, a situation arising on processing (12). It is perhaps less familiar that fully flexible chains, which on their own are not liquid crystal forming (i.e. correspond to Figure 1a), can be transformed into a mesophase by orientation and/or constraints according to the scheme in Figure 1b to be invoked explicitly for polyethylene below.

Crystal Imperfections

High (or raised) G_i (Figure 1c) means crystals which are either imperfect due to defects or, while otherwise regular, contain loosely packed molecules. The latter, i.e. loose packing, may be due to bulky side groups. The liquid crystal forming ability of the otherwise regular and fully flexible long side group containing siloxanes (2) could be relatable to this cause. In addition (or alternately) bulky side groups could, due to steric congestion, stiffen a chain which then will raise G_i contributing to (or causing) LCP formation. The possibility of chemical irregularities of course are manifold in polymers. Irregularly placed branches or side groups, whether deliberate or accidental, are common features, again to be more explicitly referred to in connection with polyethylene. The LC forming tendency will be particularly pronounced when the influence of relatively bulky side groups and irregularity

combine, as in the case of hydroxypropyl cellulose ($\underline{13}$). Here the
inherent stiffness of the chain is biasing the system in the same
direction in any case by raising G_i, while the presence and irregular disposition of the side groups suppresses crystallization by
raising G_k.
     Introduction of comonomer units, even without necessarily in-
volving branches will similarly reduce crystal perfection, hence
raise G_k, thus promoting formation of the mesophase. Again, this
will combine with the mesophase promoting tendency of stiff units
within the chain, when such are present in any case, as e.g. in the
technologically topical copolymers of hydroxybenzoic and naphthoic
acids ($\underline{14}$).
     Mesogen-flexible spacer systems are particularly amenable for
controlled adjustments in phase behavior through variation in copolymer constitution. As stated previously, introduction of flexible spacers can render otherwise rigid rod polymers fuseable. If
say, two different flexible spacer lengths are used the resulting
copolymers will have enhanced mesomorphic tendencies through raising G_k, thus lowering T_m, the result being a widened T_i-T_m interval
in the phase diagram. Varying the copolymer composition ratio for
say a given pair of flexible spacers then provides a subtle control
over T_i-T_m, hence the width of the mesomorphic zone. This situation
can be readily realized and systematically explored with polyethers
or alternating diphenyl type mesogens and alkyl spacers linked by
ether oxygen. These compounds were recently synthesized and charac-
terized ($\underline{1},\underline{15}$). They possess sufficient thermal stability to permit
the traversing of the various phase transitions without the danger
of chemical changes, such as beset the more familiar, but otherwise
analogous family of polyesters.
     Of these studies here we shall illustrate one case involving
chains with formulae such as

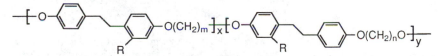

     This is a copolymer where the "mesogen" itself is a semi-
flexible moiety of diphenyl ethane, which is not LCP forming as a
homopolymer, with $(CH_2)_8$ as a sole spacer, but is turned into LCP
on copolymerization with $(CH_2)_{10}$ as the alkyl portion of the comonomer unit.   The result is shown by Figure 4. We see that a
stable mesophase (T_i>T_m) results beyond a molar fraction of $(CH_2)_{10}$
equal to 0.2 with T_i-T_m being largest at around 1:1 $(CH_2)_9/(CH_2)_{10}$
molar ratio ($\underline{15}$).
     Choosing the more rigid α-methyl stilbene, which is a familiar
mesogen, instead of diphenyl ethane, in the above family of polyethers enhances the LCP forming tendency throughout. Here T_i lies
always above T_m, hence the mesophase is the stable one over the
full comonomer composition range again with T_i-T_m being maximum at
around the 1:1 ratio ($\underline{16}$).
     The cases quoted thus provide examples for passing into the
mesophase and for the controlled widening of this mesophase through
chemical tailoring and this in a systematic manner in accordance
with basic thermodynamic principles.

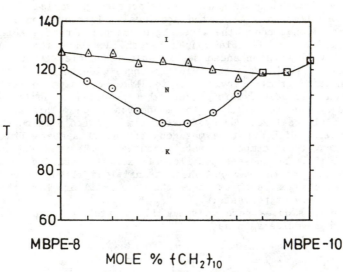

Figure 4. Crystal-isotropic (k-i), crystal-nematic (k-n), and nematic-isotropic (n-i) transition temperatures as a function of mole fraction of $-(CH_2)_{10}-$ spacer pertaining to the series of MBPE-8,10 copolyethers, containing methylbiphenyl ethane mesogen and randomly distributed $-(CH_2)_8-$ and $-(CH_2)_{10}-$ spacers (adapted from ref. 15).

Among imperfections we may count further the random placement of isomeric variants of otherwise identical chemical moieties along the chain. The mesogenic character, of α-methyl stilbene, a common mesogen (10,11) is in fact partly due to this reason. Here, the presence of the methyl group introduces nonplanarity into the otherwise planar stilbene moiety (17). Thus two enantiomers are possible, the polymer itself having no stereoregularity. The result is that G_k is raised and the melting point is suppressed; hence 4,4'-substituted α-methyl stilbenes, but not stilbenes, are nematogenic even as small molecules. When incorporated in a chain a further type of isomerism, namely head to head or head to tail (with relation to the methyl group in the mesogen) is added, the random sequence of which undoubtedly reinforces the LC forming tendency through raising the imperfection content of the crystal. All the above, relating to the methyl stilbene polymer, relies on the raising of G_k as in Figure 1c. This clearly combines with the chain stiffening effect of the stilbene group raising G_i by Figure 1b, the subject of the preceding section. The above case therefore provides an example of the mutually reinforcing interplay of the two thermodynamic factors highlighted by Figures 1b and 1c respectively.

The above merely serves to place chemical tailoring on the map within the broader framework of our scheme. This line, with the added variable of molecular weight, will be pursued further in a forthcoming publication specific devoted to it (18) and will be briefly described in the last part of this article.

EXAMPLE: POLYETHYLENE

Crystal Imperfections

In what follows we shall proceed with examples on polyethylene which illustrate the main aspects of the present scheme, and at the same time are noteworthy in their own right.

Polyethylene conforms to the situation of Figure 1a, i.e. under normal conditions it does not display a mesophase, only the familiar orthorhombic crystal form (o). Nevertheless, there exists a virtual mesophase which can be "uncovered"; the ways in which this can be achieved is the subject of what follows.

First let us discuss about the mesophase itself. This corresponds to a hexagonal packing of the straightened polyethylene chain, i.e. to the packing of rods of circular cross-section which in turn arises through cylindrical averaging of the polyethylene chain around its axis. This virtual hexagonal phase (h) in polyethylene is in direct continuity with the stable hexagonal crystal phase in n-alkanes as laid out by one of us elsewhere (6).

On the various ways of "uncovering" this virtual mesophase we consider first the route by Figure 1c, that is raising G_k. One example is provided by ethylene-propylene copolymers (19). Here for high enough polymer content, (for ca 70 $-CH_3$ branches per 1000 main chain carbon atoms) the X-ray patterns reveal the hexagonal (h) as opposed to the familiar orthorhombic (o) packing of chains even at room temperature. Here the randomly distributed (CH_3) branches disturb the regular packing of the chains thus leading to the situation as in Figure 3.

Another example is the influence of irradiation on linear PE with γ-rays or electrons (19,20) with doses large enough (more than 300 Mrads) to start influencing the a spacing of the orthorhombic lattice, but not yet sufficient to appreciably reduce, or even less destroy, crystallinity (the latter happens well beyond 1000 Mrads). When such an irradiated sample is heated the o phase changes first into an h phase before melting (Figure 5). Thus the irradiation has created a mesomorphic region in accordance with Figure 1c, the temperature interval T_i-T_m increasing with irradiation dose received (Figure 5). As familiar by now, irradiation crosslinks the polyethylene chains, preferentially in the amorphous regions (21,22), yet as the dose becomes large (approaching and exceeding 1000 Mrad) crosslinks are being increasingly created also in the crystal. The latter will produce lattice defects which will increase G_k and thus, by Figure 1c, will lead to a stable mesophase in the appropriate temperature region with lowered T_m. It is worth noting that the introduction of crosslinks also affects G_i: by reducing the configurational entropy of the melt it will raise G_i, hence lead towards a situation as in Figure 1b. (There is evidence for it through an initial increase--as opposed to the subsequent overall decrease--in the o melting point at doses which are too low, 5-20 Mrad, to have other consequences for the phase diagram (19,20).) Thus we have a situation where an increase in both G_k and G_i (Figures 1c and 1b respectively) contribute to the uncovering of the mesophase.

As laid out earlier hydrostatic pressure could promote the mesophase. Polyethylene provides a good example. As explored by Bassett et al. (5) a hexagonal phase with mesophase characteristics (no longitudinal register between chains, high chain mobility enabling ready sliding of chains) appears above 3 kbar pressure--even with the (chemical) defect-free linear PE, at suitably elevated temperature (Figure 6). Now, if the chain contains chemical defects, such as crosslinks produced by irradiation, the triple point will shift toward lower pressures with increasing crosslink content until hexagonal phase will become stable even at atmospheric pressure (20). The latter is in fact the situation in the preceding paragraph represented by Figure 5, thus highlighting the continuity between the effects of crosslinking and pressure: both increase G_i by reducing the entropy of the liquid, with the crosslinks also increasing G_k.

Dependent on whether the rise in G_i or that in G_k is dominant the crystal melting point will be higher or lower than for the corresponding linear material under atmospheric pressure. This follows from Figures 1a-c and is clearly reflected by Figures 5 and 6 where T_m values (here the o h transition) can be both elevated (Figure 6 and high pressure entries in Figure 5) and lowered (1 bar entry in Figure 5).

Figure 6 also displays the connection between the stable hexagonal phase in short n-alkanes and the hexagonal mesophase in polyethylene as a function of chain length in the p-T diagram. A full discussion is given by one of us elsewhere (6). At this place our emphasis is on the continuity, despite the fact that the pressure coefficient of the width of the temperature interval is of opposite sign in these two families of material. This continuity is apparent also by a structural feature, namely the concentration of conforma-

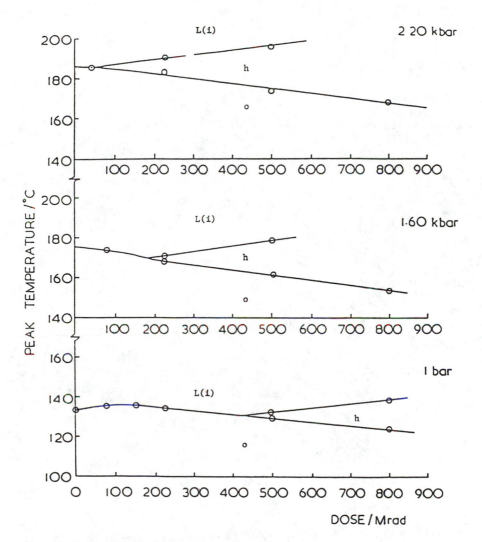

Figure 5.   Transition temperatures T_{k-lc}, T_{lc-i}, and T_{k-i} at three pressures of irradiated polyethylene as a function of radiation dose.   Here the mesophase is the hexagonal phase (h) the crystal the orthorhombic phase (o) and the liquid (L) is the isotropic (i) melt.

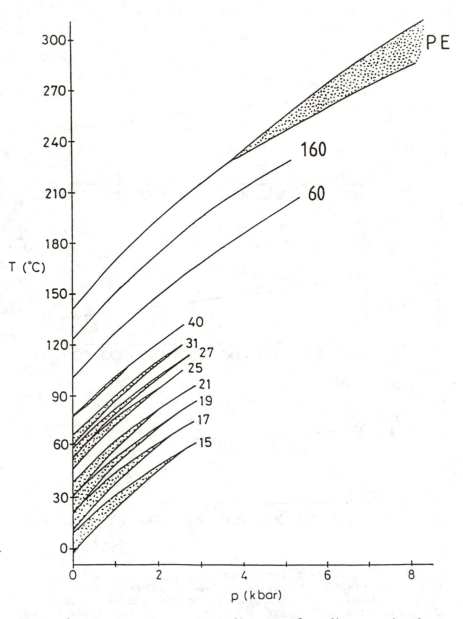

Figure 6.   Pressure-temperature diagrams of n-alkanes and poly-ethylene. Shaded areas indicate regions of stability of the meso-phase ("hexagonal" or "rotator"). The figures indicate the number of carbon atoms in the chain (from ref. 6).

tional "kink" defects, C_K. As described in ref. 6 C_K can be directly registered spectroscopically and inferred from X-ray diffraction, and is found to be a continuous function of temperature (T) and chain length L. The mesophase ("rotator" and/or "hexagonal" phase) appears in different systems under different conditions, and the corresponding C_K (L,T) values lie on a master surface given by an appropriate set of L,T coordinates. The gradual changeover from "barophobic" to "barophillic" behavior of the mesophase with increasing L and T is attributed to the steeply increasing configurational entropy associated with the introduction of "kinks", while maintaining a comparatively low volume (6).

Orientation

While still with PE, we pass on to situations conforming to Figure 1b, i.e. to the uncovering of the mesophase through raising of G_i. For a flexible chain like PE the latter is realized by stretching out of the chains and keeping them stretched (23) or otherwise constrained (24) while in the melt (or solution), thus preventing the full conformational entropy to be recovered. Since in ultra oriented (and thus ultra strong and stiff) PE fibers the molecules are virtually fully stretched out they lend themselves well to the creation of such a situation. Such fibers when held at fixed length, are heated beyond their usual orthorhombic crystal melting point, when within an appreciable temperature interval and, for a limited time, the hexagonal form is apparent. Here the DSC scans display two melting peaks corresponding to o→h and h→i transformation as verifiable by X-ray diffraction. In such a situation (realized by holding fibers clamped at their ends (23), or constrained through embedding in a resin (24) the mesophase is necessarily transient as the chain orientation will relax in time; such a situation therefore will not correspond to a final equilibrium.

An equilibrium situation, however, can be approached by light crosslinking of the fibers, where crosslinks are too few in number to significantly influence the thermodynamic stability of the solid phase (say 10 Mrad dose as compared to several hundred Mrad quoted in Figure 5), but nevertheless are sufficient in number to create a loosely connected infinite network. Such a network can be stretched out into a highly oriented system, either in the melt as an elastomer, or in the partially crystalline state through usual fiber drawing. There are several reports, past (25,26) and more recent relating to high modulus fibers (27-29), of such lightly crosslinked oriented systems converting to, or forming in the hexagonal phase above the crystallization or melting temperature of the usual orthorhombic crystal form. Here, when held at constant length the stress, and with it the hexagonal form, can be maintained over a considerable length of time, if not indefinitely, and can also be reconstituted after repeated cooling and heating into the orthorhombic and liquid regimes. While full reversibility, hence the equilibrium nature of the i→h and h→o transitions has not been purposefully followed up in existing reports, this possibility, hence the attainability of an equilibrium situation according to Figure 1b is clearly suggested inviting renewed explicit investigations to this effect.

Finally, we quote a most recent finding which seems to fall into the above scheme and is of potential practical consequence (30,31). Very high molecular weight (MW) PE-melts behave in a highly elastic manner: they do not stretch (or flow) or extrude smoothly, but fracture and spurt respectively instead. Latest work here, however, identified a narrowly and sharply defined temperature window (under existing experimental conditions between 150-152°C) where such material extrudes smoothly with surprisingly low flow resistance ("viscosity"). Going below 150°C the viscosity rises sharply until on further lowering of temperature the system blocks, while above 152° it extrudes in spurts at increased mean extrusion pressure. Figure 7 shows a pressure vs. temperature curve at a constant exit velocity (piston speed in rheometer) for a polymer of M_w = 410000, displaying the sharp minimum between 150-152°C. This is to be compared with the "normal" extrusion behavior displayed by a polymer with M_w = 280000 (Figure 8) i.e. no minimum in pressure, only a steady decrease with temperature, under otherwise identical conditions.

To our mind the existence of the above temperature window must be associated with the presence of a distinct new phase which has all the hallmarks of a liquid crystal mesophase, hence for poly-ethylene the hexagonal phase. Accordingly, the lower temperature end of the window would correspond to an o→h transformation, while the upper end (sharply defined within 0.2°C) to h→i transformation. The ensuing spurt corresponds to the usual behavior of such high molecular weight melts at conventional processing temperatures, where such materials are in practical terms unprocessable. In the above situation the reason for mesophase formation would lie in the transient alignment of the chains while passing through the conical portion (with increasing cross section) of the extrusion barrel on its way to the die exit. Here the flow field will be elonga-tional, which is known to be the requirement for stretching out chains (32). It is known from work on solutions, and from pre-ceding theoretical predictions (32) that such elongational flow induced chain stretching is critical both in strain rate $\dot{\varepsilon}$, with a critical value $\dot{\varepsilon}_c$, the coil stretch transition, and in molecular weight (MW), $\dot{\varepsilon}_c$ being a sharply decreasing function of MW. The critical dependence of the effect on MW for a given $\dot{\varepsilon}$ (defined by the polymer exit, hence piston velocity in the rheometer) is apparent from the comparison of Figure 8 with Figure 7. There is a similar effect, i.e. the appearance of a pressure minimum in the pressure-temperature curves recorded at constant piston velocity, when surpassing certain piston speeds (hence $\dot{\varepsilon}$) demonstrating the existence of a critical $\dot{\varepsilon}$ for a given grade of polyethylene (31). Clearly, the situation exemplified by Figure 7 falls within the scheme in Figure 1b with the added factor that the chain orientation, and consequent mesophase formation, is transient. It is confined to the duration of the appropriate passage time, and even then it pertains only to the appropriate portion of the flow field. The flow-enhancing effect of the mesophase with respect to the melt is attributed to self-straightening and disentangling of polymer chains promoted by the i h phase transition. Thus the number of chain entanglements, which is so high in an ultra high molecular weight polymer as to prohibit flow of the melt, is appropriately reduced in the mesophase. On the other

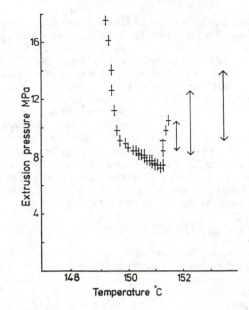

Figure 7.   Extrusion pressure plotted against extrusion temperature for extrusion of a polyethylene of $\overline{M}_w$ = 4x10^5 during heating at approximately 0.3 to 0.4°C min^{-1}.   Approximate piston velocity 5x10^{2}cm min^{-1}.   1mm diameter die.   The vertical, double headed arrow(s) represent pressure oscillations in the region of spurt (from Waddon and Keller, ref. 30).

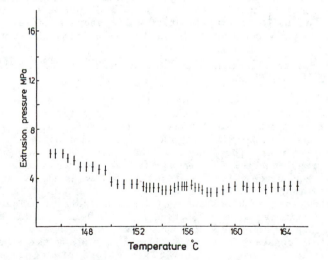

Figure 8.   Extrusion pressure plotted against extrusion temperature for extrusion of a polyethylene of $\overline{M}_w$ = 2.2x10^5 during heating at approximately 0.2 to 0.6°C min^{-1}.   Approximate piston velocity 3.3x10^{2}cm min^{-1}.   1mm diameter die (from Waddon and Keller, ref. 30).

hand, the translational chain diffusion remains high compared to
that in the ordered orthorhombic crystals accounting for the low
flow resistance.

Admittedly, the existence of the postulated transient
mesophase would still require structural confirmation. Even so,
the taking of a sharp drop in viscosity as indicator of mesophase
formation has well established precedents in the liquid crystal
field. Such is e.g. the well documented effect in Kevlar referred
to above (12) which in many respects has similarities to the
presently discussed PE, except that Kevlar is "mesogenic" and can
exist as stable liquid crystal under ambient conditions, while the
mesophase in the flexible PE is "virtual". The latter "virtual"
phase only becomes "real" transiently, which suffices to
dramatically affect the entire flow behavior of the material, the
effect through which it is being detected.

EXAMPLES: PHASE TRANSITIONS-MOLECULAR WEIGHT RELATIONSHIP

The examples discussed so far have always assumed that the polymer
molecular weight is above values which influence phase transition
temperatures. Let us discuss the relationship between phase transition temperatures and polymer molecular weight for three different
situations. Upon increasing the molecular weight from monomer to
polymer, the entropy of the liquid phase (S_i) decreases. The
decrease of the entropies of mesomorphic and crystalline phases is
lower than that of the isotropic phase. For simplicity, the decrease
of the entropy of the crystalline phase will be neglected. The decrease in S_i and S_{lc} tends asymptotically to zero with increasing
molecular weight (M). M_o from Figures 9 to 11 refers to molecular
weight of the polymer structural unit. M_1 to $M\infty$ from the same figures are arbitrary molecular weights of the corresponding polymer.
It follows that G_i and G_{lc} increase with the increase of the polymer molecular weight again asymptotically; above a certain molecular weight, we may consider both parameters as remaining constant.

Case 1. Both Monomeric Structural Unit and Polymer Display an Enantiotropic Mesophase

The first situation we will consider refers to the case in which
the monomeric structural unit displays an enantiotropic mesophase.
Upon increasing its molecular weight to dimer, trimer, etc., S_i
decreases and, therefore, G_i increases. Beyond a certain molecular
weight, G_i remains for all practical considerations constant.
Figure 9 transforms the free energy (G) versus transition temperature (T) dependence, into a transition temperature (T) versus
molecular weight (M) dependence. The T versus M plot in Figure 9
demonstrates that both melting (T_{k-lc}) and isotropization (T_{lc-i})
temperatures increase with increasing molecular weight up to a
certain range of M values beyond which T_{k-lc} and T_{lc-i} remain
approximately constant with the T_i line lying above the line for T_m
for all molecular weights. However, the slope of the increase of
T_{lc-i} is steeper than that of T_{k-lc}. The difference between these
two slopes determines the relative thermodynamic stabilities of the
mesomorphic versus that of the crystalline phases at different
polymer molecular weights.

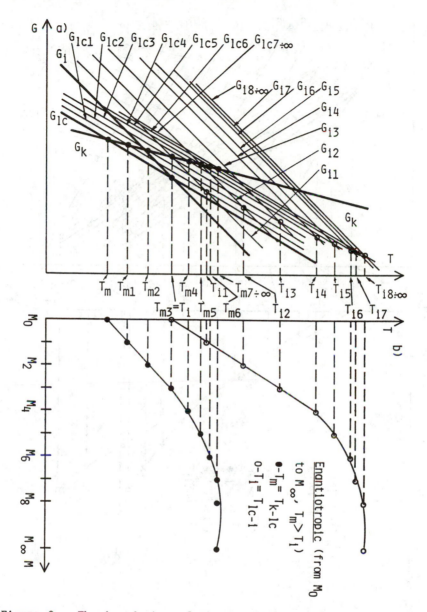

Figure 9. The broadening of the temperature range of an enantiotropic mesophase of the monomeric structural unit (M_o) by increasing the degree of polymerization. The upper part (a) describes the influence of molecular weight on the dependence between the free energies of the crystalline (G_k), liquid crystalline (G_{1c}) and isotropic (G_i) phases and transition temperatures. The translation of this dependence into the dependence phase transition temperature-molecular weight is presented in the lower part (b).

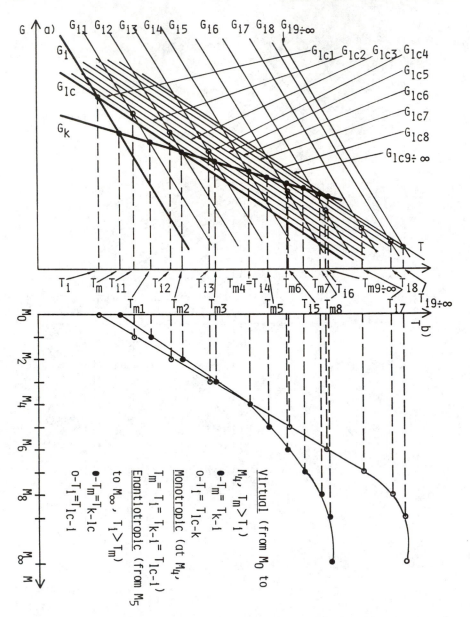

Figure 10.   Transformation of a virtual or monotropic mesophase of the monomeric structural unit (M_o) into an enantiotropic meso-phase by increasing the degree of polymerization. The upper part (a) describes the influence of molecular weight on the dependence between the free energies of the crystalline (G_k), liquid crys-talline (G_{1k}) and isotropic (G_1) phases and transition tempera-tures.   The translation of this dependence into the dependence phase transition temperature-molecular weight is shown in the lower part (b).

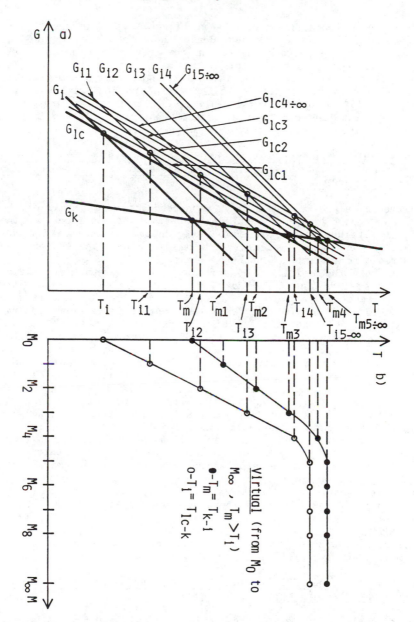

Figure 11. The narrowing of the temperature range of a virtual mesophase of the monomeric structural unit (M$_o$) by increasing the degree of polymerization. The upper part (a) describes the influence of molecular weight on the dependence between the free energies of the crystalline (G$_k$), liquid crystalline (G$_{1c}$) and isotropic (G$_i$) phases and transition temperatures. The translation of this dependence into the dependence phase transition temperature-molecular weight is presented in the lower part (b).

For this particular case, the higher slope of the T_{lc-i}-M versus that of the T_{k-lc}-M dependence leads to a widening of the temperature range between the two curves with increasing molecular weights of the polymer. This widening of the liquid crystal temperature regime with molecular weight agrees with experimental data reported for the case of both main chain (7,33,34) and side chain (35-37) liquid crystal polymers. This effect has been repeatedly labelled the "polymer effect", especially in the case of side chain liquid crystal polymers (38).

Case 2. The Monomeric Structural Unit Displays a Virtual or a Monotropic Mesophase; The Polymer Displays a Monotropic or an Enantiotropic Mesophase

The steeper slope of the $T_{lc-k(i)}$-M dependence versus that of the $T_{k-i(lc)}$ dependence has even more important implications on the molecular weight-phase transition temperature dependence for the situation when the monomer structural unit displays only a monotropic or a virtual mesophase (Figure 10).

As seen the two lines T_i versus M and T_m versus M intersect. This arises from the fact that $T_i < T_m$ for low molecular weights with a steeper slope for T_i. Specifically (for the illustration in Figure 10) the T_i (i.e. T_{lc-k}^i) values are below the corresponding T_m (i.e. T_{k-i}^m) values for M_0 to M_4, hence in the range M_0 to M_4, i.e. the monomer together with its low oligomers up to 4, display only a virtual mesophase. Beyond M_4 the mesophase becomes stable, hence the system enantiotropic. In addition to the thermodynamic criterion, kinetics also influence the phase transitions. A certain amount of supercooling of the isotropic-mesomorphic and especially much more so of the mesomorphic-crystalline transitions is possible which can lead to monotropic behavior for molecular weights slightly below and at the intersection point (i.e. at M_4). In view of the fact that T_i and T_m are expected to be continuous functions of molecular weight, the intersection point can be arbitrarily closely approached (from below), hence realization of a metastable liquid crystal phase and consequent monotropic behavior is to be expected for appropriate molecular weights in cases to which Figure 10 pertains. This effect has been observed experimentally both in the case of main chain (39-41) and side chain (42,43) liquid crystalline polymers and was labelled "transformation of a monotropic mesophase into an enantiotropic mesophase by increasing the molecular weight of the polymer".

Case 3. The Monomeric Structural Unit Displays a Virtual Mesophase; the Polymer Displays a Virtual Mesophase.

The third situation is illustrated in Figure 11 and also refers to a different case in which the monomeric unit displays only a virtual mesophase. Here as before, the slope of the T_i (i.e., T_{lc-k})-M dependence is higher than that of the T_m (i.e., T_{k-i})-M dependence, the latter lies above the former throughout hence the two curves do not intercept each other. Therefore, the resulting polymer displays also only a virtual mesophase. This thermodynamic situation was recently applied to the synthesis of virtual liquid crystal polyethers containing both flexible mesogens and flexible spacers. The most thoroughly investigated example so far refers to polyethers based on 1-(4-hydroxy)-2-(2-methyl-4-hydroxyphenyl) ethane and α,ω-dibromoalkanes ($\underline{1,15,16,44-46}$). In this last class of polymers, the transformation of the virtual mesophase of the homopolymer into a monotropic or enantiotropic mesophase was accomplished by increasing the free energy of the crystalline phase (G_k). Theoretically this situation can be envisioned by inspecting Figure 1a and c. Experimentally, the increase of G_k has been accomplished by two different techniques so far. The first involved the copolymerization of 1-(4-hydroxy)-2-(2-methyl-4-hydroxyphenyl) ethane with two ($\underline{1,15,16,44,45}$) or more than two ($\underline{46}$) flexible spacers of different length, or the copolymerization of two dissimilar flexible mesogenic bisphenols with one flexible spacer ($\underline{47}$). The resulting copolymers exhibit an increased G_k in comparison to that of their parent homopolymers since their melting temperatures are lower. The second technique is based on blends of two or more homopolymers which display virtual mesophases. When the virtual mesophases of the homopolymers are isomorphic and their crystalline phases are not, the resulting blends display either a monotropic or an enantiotropic mesophase ($\underline{48}$).

In conclusion, the steeper slope of the isotropization temperature-molecular weight dependence versus that of the melting temperature-molecular weight dependence should have the following effects on going from monomer to polymer. When the monomeric structural unit displays an enantiotropic mesophase, the resulting polymer will display a broader enantiotropic mesophase, i.e. anisotropic temperature interval. When the monomeric unit displays a monotropic mesophase, the resulting polymer will, most probably, display an enantiotropic mesophase. When the monomeric unit displays a virtual mesophase, the resulting polymer may display either an enantiotropic, monotropic or virtual mesophase. For the case of side chain ($\underline{42,43}$) and main chain ($\underline{33,40,41}$) liquid crystalline polymers containing flexible spacers and displaying a single mesophase, the nature of the mesophase displayed by the polymer is most frequently identical to that of its monomeric unit. Primarily the molecular weight-phase transition dependences are determined by the relationship between the free energies of the crystalline, liquid crystalline and isotropic phases of the monomeric structural unit yet influenced by the molecular weight of the main chain backbone through its effect on the melt entropy as in Case 1 above. Combinations of more than one mesomorphic phases of different thermodynamic stabilities in a monomeric-structural unit will follow the same molecular weight dependence as those described by the monomers displaying a single mesophase ($\underline{43,51-53}$). However, there are few examples where the nature of the mesophase displayed by the main chain ($\underline{8,19}$) and side chain ($\underline{22}$) liquid crystalline polymers is molecular weight dependent. Both in the case of the main chain ($\underline{49,50}$) and side chain ($\underline{53}$) liquid crystalline polymers it has been demonstrated that this change in the mesophase represents a continuous dependence of molecular weight. The interpretation of this dependence based on a similar thermodynamic scheme will represent the subject of further publications.

Conclusion

It has been shown how the diverse manifestations of the mesomorphic state in polymers can be encompassed by a simple unifying scheme based on the relative thermodynamic stability including metastability, of the crystalline, liquid crystalline and liquid phases. The scheme embraces not only traditional LCP's with mesogenic groups but also flexible polymers such as may not be considered liquid crystal forming a priori, but which nevertheless can be obtained as such by "uncovering" virtual mesophase regimes within their phase diagams. As shown, the latter can arise in suitable chemical variants within a given family of compounds, can be purposefully accomplished by subsequently induced chemical modifications and also by appropriate choice of physical parameters and constraints, sometimes with quite unsuspected results, (e.g. the transient mesophase and consequent narrow processing temperature window in high MW polyethylene). Since few other polymers have been studied as comprehensively as polyethylene--the polymer of most of our examples--it may well be that many more polymers, not considered to be mesogenic at present will be found to exhibit mesomorphic behavior under appropriate conditions. For example, it has been theoretically predicted (54) and recently experimentlly suggested (55) that polyethylene displays an additional virtual mesomorphic phase which may be nematic in between -150 and -198°C. The present note, among others, should hopefully provide signposts for the purposeful creation of mesophases and for their "uncovering"--in cases where they may be hidden--on the basis of simple thermodynamic considerations.

Acknowledgments

Sponsorship by U.S. Army European Office, London, and the National Science Foundation Polymer Program (DMR-86-19724) is gratefully acknowledged.

Literature Cited

1. Percec, V.; Yourd, R. Macromolecules 1989, 22, 524.
2. Godovsky, Y.K.; Papkov, V.S. Makromol. Chem., Macromol. Symp. 1986, 4, 71.
3. Schneider, N.S.; Desper, C.R.; Beres, J.J. In Liquid Crystalline Order in Polymers; Blumstein, A., Ed.; Academic Press: New York, 1978; p 299.
4. Wunderlich, B.; Grebovicz, J. Adv. Polym. Sci. 1984, 60-61, 1.
5. Bassett, D.C. In Developments in Crystalline Polymers - 1; Bassett, D.C., Ed.; Appl. Sci. Publ., 1982; p 115.
6. Ungar, G. Macromolecules 1986, 19, 1317.
7. Feijoo, J.L.; Ungar, G.; Owen, A.J.; Keller, A.; Percec, V. Mol. Cryst.,-Liq. Cryst. 1988, 155, 487-494.
8. Feijoo, J.L.; Ungar, G.; Keller, A.; Percec, V. Polymer, in press.
9. Odell, J.A.; Keller, A.; Atkins, E.D.T.; Miles, M.J. J. Materials Sci. 1981, 16, 3306.
10. Percec, V.; Schaffer, T.D.; Nava, H. J. Polymer Sci., Polym. Lett. Ed. 1984, 22, 637.

11. Ober, C.K.; Jin, J.-I.; Lenz, R.W. Adv. Poly. Sci. 1984, 59, 103.
12. Cifferi, A.; Valenti, B.   In Ultra-High Modulus Polymers; Cifferi, A.; Ward, I.M., Eds.; Appl. Sci. Pub., 1979; p 203.
13. Atkins, E.D.T.; Fulton, W.S.; Miles, M.J.   TAPPI Conference Papers: 5. International Dissolving Pulps Symposium: Vienna, 1980; p 208.
14. Blackwell, J.; Biswas, H.; Cheng, M.; Cageao, R. Mol. Cryst. Liq. Cryst. 1988, 155, 299. Jackson, W.J., Jr. Mol. Cryst. Liq. Cryst. 1989, 169, 23.
15. Percec, V.; Yourd, R. Makromol. Chem. 1990, 191, 25 and 49.
16. Percec, V.; Yourd, R.; Ungar, G.; Feijoo, J.L.; Keller, A.   To be published.
17. Young, W.R., Aviram, A.; Cox, R.J. J. Am. Chem. Soc. 1972, 94, 3976.
18. Percec, V.; Keller, A. Macromolecules; submitted.
19. Ungar, G.; Keller, A. Polymer 1980, 21, 1273.
20. Vaughan, A.S.; Ungar, G.; Bassett, D.C.; Keller, A. Polymer 1985, 26, 726.
21. Keller, A.   In Developments in Crystalline Polymers - 1; Bassett, D.C., Ed.; Appl. Sci. Publ., 1982; p 37.
22. Keller, A.; Ungar, G. Radiat. Phys. Chem. 1983, 22, 155.
23. Pennings, A.J.; Zwignenburg, A. J. Polym. Sci., Polym. Phys. Ed. 1979, 17, 1011.
24. Lemstra, P.J.; van Aerle, N.A.J.M. Polymer 1988, 20, 131.
25. Clough, S.B. J. Polym. Sci., Polym. Lett. 1970, 8, 519.
26. Clough, S.B. J. Macromol. Sci. 1970, B4, 199.
27. Hikmet, R.M.; Lemstra, P.J.; Keller, A. Colloid & Polymer Sci. 1987, 265, 185.
28. Hikmet, R.M.; Keller, A.; Lemstra, P.J. To be published.
29. Hikmet, R.M. Ph.D. Thesis, Bristol, 1987.
30. Waddon, A.; Keller, A. J. Polym. Sci. Polym. Lett. Ed.; submitted.
31. Waddon, A.; Keller, A. To be published.
32. Keller, A.; Odell, J.A. Colloid & Polym. Sci. 1985, 263, 181.
33. Blumstein, A.; Vilasagar, S.; Ponrathnam, S.; Clough, S.B.; Blumstein, R.B.; Maret, G. J. Polym. Sci., Polym. Phys. Ed. 1982, 20, 877.
34. Percec, V.; Nava, H.; Jonsson, H. J. Polym. Sci., Polym. Chem. Ed. 1987, 25, 1943.
35. Godovsky, Y.K.; Mamaeva, I.I.; Makarova, N.N.; Papkov, V.P.; Kuzmin, N.N. Makromol. Chem., Rapid Commun. 1985, 6, 792.
36. Kostromin, S.G.; Talroze, R.V.; Shibaev, V.P.; Plate, N.A. Makromol. Chem., Rapid Commun. 1982, 3, 803.
37. Percec, V.; Hahn, B. Macromolecules 1989, 22, 1588.
38. Percec, V.; Pugh, C.   In Side Chain Liquid Crystal Polymers; McArdle, C.B., Ed.; Blackie and Son Ltd.: Glasgow; Chapman and Hall: New York, 1989; p 30.
39. Majnusz, J.; Catala, J.M.; Lenz, R.W. Eur. Polym. J. 1983, 19, 1043.
40. Zhou, Q.F.; Duan, X.Q.; Liu, Y.L. Macromolecules 1986, 19, 247.
41. Percec, V.; Nava, H. J. Polym. Sci., Polym. Chem. Ed. 1987, 25, 405.

42. Stevens, H.; Rehage, G.; Finkelmann, H. Macromolecules 1984, 17, 851.
43. Percec, V.; Tomazos, D.; Pugh, C. Macromolecules 1989, 22, 2259.
44. Percec, V.; Yourd, R. Macromolecules 1989, 22, 524 and 3229.
45. Percec, V.; Tsuda, Y. Polym. Bull. 1989, 22, 489 and 497; 1990, 23, 225.
46. Percec, V.; Tsuda, Y. Macromolecules 1990, 23, 5.
47. Percec, V.; Howells, B. In preparation.
48. Percec, V.; Tsuda, Y. Polymer; in press.
49. Kumar, R.S.; Clough, S.B.; Blumstein, A. Mol. Cryst. Liq. Cryst. 1988, 157, 387.
50. Blumstein, R.B.; Blumstein, A. Mol. Cryst. Liq. Cryst. 1988, 165, 361.
51. Sagane, T.; Lenz, R.W. Polymer 1989, 30, 2269.
52. Shibaev, V. Mol. Cryst. Liq. Cryst. 1988, 155, 189.
53. Percec, V.; Lee, M. To be published.
54. Ronca, G.; Yoon, D.Y. J. Chem. Phys. 1982, 76, 3295.
55. Percec, V.; Tsuda, Y. Macromolecules. In press.

RECEIVED April 24, 1990

Chapter 24

Solubilities, Processabilities, and Head-to-Tail Polymerization of Liquid-Crystalline Polymers, Including First Super-Strong Polymers

F. Dowell

Theoretical Division, MS–B268, Los Alamos National Laboratory, University of California, Los Alamos, NM 87545

This paper presents summaries of unique new static and dynamic theories for backbone liquid crystalline polymers (LCPs), side-chain LCPs, and combined LCPs [including the first super-strong (SS) LCPs] in multiple smectic-A (SA) LC phases, the nematic (N) phase, and the isotropic (I) liquid phase. These theories are used to predict and explain new results: (1) solubilities of LCPs in various kinds of solvents, (2) the diffusion (thus, slowest rate of processability) of LCPs, and (3) head-to-tail polymerization of SS LCPs. Melt processability of some SS LCPs is also reviewed. These theories can be applied to almost any kind of organic molecule.

Polymer molecules are so long that they typically solidify too fast to order completely into the three-dimensional (3D) crystalline state in sample sizes large enough for practical applications. The molecular ordering [and thus, strength (1-5)] of a solid polymer can be significantly increased by solidifying the polymer from a LC phase. Strong polymers are solidified backbone LCPs and are used as stronger, lighter-weight replacements for metals, ceramics, and other materials in various structural applications, such as auto and airplane parts, armor, building materials, etc. Combined LCPs are polymers whose backbones and side chains both have LC ordering.

Super-strong (SS) polymers (3-11) are specially-designed combined LCPs in which the side chains of one molecule are designed to interdigitate (pack between) the side chains of neighboring molecules, thus leading to molecular self-reinforcement and enhanced molecular ordering (thus, enhanced mechanical properties) compared to backbone LCPs.

There is partial orientational ordering of the long axes of the molecules parallel to a preferred axis in the nematic (N) LC phase. In a smectic-A (SA) LC phase, there is partial orientational

0097–6156/90/0435–0335$07.00/0

ordering and partial one-dimensional (1D) positional ordering of the
centers of mass of the molecules along the preferred axis.

Theory Summary

The theories in this paper are first-principles statistical
mechanics theories used to calculate static thermodynamic and
molecular ordering properties (including solubilities of LCPs in
various kinds of solvents) and dynamic properties (diffusion from
Brownian motion). The diffusion of the LCP molecules constitutes a
lower limit for the speed of processing of the LCPs. The static
theory is used to calculate the packing of the bulky relatively
rigid side chains of SS LCPs; these calculations indicate that
head-to-tail polymerization of the monomers of these SS LCPs will be
very strongly favored. The intermolecular energies and forces
calculated from the static theory are used in the dynamic theory.

New Static Theory for Calculation of Solubilities

In this paper, an earlier theory ($\underline{12-13}$) for binary mixtures of
backbone LCPs (and/or nonpolymeric molecules) in the nematic ($\underline{N}$) LC
phase and the isotropic ($\underline{I}$) liquid phase has been extended to treat
binary mixtures in multiple smectic-A (SA) LC phases, the $\underline{N}$ LC
phase, and the $\underline{I}$ liquid phase. Either component 1 (C1) and/or
component 2 (C2) in the mixture can be a backbone LCP, a
nonpolymeric LC molecule, a polymeric non-LC molecule, or a
nonpolymeric non-LC molecule. C1 can also be a side-chain LCP or a
combined LCP (including a SS LCP). The new theory of this paper is
the mixture analogue of an earlier theory for pure
(single-component) systems derived and presented in detail in
References 3 and 7-10 (see also References 14-23). When only one
component is present, the new mixture theory of this paper reduces
to this earlier single-component theory. Therefore, only a short
summary of the new mixture theory is presented here.
  Using average bond lengths and angles for the different
functional chemical groups in each type of molecule in the mixture,
the average shape of each such molecule is calculated in continuum
space. [For example, an average bond angle of 120° is used for a
sp^2-bonded carbon atom, and an average length ($\underline{24}$) of 1.43 Å is used
for a carbon-to-nitrogen single bond (i.e., C-N) when the carbon
atom is in a benzene ring.] As discussed in References 3 and 8-10,
these average bond lengths and angles are taken from compilations
(as in Reference 24) of average values taken from experimental data
for existing materials.
  To calculate the static thermodynamic and molecular ordering
properties of a system of molecules, the configurational partition
function Q_c of the system must be derived. Q_c does not contain the
kinetic energy, intramolecular and intermolecular vibrations, and
very small rotations about molecular bonds. Q_c does contain terms
which deal with significant changes in the shapes of the molecules
due to rotations about semiflexible bonds (such as about
carbon-carbon bonds in $\underline{n}$-alkyl [i.e., $(-CH_2-)_x$] sections) in a
molecule. For mathematical tractability in deriving Q_c, the
description of the molecules in continuum space is mapped onto a

description of the molecules on a simple cubic (SC) lattice in order
to calculate the packing and interactions of all the molecules in
the system. The volume $\underline{V}$ of the system is divided into a set of
lattice sites [see Figure 1(a)]. Each molecule is divided into a
set of connected segments or sites [see Figure 1(b)], where these
sites correspond to small groups of atoms, such as methylene ($-CH_2-$)
groups. Each molecular segment (specifically, the hard-repulsive
volume of each molecular segment) occupies one lattice site. [The
hard-repulsive volume of a molecular segment is the volume of a
segment of a molecule that is completely excluded to other molecular
segments by steric (i.e., "hard", or infinitely large) repulsions.]

Analytic combinatorial lattice statistics are used to calculate
the packing and interactions of a molecule with the other molecules
in the system. [The generalized lattice statistics used in this
theory have been found to be very accurate (deviations less than 1%)
compared (20,25) with Monte Carlo computer simulations in limiting
cases presently amenable to such simulations.] Any continuum-space
orientation of a molecule or molecular part or bond can be
decomposed into its components parallel to the $\underline{x}$, $\underline{y}$, and $\underline{z}$ axes of
the system and then these components mapped onto the $\underline{x}$, $\underline{y}$, and $\underline{z}$
axes of the SC lattice [see Figure 1(c)] in a manner analogous to
normal coordinate analysis in, for example, molecular spectroscopy.

Various continuum limits of the lattice description are taken
(i.e., the number of C1 molecules $N_{m1} \to \infty$, the number of C2
molecules $N_{m2} \to \infty$, and the number of lattice sites $M \to \infty$), such that
the density and the other thermodynamic and molecular ordering
variables can vary continuously for the system of molecules. (In
these limits, the lattice statistics can treat molecules in which
the number of segments per molecule is not an integer.)

The configurational partition function $Q_c = \{\Omega \exp[-E/(k_B T)]\}$,
where k_B is the Boltzmann constant, and $\underline{T}$ is the absolute
temperature. Ω is the average number of ways to lay the
hard-repulsive volumes of all the molecules on the lattice, without
laying the hard-repulsive volumes of two molecular segments on the
same lattice site. Ω is thus the part of the partition function due
to steric repulsions between molecular segments. $\underline{E}$ is the average
sum of all the potential energies arising from other interactions
between molecules. $\underline{E}$ is thus the part of the partition function due
to other intermolecular interactions [including soft (finite-sized)
repulsions, London dispersion attractions, dipole/dipole
interactions, dipole/induced dipole interactions, hydrogen bonding,
and so forth between molecular segments].

The theory used here is a localized mean-field (LMF) theory.
That is, there is a specific mean field (or average environment) in
a given direction $\underline{i}$ in a given local region [see Figure 1(d)], where
these local regions can have more than a random probability to
contain specific parts of the molecules, such as rigid sections or
semiflexible sections or the centers of mass of the molecules. A
rigid section is formed by a sequence of conjugated aromatic,
double, and triple bonds in the molecule. The overlap of π orbitals
in the aromatic, double, and triple bonds in the section leads to
the rigidity of the section. A semiflexible section is usually
formed by an $\underline{n}$-alkyl [i.e., $(-CH_2-)_x$] chain section. Such a
section is partially flexible (semiflexible) since it costs a

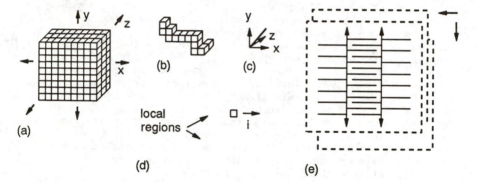

Figure 1.   Schematic illustrations: (a) SC lattice.
(b) Molecule divided into segments.   (c) Decomposition of
continuum-space orientations into x, y, and z components.
(d) Local regions (denoted with dashed lines).   (e) Packing of SS
LCPs in planes (denoted with dashed lines); the continuation of
the backbones is indicated by the small arrows; the diffusion of
a molecule in a plane between the planes of other molecules is
indicated by the bold arrows.

finite, but easily achievable, energy to make rotations about any carbon-carbon bond between methylene units in a given chain section. The net energy difference between the one trans and either of the two gauche rotational energy minima is E_g, with the gauche states having the higher energy. The compositions of the local regions in this LMF theory are determined by how the molecules actually pack and interact with lowest free energy or chemical potential. The local regions are defined such that there are no edge effects (i.e., continuity of the individual molecules and of the density is preserved from one region to another). The fact that the packing can be different for different directions $\underline{i}$ allows the treatment of partial orientational ordering along a preferred axis. The use of local regions allows the treatment of partial positional ordering of the molecules, such that similar parts of the molecules tend to pack with each other.

It is important to note that the treatment of different localized mean fields around different parts of a molecule does not break the molecule into pieces. That is, the connectivity of the molecule is preserved in this LMF theory; thus, this theory can treat molecules of any length (including high polymers).

The partition function Q_c and the resulting equations for static thermodynamic and molecular ordering properties are functions of $\underline{T}$, pressure $\underline{P}$, system composition (i.e., mole fraction of each component present), and details of the molecule chemical structures [including bond lengths and angles, net energy differences E_g between trans and gauche rotational states, hydrogen bonds, dipole moments, bond polarizabilities, site-site Lennard-Jones (12,6) potentials, degree of polymerization], and orientational and positional orderings of the different parts of the molecules. The Lennard-Jones (LJ) potentials are used to calculate repulsions and London dispersion attractions between different molecular sites, and the dipole moments and polarizabilities are used to calculate dipole/dipole and dipole/induced dipole interactions between different sites. Each interaction in the theory here depends explicitly on the intramolecular and intermolecular orientational and positional ordering of the specific molecular sites involved in the interaction.

There are no ad hoc or arbitrarily adjustable parameters in this theory. All variables used in this theory are taken from experimental data for atoms or small groups of atoms (such as benzene rings, methylene groups, etc.) or are calculated in the theory.

Here is the essence of this lattice theory: Given average variables for the orientations and positions of the different parts of the other N_{m1} C1 molecules and N_{m2} C2 molecules in the system (where $N_{m1} \to \infty$ and $N_{m2} \to \infty$) at a given $\underline{V}$ and $\underline{T}$, take a C1 test molecule with a given chemical structure and a C2 test molecule with a given chemical structure, and then (1) count the number Ω of ways that each test molecule (with its statistically weighted average of chain rotational states) can be packed into the system $\underline{V}$ and (2) sum the intermolecular interaction energies to obtain $\underline{E}$. Together, steps (1) and (2) yield Q_c. In general, step (1) is the origin of entropy and PV effects in the partition function, and step (2) is the origin of energy terms.

Ω and $\underline{E}$ in Q_c are derived by calculating probabilities that lattice sites will be vacant or occupied by molecular segments of a specific type. In the mean-field calculation of such probabilities, it is easy to see that the decomposition of the actual continuum-space orientations of the long axis of each molecule (and the semiflexible bonds in the molecule) into components parallel to the $\underline{x}$, $\underline{y}$, and $\underline{z}$ axes of the system and then the mapping of these components onto the $\underline{x}$, $\underline{y}$, and $\underline{z}$ axes of the SC lattice will yield the same probabilities that would be obtained using the continuum-space orientations. In the decomposition and mapping approach, the free energy of the system is minimized with respect to the orientations of the long axes of the molecules by minimizing the free energy of the system with respect to the $\underline{x}$, $\underline{y}$, and $\underline{z}$ components of the long axes of the molecules. A mathematically equivalent simplification in the mean field calculation of probabilities is to let the long axes of the molecules be placed parallel to the $\underline{x}$, $\underline{y}$, and $\underline{z}$ axes of the SC lattice and then minimize the free energy of the system with respect to the numbers of long axes of molecules parallel to each axis of the lattice. Since it is mathematically much simpler and much more tractable, this latter approach has been used in this theory.

$\Omega = (\Pi_k \, \Omega_k)$ and $E = (\Sigma_k \, E_k)$, where $k = 1$ refers to the packing and interactions of the C2 molecules and the backbones of the C1 molecules, $k = 2$ refers to the packing and interactions of the C2 molecules and a plate-like section of the side chains of a C1 molecule with the plate-like sections of the side chains of other C1 molecules, and $k = 3$ refers to the packing and interactions of the side chains of C1 molecules with with other side chains in the same plate or plane. [See Figure 1(e); in this schematic figure, the lines represent the long axes of the side chains and backbones of C1 molecules, where these side chains and backbones can have both rigid and semiflexible sections. The actual angles of the long axes of the side chains with respect to the long axis of the backbone are used in all the calculations for side chains in this theory. C2 molecules can also have both rigid and semiflexible sections.] For C1 backbone polymers and C1 nonpolymeric materials, $(\ln \Omega_1) \neq 0$, $E_1 \neq 0$, $(\ln \Omega_2) = (\ln \Omega_3) = 0$, and $E_2 = E_3 = 0$. For the particular C1 side-chain LCPs and C1 combined LCPs (including SS LCPs) studied (3-13) with the earlier single-component analogue of this mixture theory, it has been found that the $\underline{N}$ LC phase and the $\underline{I}$ liquid phase for these polymers involve the packing of plate-like sections of backbones and side chains; thus, $(\ln \Omega_1) \neq 0$, $(\ln \Omega_2) \neq 0$, $(\ln \Omega_3) = 0$, $E_1 \neq 0$, $E_2 \neq 0$, and $E_3 = 0$ in these phases. In calculations (3-13) with the earlier single-component analogue of this mixture theory for these side-chain LCPs and combined LCPs, it has also been found that the SA LC phase for these LCPs is a local SA phase which involves the local ordering of side chains within a plate-like section; thus, $(\ln \Omega_1) = (\ln \Omega_2) = 0$, $(\ln \Omega_3) \neq 0$, $E_1 = E_2 = 0$, and $E_3 \neq 0$ in this phase. For the C1 side-chain LCPs and C1 combined LCPs studied in this paper and in References 3-13, the physically reasonable and mathematically simplifying assumption of segregated packing of C1 backbones and C1 side chains is used, i.e., that C1 side chains pack with other side chains while C1 backbones pack with other backbones. This assumption of segregated packing permits the most efficient packing of the different parts of the C1 molecules in the available $\underline{V}$ of the system.

Now is presented a short summary of how Ω and $\underline{E}$ in Q_c are derived in this lattice theory. $\Omega_k = (\Pi_h \, \Omega_{hk})$, where $\underline{h}$ refers to the component of the mixture. To determine Ω, assume that first N_{m1} C1 molecules (where $N_{m1} \to \infty$) and then N_{m2} C2 molecules (where $N_{m2} \to \infty$) are already on the lattice, and use analytic lattice combinatorial statistics to calculate the number of ways to place a Ch test molecule on the lattice. (These lattice statistics are invariant to the order in which the test molecule of each component is laid on the lattice.) $\Omega_{hk} = (\Pi_i \, g_{Rhki} g_{Fhki})$, where $g_{\ell hki}$ is the average number of ways of placing $N_{\ell hki}$ indistinguishable Ch molecules on the lattice where ℓ refers to the local region (i.e., $\underline{R}$ and $\underline{F}$ refer to rigid-rich and semiflexible-rich local regions, respectively) and $\underline{i}$ refers to the particular axis of the SC lattice. $g_{\ell hki} = [(\Pi_{N_{\ell hki}} \, v_{\ell hki})/(N_{\ell hki}!)]$, where $v_{\ell hki}$ is the average number of ways to lay all the segments of the Ch test molecule--for $\underline{k}$-type packing--with respect to local region ℓ. $v_{\ell hki}$ is equal to the number of ways to lay the first segment of the Ch test molecule multiplied by the probabilities $\underline{p}$ to lay the second, third, etc., segments of the Ch test molecule. The number of ways to lay the first segment is the number of empty lattice sites in the local region. $p = [success/(success + failure)]$, where success is the number of empty lattice sites in the local region, and failure is the number of filled sites in the local region that would interfere with laying a segment in a given direction in the local region on the lattice. This failure term contains the variables for the average orientations and positions of the other molecules in the local region of the lattice. (For more general details about the lattice statistics used here to derive Ω, see especially Reference 20.)

$\underline{E}$ is the sum over site-site effective pair intermolecular interactions (potential energies), such as LJ potentials (for repulsions and London dispersion attractions), dipole/dipole interactions, dipole/induced dipole interactions, hydrogen bonding, etc. The sites refer to atoms or small groups of atoms (such as benzene rings or methylene groups). These potential energies depend on the molecule chemical structures, mixture composition, density of the system, and average orientations and positions of the different parts of neighboring molecules. In a manner analogous to the calculation of Ω, analytic lattice combinatorial statistics are used to calculate the probabilities that lattice sites will be occupied by molecular segments (sites) of a specific type.

The independent variables in Q_c are as follows: $\underline{T}$; ρ_k is the average density of the system (average fraction of lattice sites occupied by molecular segments) for $\underline{k}$-type packing $(0 < \rho_k \leq 1)$; v_{ok} is the volume of one lattice site (i.e., hard-repulsive volume of one molecular segment) for $\underline{k}$-type packing; dph is the degree of polymerization of the backbone of the Ch polymer molecule [i.e., is the number of times that the repeat unit is repeated to make the backbone of the polymer; dph = 1 for a nonpolymeric molecule]; r_{h1}

and f_{h1} are the number of rigid segments and semiflexible segments, respectively, in one repeat unit in the backbone of a Ch molecule; $r_{h2} = r_{h3}$ and $f_{h2} = f_{h3}$ are the number of rigid segments and semiflexible segments, respectively, in each side chain of a Ch molecule; E_{gh1} is the energy of a gauche rotational state (relative to the trans state) of a semiflexible bond in a semiflexible section in the repeat unit of the backbone of a Ch molecule; $E_{gh2} = E_{gh3}$ is the energy of a gauche rotational state (relative to the trans state) of a semiflexible bond in a semiflexible section in each side chain of a Ch molecule; x_h is the mole fraction of repeat units of Ch; $P_{2rhk} = (\langle 3 \cos^2 \theta_{hk} - 1 \rangle / 2)$ is the average orientational order of the rigid sections of the Ch molecules involved in $\underline{k}$-type packing, where θ_{hk} is the angle between the long axis of the rigid section and the preferred axis of orientation for the rigid section $(0 \le P_{2rk} \le 1)$; λ_{hk} is the average fraction of 1D positional alignment of the centers of mass of the Ch molecule parts whose rigid-section long axes are oriented parallel to the preferred axis for orientation of the rigid sections in $\underline{k}$-type packing $(0 \le \lambda_{hk} \le 1)$; a_σ is the average separation distance between segment centers at the zero of energy in the LJ pair potential for any two segments in different molecules; ϵ_{chch1} and ϵ_{thth1} are the absolute values of the minimum of potential energy between two core segments and between two tail segments, respectively, in the LJ potential for backbones of Ch molecules ("tail" refers to the semiflexible segments and the two end rigid segments in a rigid section, and "core" refers to the other rigid segments in the rigid section); $\epsilon_{chch2} = \epsilon_{chch3}$ and $\epsilon_{thth2} = \epsilon_{thth3}$ are the absolute values of the minimum of potential energy between two core segments and between two tail segments, respectively, in the LJ potential for side chains of Ch molecules; $\mu_{D\|h1}$ is the net longitudinal dipole moment for the repeat unit of the backbone of a Ch molecule; $\mu_{D\|h2} = \mu_{D\|h3}$ is the net longitudinal dipole moment for each side chain of a Ch molecule; $\mu_{D\perp h1i}$ (where $i = 1,2,3,\ldots$) are the individual transverse dipole moments for the repeat unit of the backbone of a Ch molecule; $\mu_{D\perp h2i} = \mu_{D\perp h3i}$ (where $i = 1,2,3,\ldots$) are the individual transverse dipole moments for each side chain of a Ch molecule; α_{ch1} and α_{th1} are the average polarizabilities for a core segment and a tail segment, respectively, in a repeat unit of the backbone of a Ch molecule; $\alpha_{ch2} = \alpha_{ch3}$ and $\alpha_{th2} = \alpha_{th3}$ are the average polarizabilities for a core segment and a tail segment, respectively, in each side chain of a Ch molecule; $a_{\sigma Hk}$ is the average separation distance between segment centers at the zero of energy in the pair potential for hydrogen bonding between any two segments in different molecules involved in $\underline{k}$-type packing [here a LJ (12,6) potential is used, which is different from the LJ potentials used above for London dispersion forces]; ϵ_{chchH1} is the absolute value of the minimum of potential energy for hydrogen bonding between two core segments in the backbones of Ch molecules; ϵ_{chchH2} and ϵ_{chchH3} are the absolute values of the minimum of potential energy for hydrogen bonding between two core segments in the side chains of Ch molecules for 2-type packing and 3-type packing, respectively.

$$(\ln \Omega_k)/N_m = x_{\kappa k}(\{\sum_{\ell=R,F} [(Q_{\ell1k} \ln Q_{\ell1k} + 2Q_{\ell2k} \ln Q_{\ell2k}$$

$$- C_{1k} \ln C_{1k})C_{\ell6k}/(2\rho_k)]\} - C_{2k}) ; \tag{1}$$

$$E_k/N_m = (x_{\kappa k}\rho_k/9)\{\sum_{\ell=R,F} \sum_h \sum_n [(Q_{\ell3hnk}/Q_{\ell1k})$$

$$+ (Q_{\ell4hnk}/Q_{\ell2k})]/C_{\ell6k}\} , \tag{2}$$

$$Q_{\ell1k} = 1 - [\sum_h (x_h[\rho_k/(3C_{\ell6k})]\{(1 + 2P_{2rhk})[(r_{hk} - y_{1hk})$$

$$+ q\lambda_{hk}C_{4hk}] + (1 + 2P_{2rhk}P_{2ihk})f_{hk})\}] , \tag{3}$$

$$Q_{\ell2k} = 1 - (\sum_h \{x_h[\rho_k/(3C_{\ell6k})]\{(1 - P_{2rhk})(r_{hk} - y_{1hk})$$

$$+ (1 - P_{2rhk}P_{2ihk})f_{hk} - q(1 + 2P_{2rhk})\lambda_{hk}C_{5hk}]\}) , \tag{4}$$

$$Q_{\ell3hnk} = (1 - P_{2rhk})(1 - P_{2rnk})(W_{2hnk} + W_{3hnk}) + [(1 - P_{2rhk})$$

$$\times(1 + 2P_{2rnk})W_{\ell1hnk}] + [(1 + 2P_{2rhk})(B_{1hk} + q\lambda_{hk}B_{2hk})(1 + 2P_{2rnk})$$

$$\times(B_{1nk} + q\lambda_{nk}B_{2nk})\omega_{thtnk}/2] , \tag{5}$$

$$Q_{\ell4hnk} = (1 - P_{2rhk})(1 - P_{2rnk})(W_{1hnk} + W_{3hnk} + B_{1hk}B_{1nk}\omega_{thtnk})$$

$$+ (1 - P_{2rhk})(1 + 2P_{2rnk})W_{\ell22hnk} + [(1 - P_{2rnk})(1 + 2P_{2rhk})$$

$$\times(W_{\ell21hnk} + W_{\ell4hnk})] + (1 + 2P_{2rhk})(1 + 2P_{2rnk})W_{\ell3hnk} , \tag{6}$$

$$C_{1k} = 1 - \rho_k , \tag{7}$$

$$C_{2k} = \sum_h x_h[(\{(1 + 2P_{2rhk}) \ln (1 + 2P_{2rhk}) + [2(1 - P_{2rhk})$$

$$\times \ln (1 - P_{2rhk})]\}/3) + (\sum_{\ell=R,F} \{\ln [x_h\rho_k/(3C_{\ell6k})]\}/2)] , \tag{8}$$

$$C_{3hk} = r_{hk} - y_{1hk} + P_{2ihk}f_{hk} , \tag{9}$$

$$C_{4hk} = r_{hk} - [(1 + 2P_{2ihk})f_{hk}/3] , \tag{10}$$

$$C_{5hk} = (1 - P_{2ihk})f_{hk}/3 , \tag{11}$$

$$C_{\ell6k} = \sum_h (x_h\{m_{hk} + q[(1 + 2P_{2rhk})\lambda_{hk}(r_{hk} - f_{hk})/3]\}) , \tag{12}$$

$$W_{1hnk} = B_{1nk}[2(r_{hk} - 2)\omega_{chtnk} + A_{1hk}\omega_{thtnk}] , \tag{13}$$

$$W_{2hnk} = (r_{hk} - 2)(r_{nk} - 2)\omega_{chcn\perp k} + 2(r_{hk} - 2)A_{1nk}\omega_{chtnk}$$

$$+ A_{1hk}A_{1nk}\omega_{thtnk} , \tag{14}$$

$$W_{3hnk} = (r_{hk} - 2)(r_{nk} - 2)\omega_{chcn\|k} + 2(r_{hk} - 2)A_{1nk}\omega_{chtnk}$$

$$+ A_{1hk}A_{1nk}\omega_{thtnk} , \tag{15}$$

$$W_{\ell 1hnk} = (B_{1nk} + q\lambda_{nk}B_{2nk})[2(r_{hk} - 2)\omega_{chtnk} + A_{1hk}\omega_{thtnk}] , \qquad (16)$$

$$W_{\ell 21hnk} = (r_{hk} - 2)(r_{nk} - 2)(1 + q\lambda_{hk})\omega_{chcn\perp k} + [2(r_{hk} - 2)$$

$$\times(1 + q\lambda_{hk})A_{1nk}\omega_{chtnk}] + A_{1nk}(A_{1hk} + q\lambda_{hk}A_{2hk})\omega_{thtnk} , \qquad (17)$$

$$W_{\ell 22hnk} = (r_{hk} - 2)(r_{nk} - 2)(1 + q\lambda_{nk})\omega_{chcn\perp k} + [2(r_{hk} - 2)$$

$$\times(A_{1nk} + q\lambda_{nk}A_{2nk})\omega_{chtnk}] + A_{1hk}(A_{1nk} + q\lambda_{nk}A_{2nk})\omega_{thtnk} , \qquad (18)$$

$$W_{\ell 3hnk} = (r_{hk} - 2)(1 + q\lambda_{hk})(r_{nk} - 2)(1 + q\lambda_{nk})\omega_{chcn\|k}$$

$$+ 2(r_{hk} - 2)(1 + q\lambda_{hk})(A_{1nk} + q\lambda_{nk}A_{2nk})\omega_{chtnk}$$

$$+ (A_{1hk} + q\lambda_{hk}A_{2hk})(A_{1nk} + q\lambda_{nk}A_{2nk})\omega_{thtnk} , \qquad (19)$$

$$W_{\ell 4hnk} = B_{1nk}[2(r_{hk} - 2)(1 + q\lambda_{hk})\omega_{chtnk}$$

$$+ (A_{1hk} + q\lambda_{hk}A_{2hk})\omega_{thtnk}] , \qquad (20)$$

$$A_{\xi hk} = 2 + x_\nu[f_{hk}(2 + P_{2ihk})/3] , \qquad (21)$$

$$B_{\xi hk} = y_{2hk} + x_\nu[2f_{hk}(1 - P_{2ihk})/3] . \qquad (22)$$

In Equations 21-22, $x_\nu = +1$ if $\xi = 1$; $x_\nu = -1$ if $\xi = 2$. If $\ell = R$ in Equations 1-22 above, then $q = +1$; if $\ell = F$, then $q = -1$. In Equations 3-4 and 9, $y_{1h1} = 1/dph$ and $y_{1h2} = y_{1h3} = 1$. In Equation 22, $y_{2h1} = 1/dph$ and $y_{2h2} = y_{2h3} = 1/2$. In the above equations, $x_{\kappa k} = [m_{1k}/(\Sigma_i m_{1i})]$, where $x_{\kappa k} = x_{\kappa 1} = m_{11}/m_{11} = 1$ if Cl is a nonpolymeric molecule or a backbone polymer; $x_{\kappa k} = m_{1k}/(m_{11} + m_{12})$ for $k = 1$ or 2 if Cl is a side-chain LCP or a combined LCP; and $x_{\kappa k} = x_{\kappa 3} = m_{13}/m_{13} = 1$ for $k = 3$ if Cl is a side-chain LCP or a combined LCP. $m_{hk} = (r_{hk} + f_{hk})$. $N_m = (N_{m1} + N_{m2})$. Also, $\Sigma_h = \Sigma_{h=1}^2$ and $\Sigma_n = \Sigma_{n=1}^2$, where $\underline{h}$ and $\underline{n}$ refer to the components of the mixture.

$P_{2ihk} = (\langle 3 \cos^2\psi_{hk} - 1\rangle/2) = (1 - 3u_{hk})$ is the average intramolecular orientational order of the semiflexible sections in the molecule parts of Ch involved in $\underline{k}$-type packing, where ψ_{hk} is the angle between a given semiflexible bond and the rigid section to which the semiflexible bond is attached. $0 \leq P_{2ihk} \leq 1$. For a $(-CH_2-)_x$ or a $(-CH_2-)_xCH_3$ semiflexible chain, each semiflexible bond has three choices of direction on the SC lattice which mimic the three choices [trans, gauche(+), gauche(-)] of rotational energy minima for the carbon-carbon bonds in the tetrahedral coordination in such a chain in a real molecule. For case $\underline{k}$ in Ch, a trans choice has a Boltzmann statistical weight of $\zeta_{1hk} = [1/(1 + 2\Lambda_{hk})]$, and each gauche choice has a Boltzmann statistical weight of $\zeta_{2hk} =$

$[\Lambda_{hk}/(1 + 2\Lambda_{hk})]$, where $\Lambda_{hk} = \{exp[-E_{ghk}/(k_BT)]\}$. $2u_{hk} = [(\Sigma_{h\gamma k} 2u_{h\gamma k}f_{h\gamma k})/(\Sigma_{h\gamma k} f_{h\gamma k})]$, where γ refers to semiflexible section γ. $f_{hk} = (\Sigma_{h\gamma k} f_{h\gamma k})$. For $f_{h\gamma k} = 1$, $2u_{h\gamma k} = 2\varsigma_{2hk}$. For $f_{h\gamma k} \geq 2$,

$$2u_{h\gamma k} = \{(\sum_{j=1}^{2} Y_{1jhk}) + [(f_{h\gamma k} - 2)/2][\sum_{j=1}^{2} Y_{2jhk}]\}/f_{h\gamma k}, \qquad (23)$$

$$2u_{h\gamma k} = \{(\sum_{j=1}^{3} Y_{3jhk}) + [(f_{h\gamma k} - 3)/2][\sum_{j=1}^{2} Y_{2jhk}]\}/f_{h\gamma k}, \qquad (24)$$

for even $f_{h\gamma k}$ and odd $f_{h\gamma k}$, respectively.

$Y_{11hk} = [2(\varsigma_{1hk}\varsigma_{2hk} + \varsigma_{2hk}^2)/D_{\varsigma 1hk}]$, $D_{\varsigma 1hk} = [\varsigma_{1hk}^2 + 4\varsigma_{1hk}\varsigma_{2hk} + 2\varsigma_{2hk}^2]$, $Y_{12hk} = [2(2\varsigma_{1hk}\varsigma_{2hk} + \varsigma_{2hk}^2)/D_{\varsigma 1hk}]$, $Y_{21hk} = Y_{22hk} = [2(\varsigma_{1hk}^2 + 3\varsigma_{1hk}\varsigma_{2hk} + 2\varsigma_{2hk}^2)/D_{\varsigma 2hk}]$, $D_{\varsigma 2hk} = [3\varsigma_{1hk}^2 + 2(5\varsigma_{1hk}\varsigma_{2hk} + 3\varsigma_{2hk}^2)]$, $Y_{31hk} = [2(\varsigma_{1hk}^2\varsigma_{2hk} + 3\varsigma_{1hk}\varsigma_{2hk}^2 + \varsigma_{2hk}^3)/D_{\varsigma 3hk}]$, $D_{\varsigma 3hk} = [\varsigma_{1hk}^3 + 2(3\varsigma_{1hk}^2\varsigma_{2hk} + 4\varsigma_{1hk}\varsigma_{2hk}^2 + \varsigma_{2hk}^3)]$, $Y_{32hk} = [2(2\varsigma_{1hk}^2\varsigma_{2hk} + 4\varsigma_{1hk}\varsigma_{2hk}^2 + \varsigma_{2hk}^3)/D_{\varsigma 3hk}]$, and $Y_{33hk} = [6(\varsigma_{1hk}^2\varsigma_{2hk} + \varsigma_{1hk}\varsigma_{2hk}^2)/D_{\varsigma 3hk}]$.

The intermolecular interaction energy between a y-type molecular segment and a z-type molecular segment (where y and z can each be replaced by c for "core" or t for "tail") is given by $\omega_{yhznk} = (\Sigma_{j=1}^{4} \omega_{yhznjk})$, where $j = 1$ refers to LJ interactions (for repulsions and London dispersion attractions), $j = 2$ refers to dipole/induced dipole interactions, $j = 3$ refers to dipole/dipole interactions, and $j = 4$ refers to hydrogen-bonding interactions. $\omega_{chtn\|k} = \omega_{chtn\perp k} = \omega_{chtnk}$; and $\omega_{thtn\|k} = \omega_{thtn\perp k} = \omega_{thtnk}$. (∥ and ⊥ refer to interactions between molecular segments attached to cores that are parallel and perpendicular, respectively, to each other.) $\omega_{yhzn1k} = \{4\epsilon_{yhznk}[(a_\sigma/a_k)^{12} - (a_\sigma/a_k)^6]\}$. $\omega_{chcn1\|k} = \omega_{chcn1\perp k}$. $\epsilon_{yhznk} = (\epsilon_{yhyhk}\epsilon_{znznk})^{1/2}$. If $f_{hk} = 0$, $\epsilon_{ththk} = \epsilon_{chchk}$.

a_k is the average separation distance between the centers of two first-neighbor intermolecular segments involved in k-type packing. For $k = 1$ and $k = 2$ types of packing, a_k is calculated from $\rho_k = [(\Sigma_h m_{hk}N_{mh})/M_k] = [(\Sigma_h x_h m_{hk})v_{ok}/v_k]$, with $v_1 = [\Sigma_h (x_h a_1^2\{a_1(2/dph) + v_{o1}^{1/3}[m_{h1} - (2/dph)]\})]$, and $v_2 = [\{x_1 a_2^2[a_2 + v_{o2}^{1/3}(m_{12} - 1)]\} + (x_2 a_2^2\{a_2(2/dp2) + v_{o1}^{1/3}[m_{21} - (2/dp2)]\})]$, where v_1 is the average volume associated with the repeat unit of the backbone of one C1 molecule and the repeat unit of the backbone of one C2 molecule, and v_2 is the average volume associated with

(1) one side chain of a C1 molecule as the plate of such side chains of a molecule pack with the plates containing the side chains of other molecules and (2) the repeat unit of the backbone of one C2 molecule. $a_1 = a_2$.

In the following analysis for polymers with side chains, equations are presented for the case of one side chain per repeat unit of the backbone of the C1 polymer. (Though not presented here, it is possible to perform an analogous analysis for any number of side chains per repeat unit of the backbone of the polymer.) For $k = 3$, the length a' along the backbone between side chains is given by $a' = (a_w\{r_{11} + f_{11}[(1 + 2P_{2i11})/3]\})$, where a_w is the length of a segment in the backbone of the C1 polymer. If the side chains of a C1 molecule pack on alternating sides of the backbone, $a_3 = 2a'$. If (1) all the side chains of a C1 molecule pack on one (i.e., the same) side of the backbone or (2) the side chains of a C1 molecule interdigitate on alternating sides of the backbone with the side chains of neighboring molecules, $a_3 = a'$. If the side chains of a C1 molecule interdigitate on one (i.e., the same) side of the backbone with the side chains of a neighboring molecule, $a_3 = a'/2$. Then, $v_3 = [(\{x_1 a_3{}^2[a_3 + v_{o3}{}^{1/3}(m_{13} - 1)]\}) + (x_2 a_1{}^2\{a_1(2/dp2) + v_{o3}{}^{1/3}[m_{21} - (2/dp2)]\})]$, where v_3 is the average volume associated with (1) one side chain of the C1 molecule as the side chains of the molecule pack with each other or with interdigitating side chains of other C1 molecules and (2) the repeat unit of the backbone of one C2 molecule. Then, ρ_3 is calculated.

For backbone polymers and nonpolymeric molecules (3-10), $v_{o1}{}^{1/3}$ = 1.96 Å. For side-chain LCPs and combined LCPs (3-10), $v_{o3}{}^{1/3}$ = 1.96 Å, and $v_{o1}{}^{1/3} = v_{o2}{}^{1/3} = \{[2v_{o3}{}^{1/3} + (m_{11}X_{\Gamma12}) + (m_{12}X_{\Gamma11})]/(m_{11} + m_{12})\}$, where $X_{\Gamma1k} = (ZL_{1k}/4)$ and $L_{1k} = (a_w m_{1k})$. If the side chains of a C1 molecule pack on alternating sides of the backbone, $Z = 2$. If (1) all the side chains of a C1 molecule pack on one (i.e., the same) side of the backbone or (2) the side chains of a C1 molecule interdigitate on alternating sides of the backbone with the side chains of neighboring molecules, $Z = 1$. If the side chains of a C1 molecule interdigitate on one (i.e., the same) side of the backbone with the side chains of a neighboring molecule, $Z = 1/2$.

Using Equation 13.5-3 of Reference 26(a), $\omega_{chcn2\|k} = \omega_{chcn2\perp k} = [-(\mu_{D\|hk}{}^2 + \mu_{D\perp hk}{}^2)]\alpha_{cnk}/(a_k{}^6 z_{fhk}{}^2)]$, and $\omega_{chtn2k} = [-(\mu_{D\|hk}{}^2 + \mu_{D\perp hk}{}^2)\alpha_{tnk}/(2a_k{}^6 z_{fhk}{}^2)]$, where $z_{fhk} = (r_{hk} - 2)$ for $f_{hk} \neq 0$, and $z_{fhk} = r_{hk}$ for $f_{hk} = 0$. $\mu_{D\perp hk}{}^2 = (\Sigma_{i,j} \mu_{D\perp hki}\mu_{D\perp hkj}/\Sigma_{i,j})$, where--depending on the average relative positions and orientations of the molecules--there are three choices of sums: sum over all ii pairs and ij pairs (where $i \neq j$), sum over all ii pairs only, or sum over all ij pairs only (where $i \neq j$). Using Equation 1.3-8 of Reference 26(b), $\omega_{chcn3\|k} = \{-(\mu_{D\|hk}\mu_{D\|nk})/[2a_k{}^3 z_{fhk}{}^2 z_{fnk}(1 + \chi q\lambda_{hk})]\}$, $\chi = 0$ for $\omega_{chcn3\|k}$ in

Equation 15, and $\chi = 1$ for $\omega_{chcn3\|k}$ in Equation 19. $\omega_{chcn3\perp k} = [-(\mu_{D\perp hk}\mu_{D\perp nk})/(2a_k{}^3 z_{fhk}{}^2 z_{fnk})]$. $\omega_{chcn4\|k} = \{4\epsilon_{chcnHk}[(a_{\sigma Hk}/a_k)^{12} - (a_{\sigma Hk}/a_k)^6]/[2z_{fhk}(1 + \chi q\lambda_{hk})]\}$, and $\omega_{chcn4\perp k} = \{4\epsilon_{chcnHk}[(a_{\sigma Hk}/a_k)^{12} - (a_{\sigma Hk}/a_k)^6]/(2z_{fhk})\}$, where $\chi = 0$ for $\omega_{chcn4\|k}$ in Equation 15, and $\chi = 1$ for $\omega_{chcn4\|k}$ in Equation 19. The equations here for dipole/dipole interactions and hydrogen-bonding interactions take into account (1) parallel and antiparallel orientations of the two cores and (2) fraction of positional overlap (alignment) of the two cores. $\omega_{thtn2k} = \omega_{chcn3k} = \omega_{thtn3k} = \omega_{chtn4k} = \omega_{thtn4k} = 0$.

As in an earlier version (12-13) of this mixture theory for backbone LCPs, the global mole fractions X_1 and X_2 of C1 and C2 in the mixture are defined in terms of the local mole fractions x_1 and x_2 in the mixture by $X_1 = [x_1 dp2/(x_1 dp2 + x_2 dp1)]$ and $X_2 = 1 - X_1$, where X_1 and X_2 are the mole fractions of the <u>molecules</u> of C1 and of C2, respectively; and x_1 and x_2 are the mole fractions of the <u>repeat units</u> of C1 and of C2, respectively.

From Q_c, the PVT equation of state is then derived thermodynamically, as well as equations that minimize the free energy of the system with respect to the average independent orientational and positional order variables of the different parts of the molecules. [These variables are P_{2rhk} and λ_{hk}, where (as discussed earlier) <u>h</u> refers to the particular component in the mixture and <u>k</u> to each type of packing of the different parts of the molecules.] This paper presents results calculated at constant <u>P</u>. <u>V</u> can be varied by varying the number of empty lattice sites in the system. It is easy to show (9-10,19,21-23,27) that the equations that minimize the configurational Helmholtz free energy A_c with respect to the order variables at constant <u>V</u> and <u>T</u> also minimize the configurational Gibbs free energy G_c with respect to the order variables at constant <u>P</u> and <u>T</u>. The most stable state is the state of lowest free energy or lowest chemical potential at constant <u>P</u> and <u>T</u>.

The density ρ and the independent average order variables are calculated at a given <u>P</u> and <u>T</u> by simultaneously solving the PVT equation of state and the equations minimizing the free energy with respect to these order variables. Then, the dependent average order variables and the other static thermodynamic properties of the system can be calculated.

See References 3, 8, and 10 for examples of how the input variables [such as bond lengths and angles, trans-gauche rotational energy difference E_g, ϵ and a_σ (for LJ potentials), dipole moments, bond polarizabilities, and hydrogen bonding] for this theory are calculated for complex molecules from compilations of existing experimental data for functional chemical groups (such as benzene rings and methylene groups) and simple molecules. Once each input variable has been determined for a particular functional chemical group, that variable has then been held constant in all of the calculations with this mixture theory and with the single-component theory (3-10).

Agreement of Static Theory with Experiment. As shown in
References 3-10, 12-21 and 23, the single-component analogue of this
new mixture theory (and earlier versions of the analogue) have
predicted and reproduced experimental trends in thermodynamic and
molecular ordering properties [such as phase transition
temperatures, phase stabilities (including of $\underline{N}$ and multiple SA LC
phases and of the $\underline{I}$ liquid phase), orientational order P_2, parallel
and perpendicular radii of gyration, and odd-even effects (i.e.,
alternations in the magnitudes of various thermodynamic and
molecular ordering properties as the number of carbon atoms in a
$\underline{n}$-alkyl chain varies from odd to even)] for various backbone LCPs,
side-chain LCPs, and nonpolymeric LC materials as a function of $\underline{T}$,
$\underline{P}$, and molecule chemical structure. This theory has also given
(3,7-10) very good quantitative predictive and reproductive
agreement (relative deviations between 0% and less than about 6.4%)
with available experimental data for these properties.
        Earlier versions (12-13,22) of this new mixture theory have
predicted and reproduced experimental trends in various
thermodynamic and molecular ordering properties [such as phase
transition temperatures, phase stabilities (including of the $\underline{N}$ LC
phase and the $\underline{I}$ liquid phase), curvatures of lines of coexisting
phases, and orientational order P_2] in binary mixtures as a function
of $\underline{T}$, $\underline{P}$, system composition, and molecule chemical structures.
These earlier versions (12-13,22) of the new mixture theory have
been applied to systems in which C1 and/or C2 can be a backbone LCP,
a nonpolymeric LC molecule, or a nonpolymeric non-LC molecule.
        Figure 2(a) shows an example of the good agreement of the new
mixture theory of this paper with experiment. This figure compares
the results calculated from this theory with experimental results
(28) for the range of stability of the $\underline{N}$ LC phase with respect to
the $\underline{I}$ liquid phase as the weight % of a backbone LCP in a mixture is
varied. The backbone LCP is PBA, and the solvent is DMAc with 4
weight % LiCl. [See Figure 2(b) for the molecule chemical
structures of PBA and DMAc.] The theoretical calculations have been
performed for T = 303 K and P = 1 atm and have used values from pure
DMAc as the input variables for the solvent in the calculation. The
theoretical calculations were done for different dp of PBA and were
related to the number average molecular weight M_n, the weight
average molecular weight M_w, and the intrinsic viscosity $[\eta]$ (in
H_2SO_4) of PBA using both the data in Table I of Reference 29 and the
relation $[\eta] = 1.9 \times 10^{-7} M_w^{1.7}$ for PBA in H_2SO_4 given in
Reference 29. [The experimental data in Figure 2(a) is given for
inherent viscosity. While the data is not available now to convert
these inherent viscosities to intrinsic viscosities, the difference
between these two types of viscosities for the experimental data in
References 28 and 29 is very small (30-31) and thus can be
ignored.]

New Dynamic Theory for Calculation of Lower Limits of Processability

Here is summarized a recent new dynamic theory (9-11) for Brownian
motion and diffusion for almost any organic polymeric or
nonpolymeric molecule in the $\underline{I}$ liquid phase and in various LC

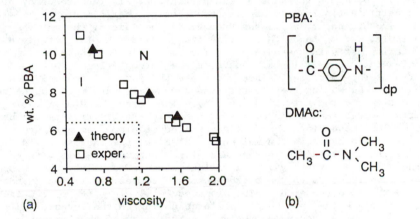

Figure 2. (a) Plot of theoretical results and experimental results (28) of weight (wt.) % of PBA vs. viscosity (in H$_2$SO$_4$) of PBA, showing the limit of stability of the N LC phase with respect to the I liquid phase at T = 303 K and P = 1 atm for mixtures of PBA in the solvent DMAc (with 4 weight % LiCl). (b) Molecule chemical structures of the backbone LCP PBA and of DMAc.

phases. The diffusion of the molecules constitutes a lower limit (i.e., slowest rate) for the speed at which a material can be processed.

The basic idea of this dynamic theory is to use the site-site intermolecular energies and forces from the static theory presented earlier in this paper to calculate friction coefficients, etc., for analytic Brownian motion calculations for the molecules. From References 32-33, the aperiodic case of Brownian motion of a harmonically-bound particle is given by

$$x_b^2 = H_1 + (x_{b0}^2 - H_1)[1 + (H_2/2)]^2 \exp(-H_2) , \qquad (25)$$

where $H_1 = [k_BT/(m_b\omega_b^2)]$, $H_2 = (\beta t)$, $\beta = (f_b/m_b)$, where $\underline{t}$ is the time, x_b is the average position at time $\underline{t}$, x_{b0} is the position at time $t = 0$, m_b is the mass of the particle, ω_b is the frequency of the particle (in 2π), and f_b is the friction coefficient.

In this paper, m_b is assigned to be the mass of the entire molecule, and ω_b to be the average frequency of the largest rigid vibrating parts in the molecule. The rationale for these assignments is as follows: Focus on the largest rigid vibrating parts of the molecules, since these are the parts of the molecules that will be moving the slowest (and thus will most limit the degree of motion and the rate of diffusion). The largest rigid vibrating parts undergo random vibrations with respect to other parts of the molecule; these vibrations are somewhat decoupled from each other by semiflexible bonds in the molecule. However, the largest rigid vibrating parts must drag the mass of the entire molecule with them. [There is one largest rigid vibrating part for each $\underline{k}$-type packing, where $k = 1$ refers to the backbones of molecules, while $k = 2$ and $k = 3$ refer to the side chains of molecules.] $f_b = (F_b/v_b)$, where F_b is the average intermolecular force acting on the largest rigid vibrating parts, and v_b is the average velocity of the largest rigid vibrating parts. v_b is calculated by equating the kinetic energy of the molecule to the potential energy $\underline{U}$ of the largest rigid vibrating parts: $(m_b v_b^2/2) = U$, where $U = \{\Sigma_k [(\Sigma_{y,z} \omega_{yzk})/m_k]\}$, $F_b = (\Sigma_k \{[\Sigma_{y,z} (\partial \omega_{yzk})/(\partial a_k)]/m_k\})$, and $(\Sigma_k \omega_{yzk})$ is the sum of all site-site intermolecular potential energies for the largest rigid vibrating parts in all the $\underline{k}$-type packings. ω_{yzk}, a_k, and m_k are calculated in the static theory presented earlier in this paper.

Note that $\underline{U}$ and F_b are calculated from the intermolecular potential energies ω_{yzk} and separation distances a_k which are calculated in the static theory for the phase of lowest free energy as a function of $\underline{T}$, $\underline{P}$, system composition, and molecule chemical structures. Thus, $\underline{U}$ and F_b are determined by the details of the molecule chemical structures and the orientational and positional ordering of the molecules. $\omega_b^2 x_b = F_b/m_b$, from the harmonic oscillator equation.

Then solve Equation 25 for x_b. t_a is the smallest value of $\underline{t}$ at which the value of x_b has increased to its asymptotic value. x_{ba} is the value of x_b at $t = t_a$. The macroscopic diffusion coefficient $D_b = (x_{ba}^2/t_a)$.

<u>Agreement</u> <u>of</u> <u>Dynamic</u> <u>Theory</u> <u>with</u> <u>Experiment</u>. This theory reproduces
(9-11) the experimental trends in diffusion coefficients for
polymeric and nonpolymeric molecules in the pure state or in binary
mixtures as a function of $\underline{T}$, $\underline{P}$, and molecule chemical structure.
This theory also gives good quantitative agreement with available
experimental data for these properties. For example, for the non-LC
backbone polymer polyisoprene [see Figure 3(a)] at infinite dilution
in hexane $[CH_3-(CH_2)_4-CH_3]$ in the $\underline{I}$ liquid phase at T = 293 K, the
infinite dilution diffusion coefficient D_{b0} (in units of

10^{-7} cm^2 sec^{-1}) of the polyisoprene is 2.93 and 1.13 calculated from
this theory [compared with 3 and 1.01 from experiment (34)] for
polyisoprene molecular weights (in units of 10^4) of 27 and 166,
respectively. Also, for example, for the nonpolymeric LC material
PAA [see Figure 3(a)] in the $\underline{N}$ LC phase at T = 400 K, the

self-diffusion coefficient D_b (in units of 10^{-6} cm^2 sec^{-1}) is 3.47
calculated from this theory, compared with ~3.4 from experiment
(35). [For the PAA molecule, the (largest) rigid vibrating part
(which is used in this theory to calculate D_b) is $O-\phi-(NO)=N-\phi-O$,
where ϕ is a para-bonded phenyl group (benzene ring).] In the above
calculations, P = 1 atm.

Further Results and Discussion

For some of the first theoretically designed candidate SS LCPs, new
results are now presented for (1) enhanced solubilities of the
molecules (compared with backbone LCPs) in nonpolymeric LC solvents
calculated using the static theory, (2) diffusion (i.e., lower limit
of processability) of the molecules calculated using the dynamic
theory, and (3) head-to-tail polymerization of the monomers
predicted from the packing of the bulky relatively rigid side chains
of the molecules as calculated using the static theory. Melt
processability of some SS LCPs is also discussed.

In earlier papers (3-7,9-10), the single-component analogue of
the static mixture theory was used to predict and design (atom by
atom, bond by bond) the first candidate SS LCPs, i.e., the first
polymers designed to have good compressive strengths as well as to
have tensile strengths and tensile moduli significantly larger than
existing strong polymers (backbone LCPs, such as PBO or DuPont's
Kevlar). [See Figure 3(a).] SS LCPs are combined LCPs designed
such that the state of lowest free energy is the state in which the
side chains of one molecule physically interdigitate with (pack
between) the side chains of neighboring molecules, as shown in
Figure 1(e). The interdigitation of the side chains keeps the
molecules from being pushed past each other or peeled apart and is
thus the origin of the good compressive strength. Some SS LCPs have
been designed (4) with side chains that pack on alternating sides of
the backbones [see Figures 1(e) and 3(b)-(c)], and some SS LCPs
[such as b-PBO-1/s-PBO-1, b-polest-1/s-polest-2, and
b-PBA-3/s-polest-2 in Figure 3(a)] have been designed with side
chains that pack on one side of the backbone [see Figures 3(d)-(e)].
See especially Reference 4 for a summary of specific details of
molecule chemical structure that are necessary for combined LCPs to
be SS LCPs. A theoretical design "cookbook" (36) has been developed

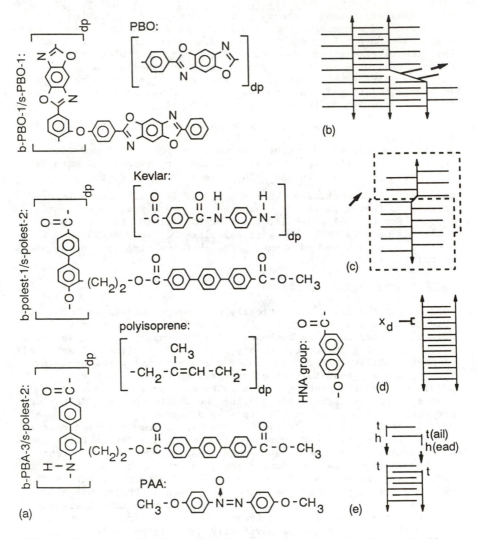

Figure 3.　(a) Molecule chemical structures for the theoretically designed candidate SS LCPs b-PBO-1/s-PBO-1, b-polest-1/s-polest-2, and b-PBA-3/s-polest-2; for the backbone LCPs PBO and Kevlar; for the non-LC polymer polyisoprene; for the nonpolymeric LC molecule PAA; and for the HNA group.　Schematic illustrations:　(b) and (c):　Small defects in the packing of the SS LCP molecules generate effective long-range 3D LC ordering; the bold arrows and molecular parts indicate packing in the third dimension.　(d) Spacing distance x_d (along the backbone) between interdigitated side chains of SS LCP molecules.　(e) Head-to-tail polymerization of SS LCP molecules.

for SS LCPs. The candidate SS LCPs presented in this paper and earlier papers (3-7,9-10) contain known functional chemical groups and thus have been designed to be easy to chemically synthesize.

As illustrated in Figure 1(e), the theory predicts (3-7,9-10) that the backbones and side chains of SS LCPs tend to pack in planes, such that backbones tend to orientationally order with other backbones in the same plane, and similarly for side chains. There is thus $\underline{N}$ LC ordering of the backbones $\underline{in}$ a plane and also $\underline{local}$ interdigitated SA LC ordering of side chains $\underline{in}$ a plane. There is also orientational ordering (alignment) of planes, such that backbones in one plane orientationally align with backbones in other planes, and similarly for side chains. There is thus $\underline{N}$ LC ordering of the planes in 3D. At finite temperatures, small naturally occurring defects in the packing of the molecules lead to an effective long-range 3D LC orientational and positional ordering of the molecules [see Figures 3(b)-(c)] and thus to effective long-range 3D strength in the system.

The chemical syntheses of several of the theoretically designed candidate SS LCPs have begun at multiple research institutions.

Enhanced Solubilities of SS LCPs in Nonpolymeric LC Solvents. Figure 4(a) shows calculated results for the theoretically designed candidate SS LCP b-PBO-1/s-PBO-1 at dp = 100 in the nonpolymeric LC solvent PAA, while Figure 4(b) shows calculated results for the backbone LCP PBO at dp = 100 in the solvent PAA. Both figures show plots of values of the average orientational order variable P_2 for the backbones of each LCP and for PAA as a function of weight % of the LCP in the mixture in the $\underline{N}$ LC phase at T = 400 K and P = 1 atm. The P_2 values of the LCPs and of PAA increase as the concentration of the LCP in the mixtures increases (i.e., the LCP enhances the ordering of the nonpolymeric LC solvent). The P_2 values for both the LCP and PAA are greater in the mixture with the SS LCP than in the mixture with the backbone LCP. As calculated (3-7,9-10) for pure SS LCPs, the ordering of the side chains of SS LCPs in mixtures enhances the ordering of the backbones of the SS LCPs, and vice versa. The enhanced ordering of the SS LCP molecules also enhances the ordering of the nonpolymeric LC solvent PAA. The enhanced LC ordering of SS LCP molecules is consistent with calculations (5) of enhanced mechanical properties (tensile strength, tensile modulus, and compressive strength) of SS LCPs over backbone LCPs.

In each of these figures, the smallest value of the weight % of LCP for which P_2 values are plotted corresponds to the lowest limit of solubility of the LCP in the $\underline{N}$ LC phase of the PAA solvent. The SS LCP is more soluble (by almost an order of magnitude in weight %) in PAA than is the backbone LCP. The side chains of the SS LCP molecules enhance the solubility of the backbones of the SS LCP molecules; that is, the side chains of the SS LCP molecules are more soluble (than the backbones) in PAA and help pull the backbones of the SS LCP molecules into solution. (In these figures, results are plotted up to about 40 weight % of the LCP.)

Opposing Trends in Strength vs. Processability Limits for SS LCPs. Table I shows how the backbone P_2, side-chain P_2, and diffusion coefficient D_b of the theoretically designed candidate SS LCP

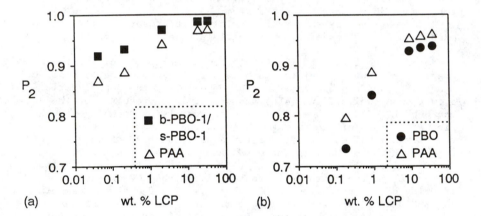

Figure 4. Plot of theoretical results of P_2 vs. weight (wt.) %
of LCP, showing the backbone P_2 values for each component of the
binary mixture of a LCP in the N̲ LC phase of PAA at T = 400 K and
P = 1 atm, where the LCP dp = 100:   (a) SS LCP b-PBO-1/s-PBO-1,
(b) backbone LCP PBO.

b-polest-1/s-polest-2 varies as the interdigitated side-chain spacing distance x_d [see Figure 3(d)] is varied along the backbone of the SS LCP. In Table I, the values of P_2 and D_b are the values for the SS LCP at dp = 100 in the limit of removal of the last drop of solvent just before forming a polymer solid (glass) at T = 400 K and P = 1 atm. P_2 is calculated using the static theory, and D_b using the dynamic theory (with input from the static theory). (To isolate the effect of varying x_d in Table I from the effects of different molecule chemical structures, start with the SS LCP structure b-polest-1/s-polest-2 with dp = 100 at T = 400 K and P = 1 atm. Then, scale the length of the repeat unit along the backbone, all intermolecular interactions, etc., of b-polest-1/s-polest-2 by the ratio r_d of a new interdigitated side-chain separation distance x_{di} to the original interdigitated side-chain separation distance x_{d1}.)

Table I. Theoretical values of the backbone P_2, side-chain P_2, and D_b of SS LCPs that are scaled versions of b-polest-1/s-polest-2

r_d	x_d (Å)	back P_2	side P_2	D_b ($\times 10^{-6}$ cm^2 sec^{-1})
1.000	5.32	0.991	0.886	2.56
0.846	4.50	0.965	0.980	0.14

As x_d decreases in Table I, the side-chain P_2 (and thus, the compressive strength) increases, but the diffusion coefficient D_b decreases (that is, the SS LCP molecules are slower to rearrange, and thus the SS LCP is harder to process). Thus, as x_d varies, there are opposing trends in (1) side-chain ordering (thus, compressive strength) and (2) diffusion coefficient (thus, ease of molecular rearrangement for processing). That is, the compressive strength is increased at the expense of the ease of molecular rearrangement needed for processing.

Note that although there is considerable variation in the magnitudes of the diffusion coefficients for the SS LCPs in Table I, the diffusion coefficients for these SS LCPs are reasonable compared with the diffusion coefficients for nonpolymeric LC molecules and non-LC polymeric molecules presented earlier in this paper.

The calculations with the dynamic theory of this paper also predict that an SS LCP molecule diffuses faster when it diffuses in a plane between (and parallel to) the planes of the other molecules, as indicated in Figure 1(e).

As x_d decreases at constant dp in Table I, the backbone P_2 decreases; the backbone is shorter and thus orientationally orders less.

Substituting the group -NH- for the group -O- in the backbone of the SS LCP b-polest-1/s-polest-2 to form the theoretically designed candidate SS LCP b-PBA-3/s-polest-2 [see Figure 3(a)] has almost no effect on the backbone P_2 and side-chain P_2, as can be seen by comparing the results in Table I for $r_d = 1$ with the following results: For b-PBA-3/s-polest-2 with dp = 100 at T = 400 K and P = 1 atm, the backbone $P_2 = 0.992$ and the side-chain $P_2 = 0.886$.

Melt Processability of Some SS LCPs. As discussed in earlier papers (4-7,10), some of the theoretically designed candidate SS LCPs are designed to be processed in solution, and some to be processed in the melt (pure state). As discussed in References 4 and 10, the inclusion of semiflexible sections [such as $(-CH_2-)_x$, with $x \leq 6$] and/or of slightly nonlinear rigid sections [such as the HNA group, meta-bonded (instead of para-bonded) phenyl rings, etc.] in SS LCPs will tend to disrupt the ordering of linear rigid sections somewhat and thus will lower the melting temperature below the temperature of the onset of chemical decomposition. [Recall that Celanese's Vectra (37) is an example of including the HNA group to obtain a melt-processable backbone LCP.]

Head-to-Tail Polymerization Predicted for SS LCPs. In polymerizing monomers of the theoretically designed candidate SS LCPs [presented in this paper and in earlier papers (3-7,9-11)] to form the polymer molecules, the state of lowest free energy occurs if the monomers interdigitate as they polymerize and if the spacing between the chains is regular. [Note that the minimization of the free energy includes the optimization of entropy effects, pressure-volume effects, and energetic effects (including repulsive and attractive forces).]

The presence of the bulky side chains with relatively rigid sections having long axes more or less perpendicular to the long axis of the backbone in SS LCP monomers will very strongly favor interdigitation of the side chains and head-to-tail polymerization of the monomers in order to achieve the interdigitated state of lowest free energy for the polymer molecules [see Figure 3(e)]. Thus, head-to-head and tail-to-tail defects in the polymerization of the SS LCP monomers are very unfavorable from the standpoint of minimizing the free energy and thus are very unlikely to occur in SS LCPs. Although such defects can reduce the regularity of spacing of the side chains and thus the uniformity of compressive strength on a submolecular level, some head-to-head and tail-to-tail defects can be tolerated in a SS LCP material.

Acknowledgments

This research was supported by the U. S. Department of Energy, Office of Basic Energy Sciences, Division of Materials Sciences.

Literature Cited

1. See, for example, Haward, R. N. The Strength of Plastics and Glass; Interscience: New York, 1949; pp. 12-16.
2. See, for example, Houwink, R.; de Decker, H. K.; van den Tempel, M. In Elasticity, Plasticity, and Structure of Matter; Houwink, R. R.; de Decker, H. K., Eds.; Cambridge University Press: London, 1971, 3rd ed.; p. 12.

3. Dowell, F. J. Chem. Phys. 1989, 91, 1316.
4. Dowell, F. J. Chem. Phys. 1989, 91, 1326.
5. Dowell, F. Polymer Preprints 1989, 30 (2), 532.
6. Dowell, F. In Industry-University Advanced Materials Conference II; Smith, F. W., Ed.; Advanced Materials Institute, Colorado School of Mines: Golden, 1989; p. 605.
7. Dowell, F. In Materials Science and Engineering of Rigid Rod Polymers, Proceedings of the Materials Research Society, Vol. 134; Adams, W. W.; Eby, R. K.; McLemore, D. E., Eds.; Materials Research Society: Pittsburgh, 1989; p. 33.
8. Dowell, F. In Materials Science and Engineering of Rigid Rod Polymers, Proceedings of the Materials Research Society, Vol. 134; Adams, W. W.; Eby, R. K.; McLemore, D. E., Eds.; Materials Research Society: Pittsburgh, 1989; p. 47.
9. Dowell, F. In Nonlinear Structures in Physical Systems: Pattern Formation, Chaos, and Waves; Lam, L.; Morris, H. C., Eds.; Springer-Verlag: New York, 1990; p. xx.
10. Dowell, F. Adv. Chem. Phys. 1990, xx, xxx.
11. Dowell, F. J. Stat. Phys. 1990, xx, xxx.
12. Dowell, F. Mol. Cryst. Liq. Cryst. 1988, 157, 203.
13. Dowell, F. Mol. Cryst. Liq. Cryst. 1988, 155, 457.
14. Dowell, F. Phys. Rev. A 1988, 38, 382.
15. Dowell, F. In Competing Interactions and Microstructures: Statics and Dynamics; LeSar, R.; Bishop, A.; Heffner, R., Eds.; Springer-Verlag: Berlin, 1988; p. 177.
16. Dowell, F. Phys. Rev. A 1987, 36, 5046.
17. Dowell, F. Phys. Rev. A 1985, 31, 3214.
18. Dowell, F. Phys. Rev. A 1985, 31, 2464.
19. Dowell, F. Phys. Rev. A 1983, 28, 3526.
20. Dowell, F. Phys. Rev. A 1983, 28, 3520.
21. Dowell, F. Phys. Rev. A 1983, 28, 1003.
22. Dowell, F. J. Chem. Phys. 1978, 69, 4012.
23. Dowell, F.; Martire, D. E. J. Chem. Phys. 1978, 68, 1094.
24. Tables of Interatomic Distances and Configuration in Molecules and Ions, Supplement 1956-1959, Special Publication No. 18; Sutton, L. E.; et. al., Eds.; Chemical Society: London, 1965; pp. S3s-S23s.
25. McCrackin, F. L. J. Chem. Phys. 1978, 69, 5419.
26. Hirschfelder, J. O.; Curtiss, C. F.; Bird, R. B. Molecular Theory of Gases and Liquids; Wiley: New York, 1964; (a) p. 984 and (b) p. 27.
27. Alben, R. Mol. Cryst. Liq. Cryst. 1971, 13, 193.
28. Kwolek, S. L.; Morgan, P. W.; Schaefgen, J. R.; Gulrich, L. W. Macromolecules 1977, 10, 1390.
29. Schaefgen, J. R.; Foldi, V. S.; Logullo, F. M.; Good, V. H.; Gulrich, L. W.; Killian, F. L. Polymer Preprints 1976, 17 (1), 69.
30. Kwolek, S. L. Private communication.
31. Schaefgen, J. R. Private communication.
32. Uhlenbeck, G. E.; Ornstein, L. S. Phys. Rev. 1930, 36, 823.
33. Chandrasekhar, S. Rev. Mod. Phys. 1943, 15, 1.
34. Polymer Handbook; Brandrup, J.; Immergut, E. H., Eds.; Wiley: New York, 1975, 2nd ed.; p. IV-68.

35. Kruger, G. J. Physics Reports 1982, 82, 230.
36. Dowell, F. Patents filed.
37. East, A. J.; Charbonneau, L. F.; Calundann, G. W.
     Mol. Cryst. Liq. Cryst. 1988, 157, 615.

RECEIVED April 24, 1990

Chapter 25

Molecular Motion in the Homopolyester of 4-Hydroxybenzoic Acid

J. R. Lyerla[1], J. Economy[2], G. G. Maresch[1,3], A. Mühlebach[4], C. S. Yannoni[1], and C. A. Fyfe[5]

[1]IBM Research Division, Almaden Research Center, 650 Harry Road, San Jose, CA 95120–6099
[2]University of Illinois, Urbana, IL 61801
[3]Max Plank Institute für Polymerforschung, Postfach 3148, D–6500 Mainz, Federal Republic of Germany
[4]CIBA–Geigy AG, Forchungszentrum, 180.053, CH–1701, Fribourg, Switzerland
[5]Department of Chemistry, University of British Columbia, Vancouver, British Columbia V6T 1Y6, Canada

Proton and ^{13}C NMR have been utilized to characterize the motional processes occurring in the homopolymer of 4-hydroxybenzoic acid, PHBA, over the temperature range 27-400°C. Changes in the line shape of the ring protons and of the chemical shielding anisotropy pattern of the carboxyl carbon provide a picture of motion in PHBA in which: 1) phenyl ring motion in the polymer occurs about the C_1 - C_4 axis well-below the 350°C phase transition and thus is not tied to the transition; 2) the onset of motion for some of the rings occurs at ca. 120°C while all the rings show motion (180° ring flipping) by 300°C; 3) the motion of the ester group only ensues with the phase transition and involves occurrence of a second motion about the chain axis, again involving jumps by ca. 180° but with the entire repeat unit participating in the motion.

The homopolyester of 4-hydroxybenzoic acid and its copolyesters with 6-hydroxy-2-naphthoic acid and with 4,4'-biphenol and terephthalic acid are of continuing commercial and academic interest. (1,2) From the applications side, interest stems from the high moduli and high temperature properties of these polymers, while on the fundamental side, interest arises from the main-chain thermotropic liquid crystalline

0097–6156/90/0435–0359$06.00/0

character of these polymers. The homopolymer, PHBA, is itself a highly crystalline material that has been the subject of a number of structural investigations by X-ray and electron-diffraction. (3-8) Of particular interest has been the nature of the reversible DSC transition at ca. 350°C (Figure 1). The diffraction data, taken below and above the transition, are described in terms of a transition from an orthorhombic unit cell to a "pseudo-hexagonal" structure. Recent high-temperature X-ray diffraction data[6] show that the unit cell expands significantly along one direction (a-axis) perpendicular to the chain axis (c-axis) between -100°C and 340°C. At the transition, the unit cell volume increases sharply (ca. 9.5%) (7) due to further expansion perpendicular to the chain axis. The reduced density of the high temperature phase suggests that torsional degrees of freedom may be introduced for the phenyl and ester groups at the transition. Indeed, the interpretation derived from recent X-ray studies (6-8) suggests that the "pseudo-hexagonal" phase has a 180° rotationally degenerate, disordered structure. However, the X-ray results do not distinguish whether the disorder is dynamic or static. Thus, to examine directly the question of whether molecular motion accompanies the 350°C phase transition in PHBA, we have carried out solid state proton and ^{13}C NMR lineshape measurements to probe the motions of the phenyl ring and the carboxyl carbon units of this homopolyester in the temperature range of 27 - 400°C.

Experimental

The homopolyester, enriched to 60% in ^{13}C at the carboxyl carbon, was prepared from labelled acetoxy-benzoic acid monomer as described by Mühlebach, et al.. (9) The polymer was characterized by DSC, TMA, and X-ray diffraction. Number-average molecular weight was determined to be >30K from the ^{1}H - NMR spectrum of hydrolyzed polymer (method of Kricheldorf and Schwarz). (10)

    Solid state NMR experiments were carried out on a spectrometer (described previously) (11) operating at 60 MHz for proton observation and 15.1 MHz for ^{13}C observation. Broad-line proton spectra were obtained using a standard $\pi/2$ pulse sequence while proton-decoupled ^{13}C spectra were obtained by standard cross-polarization (CP) techniques. Temperature dependent spectra were obtained in a range of 27 - 400°C using an inductive heating method that employs NMR tubes coated with a thin film of metal (Pt) as described by Maresch et al.. (12) Temperature was measured with thermocouples placed inside the sample. Simulation of the chemical shielding anisotropy (CSA) patterns in PHBA was carried out to derive the values of the principal elements of the shielding tensor at each temperature. Because of the magnitude of the [13]C - [13]C dipolar interactions in the enriched sample, this source of non-averaged dipolar broadening affects the CSA pattern. A nutation experiment (13) was

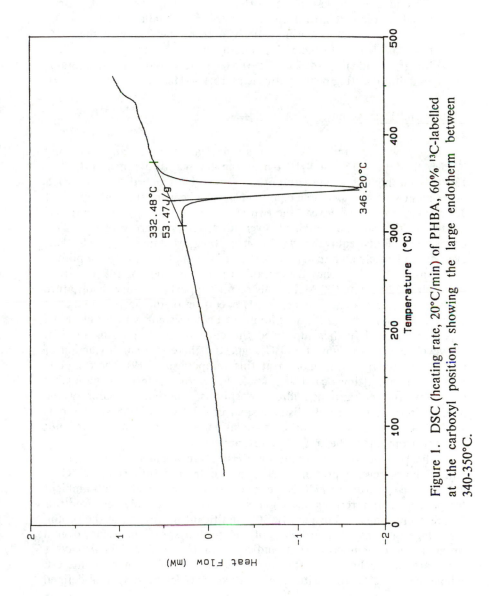

Figure 1. DSC (heating rate, 20°C/min) of PHBA, 60% ^{13}C-labelled at the carboxyl position, showing the large endotherm between 340-350°C.

carried out at 27°C to determine the magnitude of the broadening which was found to be ca. 440 Hz. Only upon inclusion of the broadening is satisfactory fitting of the patterns obtained. A CSA pattern has been obtained at high field (100 MHz for ^{13}C observation) (14) where the dipolar broadening has less effect on the pattern shape. An axial pattern is observed at room temperature having the same values found for the elements of the shielding tensor as derived from the fitting of the 27°C pattern at 15.1 MHz. In the high temperature phase, inclusion of a ^{13}C - ^{13}C dipolar broadening of 240 Hz (having been reduced by motion) provides suitable fitting of the patterns at 15.1 MHz.

Results and Discussion

Previously, we have reported ^{13}C cross-polarization, magic-angle spinning (CPMAS) data (9,15) on PHBA over the temperature range -196°C to 130°C (the highest temperature that could be reached in the MAS experiment with the available equipment). A chemical shift difference of 3.2 ppm is observed between the two ring carbons ortho to the carboxyl group in the spectra obtained over this temperature range. This persistence of non-equivalency of ring carbon resonances suggests that, even at moderately high temperature, there is no motion of large amplitude occurring with a frequency comparable to the splitting, ca. 48 Hz at 15 MHz and 320 Hz at 100 MHz. This result supports the very rigid nature of this highly crystalline polymer. A ^2H spectrum of the polymer (labelled at the 3,5 positions) at 25°C was found to be consistent with a rigid Pake spectrum (9) and further supported the carbon results on ring motion. However, a ^2H spectrum at 140°C suggested there is a small fraction of rings that undergo rapid motion at this temperature. (9) These mobile rings (correlation times ca. 10^{-7}s), could arise from defects in the crystal structure, from end groups and/or amorphous material. Apparently, the mobile rings are not readily distinguished in the ^{13}C-CPMAS spectrum at 130°C because they do not cross-polarize efficiently and/or are not resolved from under the crystalline resonances.

The previous NMR results on PHBA are augmented in Figure 2 which shows several proton spectra in the temperature range 27°C to 400°C. The spectrum at 27°C is the same as those observed at sub-ambient temperatures and reflects the rigid structure of the polymer over this temperature range. Above ca. 120°C, significant narrowing of the proton resonance line ensues. Narrowing of the proton spectrum occurs over a broad range of temperatures as indicated in Figure 3 which displays a summary of the linewidth data over the temperature range of interest. Above ca. 260°C, the features of a Pake doublet become well-defined

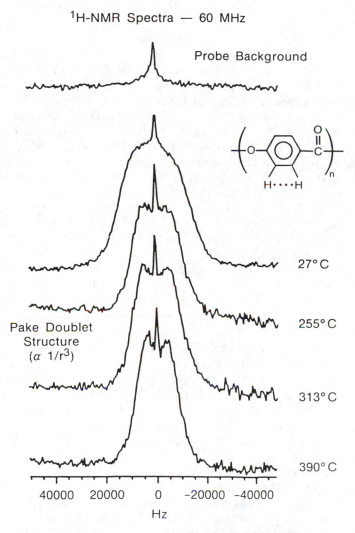

Figure 2. Proton spectra at 60 MHz of PHBA at various temperatures. Note, the probe has a background proton signal (as shown) and this gives rise to the sharp center line seen in the spectra.

(16) (e.g. in the 313° C spectrum in Figure 2). The observed splitting of the doublet is ca. 12 kHz and is consistent with a spatial separation between interacting protons of .25 nm. This value is the separation of adjacent protons on the phenyl ring of the repeat unit which are also the protons that give rise to the dominant ^{1}H - ^{1}H dipolar interaction in the polymer. Since the Pake pattern persists in the presence of the line narrowing, the molecular motion giving rise to line narrowing must occur about an axis (nearly) parallel to the two-fold axis of the phenyl unit. This motion accounts for the observed overall narrowing (Figure 2) in that it reduces intermolecular dipolar interactions and cross-ring and ring-ring intramolecular dipolar interactions as the time-scale of the motion becomes fast relative to the magnitude of these interactions (ca. 1 - 10 kHz). The proton data establish directly that the phenyl rings in PHBA have large amplitude motion in the mid-kHz region well below the 350°C DSC transition; however, the data cannot be used to distinguish whether the motion is in the form of 180° jumps or continuous rotation. Preliminary high temperature ^{2}H-NMR results (9,14) demonstrate that the motion is in the form of 180° flips about the C_1 - C_4 axis of the ring and is augmented by ring oscillations. Also, the ^{2}H data provide definite evidence of heterogeneity in the ring motion in that the spectra consist of composites of lineshapes from rings both fast and slow on the ^{2}H timescale. These data on motional heterogeneity support a similar finding in the proton data - i.e. while we have characterized the proton spectra by their fullwidth at half-height, the spectra in the temperature range 120° - 260°C are best represented by a composite of rings whose motion appears as rigid, intermediate, or fast on the relevant proton dipolar timescale (ca. 100 μsec). However, by 300°C, the entire population of rings are in rapid motion on the proton dipolar timescale.

The question arises as to whether the carboxyl group moves in concert with the phenyl ring reorientation. Insight is provided by the carboxyl carbon chemical shift anisotropy (CSA) patterns shown in Figure 4. ^{13}C enrichment of this carbon allows it to be isolated in the carbon spectrum - the ring carbon resonances, at natural abundance, being at the level of the noise in the spectrum. The spectrum observed at 27°C (Figure 4) is characteristic of the CSA pattern below the transition. In fact, the same spectrum is observed from -196 to 320°C and such a large range of temperature independence indicates that the axial symmetry observed is not due to any motional averaging, but to a fortuitous coincidence of two elements of the chemical shielding tensor. Above 320°C and through the phase transition, narrowing of the CSA pattern occurs and deviation from axial symmetry develops until above the transition a narrowed, non-axial pattern evolves (Figure 4) which is unchanged up to 400°C. The onset of

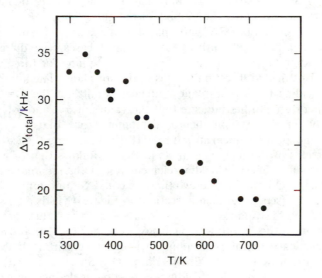

Figure 3. Proton linewidths (full-width at half-maximum) in PHBA as a function of temperature.

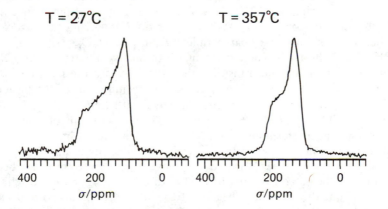

Figure 4. ^{13}C spectrum at 15.1 MHz of the carboxyl carbon in PHBA at 27°C and 357°C obtained by cross-polarization techniques. Note the change of width and shape of the CSA pattern.

significant narrowing of the CSA pattern occurs above 340°C which provides direct evidence that motion of the carboxyl unit only occurs with the phase transition (in accord with recent dielectric data in the region of the 350°C DSC transition) (7) and that the motion, at much lower temperatures, involves only the flipping of the phenyl ring.

Literature data on CSA patterns of static carbonyl and carboxyl carbons (17-19) have established that, in most cases, the direction of greatest shielding (σ_{33}) is perpendicular to the sp^2 plane for these carbons while the least shielded direction (σ_{11}) is in the sp^2 plane. The intermediate value of shielding (σ_{22}) is in-plane and near-parallel to the C=O bond axis. The literature studies indicate that the values of σ_{33} and σ_{11} do not vary greatly for a large number of compounds but that σ_{22} is very sensitive to the local chemical structure. (18,19) Assuming the assignment of the directions of the tensor elements in PHBA follows that of a typical carboxyl carbon, then the shielding element out of the sp^2 plane and the one near the C=O bond are the two that are equal. Simulation of the CSA pattern at 27 °C (see Experimental section) in Figure 4 yields a tensor with elements: $\sigma_{11} = 258$ ppm; $\sigma_{22} = \sigma_{33} = 116$ ppm. The values are typical of those found in the literature for carboxyl carbons (17-19) and yield an isotropic chemical shift of 163.3 ppm, in good agreement with the value of 162.4 ppm reported in the CPMAS data. (15) Since ($\sigma_{33} - \sigma_{22}$) < ($\sigma_{22} - \sigma_{11}$), by convention, $\Delta\sigma$, the shielding anisotropy, is defined as $\Delta\sigma = \sigma_{11} - 0.5(\sigma_{22} + \sigma_{33})$. This gives a value of 142 ppm for $\Delta\sigma$ below the transition.

The fact that: 1) motion of the carboxyl unit independent of the phenyl ring is unlikely and 2) the proton Pake pattern shows only a small amount of narrowing (ca. 10% in the magnitude of the splitting) above the 350°C transition strongly suggests that the motion involving the carboxyl unit occurs about an axis near-parallel to the phenyl C_1 - C_4 axis and thereby the chain axis. The change in width and symmetry of the CSA pattern accompanying the transition allows insight into the nature of the motion involving the carboxyl unit. Satisfactory simulation of the pattern at 357°C (Figure 4) requires a non-axial tensor with elements: $\sigma'_{11} = 227$ ppm; $\sigma'_{22} = 145$ ppm; $\sigma'_{33} = 118$ ppm. The development of non-axial symmetry in the pattern suggests that the motion occurs in the form of jumps rather than rapid rotation which would maintain the axial symmetry. Indeed, the X-ray data above the transition have been interpreted as being consistent with a structure involving motion of chain segments about the chain axis between two sites involving a two-fold axis of reorientation through an angle of 180°. A 180° jumping motion about the chain axis would leave any shielding element perpendicular to this axis unchanged in value with the onset of motion - which is the result observed (within experimental error of the fitting of the CSA pattern) for σ_{33}, the element perpendicular to the carboxyl plane. The values of σ'_{11} and σ'_{22} combined with the static values of the tensor can be used to calculate the angle, α, between the jump (chain) axis and σ_{11}. The calculation yields a value of 27°. Using the geometries for PHBA from Coulter and Windle (20) and the assignment of the carboxyl shielding tensor due to Pines et al., (18), this value of α implies that σ_{22} would lie in the sp^2 plane ca. 9° off the parallel to the C=O bond axis. This value compares very well with those found for this geometric relationship of σ_{22} in esters (18) and amides (peptides) (21) which range from 0 to 14°.

The expression for the narrowing of the CSA pattern when motion is in the form of rapid rotation is given by (17)

$$\Delta\sigma' = 0.5(3\cos^2\beta - 1)\Delta\sigma \qquad (1)$$

where β is the angle between the rotation axis and the unique axis of the axially symmetric shielding tensor and $\Delta\sigma$ is the static anisotropy. Using the assignment of the tensor given above, σ_{11}, the unique tensor element, would lie at ca. 27° relative to the chain axis. For this angle, eq. 1 yields a value of $\Delta\sigma'$ of 98 ppm in the fast motion limit. While this value is in reasonable agreement with the experimental value of width of 94 ppm, the axial symmetry of the CSA pattern for rotation is, as mentioned, inconsistent with the observed pattern. In addition, free rotation is inconsistent with the interchain packing of phenyl groups as edge to face in the "pseudo-hexagonal" structure above 350°C based on the X-ray data. Thus, the reorientational motion accompanying the phase transition is best described as a jumping motion, not a free rotation, about an axis near-parallel to the chain axis. (7)

Conclusions

In summary, the NMR lineshape data provide the following insight into the details of motion accompanying the 350°C DSC transition in PHBA: 1) phenyl ring motion in the polymer occurs about the C_1 - C_4 axis well-below this transition and thus is not associated with the transition; 2) the onset of motion for some of the rings occurs at ca. 120°C while all the rings show motion (180° ring flipping) rapid on the proton dipolar timescale by 300°C; 3) the motion of the ester group only ensues with the transition and involves occurrence of a second motion about the chain axis, again involving jumps by ca. 180° but with the entire repeat unit participating in the motion. Below the 350°C transition in PHBA, the lack of motion except for phenyl ring flips supports the existence of strong interchain interactions. These interactions are then reduced with the lattice expansion accompanying the transition so that a 180° jumping motion about the chain axis, rather than free or random rotation, is facilitated for chain segments above the transition. (7,8)

Literature Cited

1. Economy, J. J. Macromol. Sci.-Chem. 1984, A21, 1705.
2. Davies, G. R.; Ward, I. M. High Modulus Polymers, Zachariades, A. E.; Porter, R. S. Eds. Marcel Dekker, Inc., N. Y., 1988, Chapter 2.
3. Lieser, G. J. Polym. Sci., Polym. Phys. Ed. 1983, 21, 1611.
4. Economy, J.; Volksen, W.; Geiss, R. H. Mol. Cryst. Liq. Cryst. 1984, 105, 289.
5. Geiss, R. H.; Street, G. B.;Volksen, W.; Economy, J., IBM J. Res. and Develop. 1983, 27, 321.
6. Hanna, S.; Windle, A. H. Polymer Comm. 1988, 29, 236.
7. Yoon, D. Y.; Masciocchi, N.; Depero, L. E.; Viney, C.; Parrish, W. Macromolecules. 1990, 23, 1793.
8. Coulter, P. D.; Hanna, S.; Windle, A. H. Liquid Crystals. 1988, 5, 1603.
9. Mühlebach, A.; Economy, J.; Lyerla, J. R.; Yannoni, C.; Facey, G.; Fyfe, C. A.; Geis, H. Polym. Preprints. 1988, 29(1), 40.
10. Kricheldorf, H. R.; Schwarz, G. Makromol. Chem. 1983, 184, 475.
11. Lyerla, J. R.; Yannoni, C. S.; Fyfe, C. A. Acc. Chem. Res. 1982, 15, 208.
12. Maresch, G.; Kendrick, R. D.; Yannoni, C. S. Rev. Sci. Instru. 1990, 14, 2207.
13. Yannoni, C. S.; Kendrick, R. D. J. Chem. Phys. 1982, 74, 747.
14. Fyfe, C. A.; Geis, H.; Economy, J.; Muehlebach, A.; Neissner, N.; Lyerla, J. R.; to be published.
15. Fyfe, C. A.; Lyerla, J. R.; Volksen, W.; Yannoni, C. S. Macromolecules. 1979, 12, 759.
16. Pake, G. E. J. Chem. Phys. 1948, 16, 327.
17. Mehring, M. High Resolution NMR in Solids. 1983, 2nd ed. Springer, Berlin, Chapter 7.
18. Pines, A.; Chang, J. J.; Griffin, R. G. J. Chem. Phys. 1974, 61, 1021 and references therein.
19. Jagannathan, N. R. Magn. Reson. in Chem. 1989, 27, 941.
20. Coutler, P.; Windle, A. H. Macromolecules. 1989, 22, 1129.
21. Oas, T. G.; Hartzell, C. J.; McMahon, T. J.; Drobny, G. P.; Dahlquist, F. W. J. Am. Chem. Soc. 1987, 109, 5956.

RECEIVED April 24, 1990

APPLICATIONS OF LIQUID-CRYSTALLINE POLYMERS: RHEOLOGY AND PROCESSING BEHAVIOR

Chapter 26

Effect of Shear History on the Rheological Behavior of Lyotropic Liquid Crystals

P. Moldenaers, H. Yanase[1], and J. Mewis

Department of Chemical Engineering, Katholieke Universiteit Leuven, de Croylaan 46, B–3030 Leuven, Belgium

Two liquid crystalline polybenzylglutamate solutions, adjusted to the same Newtonian viscosity, have been investigated rheologically. The steady state shear properties and the transient behaviour are measured. For the same kind of polymer, the dynamic moduli upon cessation of flow can either increase or decrease with time. This change in dynamic moduli shows a similar dependency on shear rate as the final portion of the stress relaxation but no absolute correlation exists between them. By comparing the transient stress during a stepwise increase in shear rate with that during flow reversal the flow–induced anisotropy of the material is studied.

During the last ten years the interest in polymeric liquid crystals (PLCs) has been growing rapidly. Nevertheless our fundamental understanding of their flow behaviour is still rather limited. This is due to the fact that PLC rheology is much more complicated than that of ordinary isotropic polymeric fluids (1). Systematic and reliable data are lacking so far although this is the kind of information needed for the development and assessment of theoretical models for these unusual fluids.

The purpose of this paper is to explore various aspects of the rheological behaviour of lyotropic liquid crystalline systems. Lyotropics are often used as model systems for thermotropics because their viscoelastic behaviour seems to be quite similar (1) and solutions are much more easier to handle and can be studied more accurately than melts. The emphasis is on transient data as these are essential for verifying viscoelastic models but are hardly available in the literature. Transient experiments can also provide insight in the development of flow–induced orientation and structure. The reported experiments include relaxation of the shear stress and evolution of

[1]Current address: Kyoto University, Kyoto 606, Japan

0097–6156/90/0435–0370$06.00/0

the dynamic moduli upon cessation of flow as well as flow reversal and stepwise changes in shear rate.

Experimental

Two samples of poly(γ–benzylglutamate) in *m*–cresol have been used. Polybenzylglutamates (PBGs) have proven to be useful as model systems to demonstrate characteristic aspects of PLC rheology (2–4). The first sample under investigation consists of a solution of 12%, by weight, of PBLG (MW = 250000). The second one is a solution of PBDG (concentration 25%, MW = 310000). In both cases the concentration is high enough to ensure a fully liquid crystalline phase. At rest the materials are cholesteric, but during flow they develop into a nematic structure. For convenience they will be referred to as the PBLG and the PBDG sample although the relevant differences between the two samples are their concentration and molecular weight. No difference in the rheological behaviour, related to the L and the D form of the glutamates, is expected. The basic rheology of the PBLG sample under investigation has already been studied extensively (4–6).

The rheological experiments were performed on a Rheometrics Mechanical Spectrometer RMS 705F. Cone and plate geometry was used in order to generate a homogeneous shear history throughout the sample, a prerequisite in order to analyse transient behaviour. All experiments have been performed at 293K.

With liquid crystalline materials the flow can always be affected by the gap size or the measurement geometry. This has been shown not to be the case for equilibrium viscosities of PLCs (2). For transients the situation could be different because of propagation effects from the wall. Here two types of transients have been investigated for this purpose. Figure 1 indicates that the stress transients for stepwise changes in shear rate remain identical when the cone angle is doubled. Results independent of gap size have also been obtained for the evolution of the dynamic moduli after cessation of flow (7) (see below). The latter result is in contrast with available results for the relaxation of birefringence after cessation of flow. The time scale for the final part of this curve has been reported to depend on the gap size (3). For smaller gaps a slower optical relaxation was measured. Ar any rate, it can be concluded that, for the range of gap sizes used here, the rheological data are not affected. Hence an Ericksen number based on the gap size as the characteristic length scale will fail to scale the results (Burghardt, W. R.; Fuller, G. G., in press).

Equilibrium Results

Due to their intrinsic physical nature one expects liquid crystals to exhibit a yield stress in the zero shear limit. This yield stress, associated with region I in the three–region flow curve of Onogi and Asada (8), has been reported for various types of liquid crystalline materials. Figure 2 shows the steady shear flow results for the two samples. Because of the possibility of a yield stress, shear stresses were determined here by taking the average of a clockwise and a counterclockwise experiment. For the PBLG sample under investigation no indication of an upturn of the viscosity curve at the low shear rates could be detected. For the PBDG sample on the other hand, a small increase of the viscosity seems to occur in this region. However, an eventual yield stress will be very small for both samples and does not interfere with the experiments reported here.

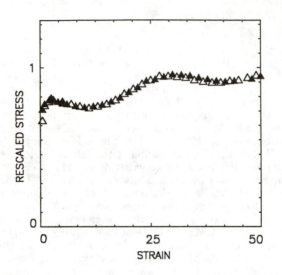

Figure 1. Effect of gap size on step–up transients for the PBLG sample;
(cone angle: $\triangle$: 0.02 rad; $\blacktriangle$: 0.04 rad)
($\dot{\gamma}_i = 0.05$ (1/s); $\dot{\gamma}_f = 0.5$ (1/s))

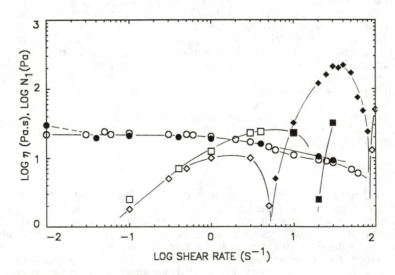

Figure 2. Steady shear flow results at 293 K;
(Viscosity: $\bigcirc$= PBLG sample; $\bullet$= PBDG sample),
(Positive N_1 : $\diamond$= PBLG; $\square$= PBDG; negative N_1 : $\blacklozenge$= PBLG; $\blacksquare$=PBDG)

The two polybenzylglutamates differ in molecular weight. Their concentrations have been adjusted to give the same Newtonian viscosity. In the Newtonian shear rate region their normal stresses turn out to be identical as well. A comparison of the two samples might shed some light on the role of the viscosity as such in controlling the time scales of some transient phenomena. For the PBLG sample the concentration is about 1.5 times the critical concentration c*. For the PBDG the concentration reaches about 4c*. This means that the PBLG sample is in a concentration region where viscosity decreases with concentration whereas it might be increasing again for the PBDG (9). The first normal stress difference has also been reported to be a decreasing and subsequentely increasing function of concentration for PLCs in the low shear limit (2). The equality of viscosity and first normal stress difference does of course not imply that the order parameter is the same for the two samples.

The first normal stress difference exhibits a linear dependency on the shear rate in the region of constant viscosity for the two solutions in Figure 2. This proportionality is predicted by the Doi theory (10) and the Leslie–Ericksen theory (11) although the basic assumption in these theories, i.e. a monodomain structure, is not satisfied.

Another prominent feature in Figure 2 is the occurrence of a negative first normal stress difference in an intermediate shear rate region for both samples. Negative normal stresses have been reported repeatedly for various liquid crystalline systems, including thermotropic ones (2, 12, 13). The critical shear rate and the critical shear stress at the transition from a positive to a negative first normal stress difference (Figure 2) are considerably larger for the PBDG sample than for the PBLG one. This is in agreement with the data of Kiss and Porter (2, 14) who reported an increasing critical shear rate and critical shear stress with increasing concentration and molecular weight. The molecular theory, as originally presented by Doi (10), does not include the possibility of negative normal stresses in a shear flow. However, recently Marrucci and Maffettone (15) demonstrated with a two–dimensional analysis the existence of a shear rate region with negative normal stresses by using the total orientational distribution function in their calculations.

Stress Relaxation upon Cessation of Flow

For isotropic polymer fluids, stress relaxation upon cessation of flow reflects the relaxation time scales during the previous flow. As the relaxation spectrum is determined by the microstructure such measurements can be used to probe the effect of shear on the structure. This turns out to be a rather insensitive technique in polymer fluids because of changes which already occur during the relaxation (16).

The liquid crystalline PBLG sample is characterized by a stress relaxation which depends on shear rate, even in the Newtonian region (Moldenaers, P.; Mewis, J. J. Non–Newtonian Fluid Mech., in press). This proves the existence of a shear rate dependent structure in the linear shear rate region. It was also found that the stress relaxation curve could be divided in two different sections. The temperature dependence of the initial part scales with the viscosity and does not depend on shear rate in the Newtonian region. The second part does not depend on temperature but scales with the inverse of the previous shear rate.

The effect of the shear rate on the stress relaxation for the PBLG sample is shown in Figure 3 for the Newtonian as well as the non–Newtonian

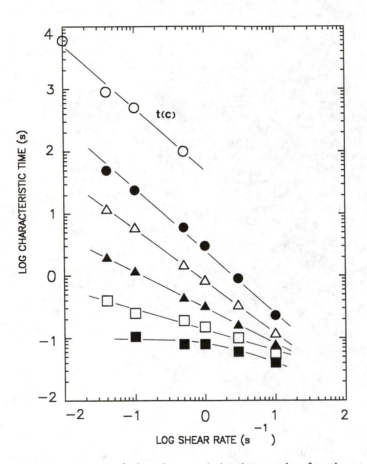

Figure 3. Comparison of the characteristic time scales for the stress relaxation and the evolution of the dynamic moduli (t(c)) upon cessation of flow for the PBLG sample;
(% stress relaxation: ■: 50%; □: 60%; ▲: 70%; △: 80%; ●: 90%)

shear rate region. The relaxation behaviour of the more concentrated PBDG solution is displayed in Figure 4. Qualitatively the relaxation behaviour of the two mesomorphic materials is similar. In particular the tail of the relaxation curve changes inversely proportional to the previous shear rate, whereas the initial part tends to become independent of shear rate. However, the final relaxation takes much longer in the more concentrated PBDG sample than for the PBLG one, notwithstanding their identical Newtonian viscosities. The data could be explained by a molecular reorientation mechanism for the fast part and a change in supermolecular structure, domains or defects, for the slow part.

A comparison of the present results with the rheo–optical relaxation data, reported by Asada et al. for similar materials (3) seems indicated. These authors found that the relaxation time of the birefringence increases inversely proportional to the previous shear rate and proportionally to the third power of the molecular weight. Concentration did hardly affect the rheo–optical time scale at identical stress levels. For the molecular weight of the two samples under investigation here, one would thus expect a time constant for the PBDG sample which is twice that of the PBLG sample. The tail of the stress relaxation curves in Figures 3 and 4 yields a ratio of about 3, quite comparable with that of the optical relaxation. Thus both phenomena could be governed by the same mechanisms.

Oscillatory Testing upon Cessation of Flow

The evolution of the structure of a material upon cessation of flow can be probed by several techniques. Measuring the variation of the linear dynamic moduli has proven to be useful in this respect. For the PBLG sample both G' and G" decrease monotonically upon cessation of flow (4). Moreover, the time scale of this effect changes inversely proportional to the previous shear rate, even in the Newtonian region. Figure 5 compares the evolution of G" for the PBLG and the PBDG sample at a frequency of 10 rad/s. The previous shear rate amounted to 3 reciprocal seconds for both materials. The moduli continue to change over an extensive period of time, suggesting a slow evolution from the flow–induced structure to the zero–shear one. The inverse proportionality with shear rate was also found in elastic recovery following cessation of flow (17).

There are also some striking differences between the two samples. For the PBLG solution, having the lowest molecular weight and concentration, the moduli decrease monotonously. On the contrary, the PBDG sample, with the higher concentration and molecular weight, displays an initial increase of the moduli, followed by a subsequent decrease. This difference was found to persist over a wide range of shear rates but the maximun for the PBDG solution tends to become less prominent at low shear rates. It can not be excluded that the moduli of the PBLG sample might also display a maximum for the dynamic moduli but then after an immeasurably short time, although measurements at low shear rates, where the changes become sufficiently slow, show a flat plateau rather than a maximum.

A second difference is more quantitative. The time over which the moduli evolve is considerably larger for the PBDG sample than it is for the PBLG sample. In Figure 3 the characteristic time (t(c)) for the change of the moduli in the PBLG sample as a function of shear rate is also included. This kinetic factor has been defined as the time after which the moduli have completed one third of their total decay. It is concluded from Figure 3 that t(c) is larger than the average relaxation time but displays the same shear

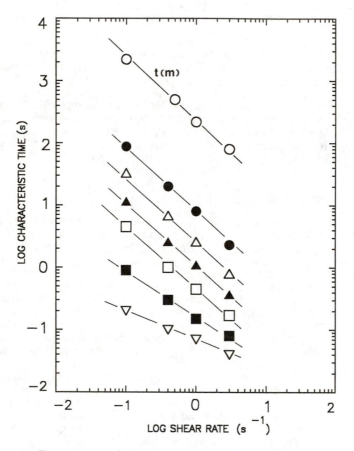

Figure 4. Comparison of the characteristic time scales for the stress relaxation and the evolution of the dynamic moduli (t(m)) upon cessation of flow for the PBDG sample;
(% stress relaxation:▽ : 40%; other symbols as in Figure 3).

rate dependency. The decay time could not be accurately determined for the PBDG sample as the scatter on the data was larger, the time for the moduli to reach their maximun value (t(m)) is included instead in Figure 4. It can be seen in Figure 5 that the characteristic time t(c) for the evolution of the moduli of the PBDG sample will be much larger than t(m). Hence, comparing the two samples shows that the difference between the time scales for the stress relaxation and the evolution of the dynamic moduli is much greater for the more concentrated solution. The effect of temperature and shear rate suggests that the final relaxation is associated with structural changes as detected by the dynamic moduli. However the present lack of correlation is counterindicative of an absolute correspondence between the two time scales.

Isotropic polymeric systems as well as particulate systems might also show time—dependent moduli after cessation of flow. As long as the shear does not induce structure growth, the moduli always increase with time after flow. An increase of the moduli upon cessation of flow has also been reported for thermotropic PLCs (18) as well as for lyotropic solutions of hydroxypropyl cellulose in water (19) and in acetic acid (20). The possibility of changing in either direction seems to be characteristic for mesomorphic materials. A fundamental theory for describing complex moduli does not exist for such materials. The present results, combined with the information about optical relaxation mentioned above, could be explained on the basis of reorientation of domains or defects. The different domains orient differently, even randomly, at rest whereas flow causes an overall orientation. Depending on the molecular interaction the flow could then cause an increase or decrease in moduli as recently suggested by Larson (21).

Flow Reversal and Stepwise Changes in Shear Rate

A change in flow direction might be a useful test method for picking up the anisotropy expected during flow in liquid crystals. Figures 6 and 7 display the response of the shear stress to a sudden reversal in flow direction for the two samples. All shear rates are in the Newtonian region and the shear stress is scaled with the equilibrium value, in order to facilitate comparison. In all cases a complex, damped oscillatory pattern is registered. In addition, the transients for different final shear rates could be superimposed for each solution by scaling the curves with strain (Moldenaers, P; Mewis, J. J. Non—Newtonian Fluid Mech., in press). The frequency of the damped oscillation does however depend on the actual solution under consideration. According to figures 6 and 7 the period is approximately 50% higher for the system with the highest molecular weight and concentration, possibly reflecting a stronger coupling between the individual molecules. Published transients for various PLCs all seem to have rather similar frequencies (22, 23). The separate effect of concentration and molecular weight still remain to be investigated.

In a second experiment the shear rate is suddenly changed, without altering the direction of flow., i.e. a stepwise increase in shear rate. This also causes a stress transient with a damped oscillatory component, scaling with strain. It is possibly caused by director tumbling as a means to readjust to the new conditions (Burghardt, W.R.; Fuller, G. G., in press). A comparison of the step—up and flow reversal experiments (Figure 7).indicates that the oscillatory components have a phase shift of nearly 180°. This experiment is one of the very few rheological tests which provide a direct proof of the anisotropic behaviour of PLCs during flow. It also indicates, together with

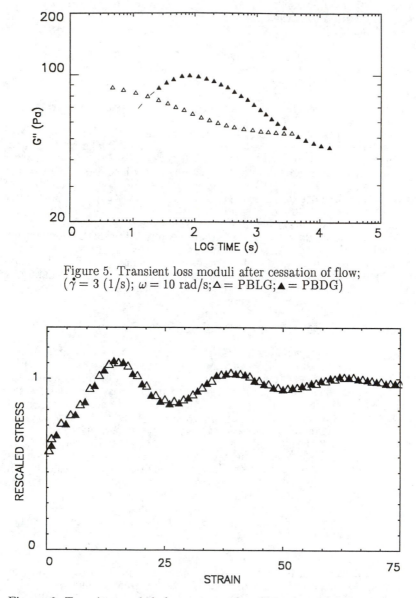

Figure 5. Transient loss moduli after cessation of flow;
($\dot{\gamma} = 3$ (1/s); $\omega = 10$ rad/s; $\triangle =$ PBLG; $\blacktriangle =$ PBDG)

Figure 6. Transient scaled shear stress after flow reversal for the PBLG sample;
($\dot{\gamma}$: $\blacktriangle = 1$ (1/s); $\triangle = 0.5$ (1/s))

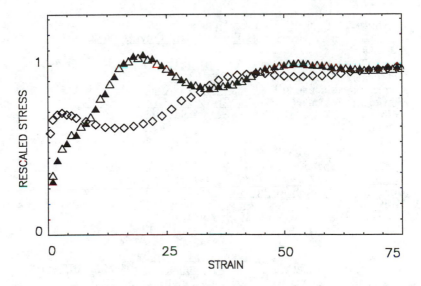

Figure 7. Comparison of the scaled shear stress after flow reversal and stepwise increase in shear rate for the PBDG sample;
(flow reversal: $\dot{\gamma}$: ▲ = 1 (1/s); △ = 0.4 (1/s);
stepwise increase in shear rate: $\dot{\gamma}_i = 0.1$ (1/s); $\dot{\gamma}_f = 1$ (1/s) ◇)

the strain scaling, that the oscillatory part of the shear stress is associated with an oriented structural feature, which forms a small angle with the flow direction. The director itself comes to mind first. Theoretically it could satisfy the experimental requirements, according to the standard theories. However experimental evidence seems to indicate an orientation of the director in the flow direction itself (24). Further experimentation is required to solve this problem.

Acknowledgments

This work was supported by a grant from AKZO International Research, Arnhem, The Netherlands.

Literature Cited

1. Wissbrun, K. F. J. Rheol. 1981, 25, 619–662.
2. Kiss, G.; Porter, R. S. J. Polym. Sci.: Polym. Phys. Ed. 1980, 18, 361–388.
3. Asada, T.; Onogi, S.; Yanase, Y. Polym. Eng. Sci. 1984, 24, 355–360.
4. Moldenaers, P.; Mewis, J. J. Rheol. 1986, 30, 567–584.
5. Mewis, J.; Moldenaers, P. Chem. Eng. Commun. 1987, 53, 33–47.
6. Mewis, J.; Moldenaers, P. Mol. Cryst. Liq. Cryst. 1987, 153, 291–300.
7. Moldenaers, P. Ph. D. Thesis, Katholieke Universiteit Leuven, Leuven, Belgium, 1987
8. Onogi, S.; Asada, T. in Rheology; Astarita, G.; Marrucci. G.; Nicolais, L., Eds; Plenum: New York, 1980; Vol. 1, p.127.
9. Asada, T,; Tanaka, T.; Onogi, S. J. Appl. Polym. Sci.: Appl. Polym. Symp. 1985, 41, 229–239. 1985, 41, 229–239.
10. Doi, M. J. Polym. Sci.,: Polym. Phys. Ed. 1981, 19, 229–243.
11. Leslie, F. M. in Advances in Liquid Crystals; Brown, G. H., Ed.; Academic Press: New York, 1979; Vol. 4, p. 1.
12. Navard, P. J. Polym. Sci.: Polym. Phys. Ed. 1986, 24, 435–442.
13. Gotsis, A. D.; Baird, D. G. J. Rheol. 1986, 25, 275.
14. Kiss, G.; Porter, R. S. J. Polym. Sci.: Polym. Symp. 1978, 65, 193–211.
15. Marrucci, G.; Maffettone, P. L. Macromolecules 1989, 22, 4076–4082.
16. De Cleyn, G.; Mewis, J. J. Non–Newtonian Fluid Mech. 1981, 9, 91–105.
17. Larson, R. G.; Mead, D. W. J. Rheol., 1989, 33, 1251–1281.
18. Wissbrun K. F.; Griffin, A. C. J. Polym. Sci.: Polym. Phys. Ed., 1982, 20, 1835–1845.
19. Ernst, B.; Navard, P.; Haudin, J. M. J. Polym. Sci.: Part B: Polym. Phys., 1988, 26, 211–219.
20. Moldenaers, P.; Mewis, J. in Proc. Xth Int. Cong. on Rheol., 1988, Sydney, Vol. 2, p. 134.
21. Larson, R. G.; Mead, D. W. J. Rheol., 1989, 33, 185–206.
22. Doppert, H. L.; Picken, S. J. Mol. Cryst. Liq. Cryst., 1987, 153, 109–116.
23. Viola, G. G.; Baird, D. G. J. Rheol., 1986, 30, 601–628.
24. Yanase, H. Ph. D. Thesis, Kyoto University, Kyoto, Japan, 1988.

RECEIVED March 16, 1990

Chapter 27

Shrinkage in Parts Molded from Thermotropic Liquid-Crystalline Polymers

Dependence upon Part Geometry and Filler Content

Paul D. Frayer and Paul J. Huspeni

Amoco Performance Products, Inc., 3702 Clanton Road, Augusta, GA 30916–5867

This analysis reveals that measurement of shrinkage or linear coefficients of thermal expansion (CTE's) in just flow and width directions, as is done for conventional polymers, is not sufficient for liquid crystal polymers (LCP's) and can lead to erroneous shrinkage predictions. This is a consequence of inherent LCP anisotropy, resulting in a relatively large linear CTE and shrinkage in the thickness direction of associated molded LCP parts. Linear and volumetric CTE data for neat and filled LCP molded parts of different geometries are presented. Volumetric CTE appears to be preserved for a specified formulation, independent of gate or part geometry. Analysis of CTE behavior for the filled LCP compositions indicates some anisotropy remains even at relatively high loadings of filler.

A method for predicting LCP part shrinkage from known linear CTE measurements is suggested. This method includes part shrinkage predictions based upon the extremes of cooling from melt and mold temperatures. For an edge-gated flex bar the results of this analysis indicated that shrinkage in the flow direction is small and can probably be modeled as shrinkage from the mold temperature, while width and thickness direction shrinkages can be modeled as shrinkage from temperatures closer to the melt temperature.

In conventional amorphous polymers, molded part shrinkage can usually be predicted using a single measured coefficient of thermal expansion since the CTE values for the flow, transverse, and thickness directions are very similar. Knowledge of CTE or actual shrinkage behavior in the flow and width directions for a

0097–6156/90/0435–0381$06.00/0

semi-crystalline polymer is usually sufficient for mold design purposes. Reference (1) provides comprehensive information on how injection molding processing parameters such as mold temperature, injection pressure, and mold gate design affect part shrinkage. These parameters may affect LCP molded part shrinkages in a similar manner, but such effects are not well understood at present and are likely to be confounded by the anisotropic nature of these materials.

Liquid crystalline polymers exhibit anisotropy in extruded and molded articles as a result of preferential orientation of LCP domains or individual chains. Reference (2) highlights some of the molecular structural features of LCP's that account for their fundamental anisotropy. These include the large aspect ratio of the individual polymer chains and their tendency to form aligned, highly crystalline domains.

The "fountain flow" effect in an LCP neat resin molding can translate into a complex skin/core structure exhibiting significantly different flow, transverse, and thickness directional shrinkages. References (2), (3), (4), (5), and (6) detail the complex nature of this skin/core morphology. These studies indicate that the skin layer is highly oriented in the flow direction, while the core can exhibit either no or 90 degree orientation to the flow direction. This situation complicates shrinkage analysis further, since part shrinkage will not be affine throughout its thickness. The absolute part thickness will affect shrinkage because of the attendant effect on the relative amounts of skin and core layers, the core layer being a larger percentage of the overall thickness as part thickness is increased.

The geometry of the molded part and the associated gating arrangement will affect molecular orientation within a part. A center gated part will reflect a biaxial orientation in the radial and circumferential directions. The skin layer of parts edge-gated along their entire width would be expected to exhibit preferential orientation in the flow direction.

Reference (2) indicates that LCP molded part anisotropy is reduced by the addition of most fillers, including glass fibers. This is opposite to the effect observed with more conventional polymers where anisotropy increases, especially upon addition of glass fibers or other fillers with a significant aspect ratio. In LCP's the reduction in anisotropy with filler addition may be a result of interference with the self-reinforcing nature. It is not clear exactly how the skin/core morphology of molded LCP parts changes with the addition of filler. Some residual skin/core structure appears to be present even at very high filler loadings, however (6).

This paper attempts to treat LCP molded part shrinkage first from a theoretical viewpoint related to directional and volumetric CTE's, and then from a more empirical perspective, where both actual shrinkage and CTE data are examined. Differences between LCP and conventional amorphous polymer molded part shrinkages are highlighted. A method is suggested for predicting potential shrinkage in a molded LCP article.

THEORY

Figure 1 details the 3-dimensional nature of the linear CTE's for an LCP molded part. These linear CTE's can be very dependent upon part geometry, gate location, and molding conditions. From reference (7) and Figure 1, the following definitions can be made:

$$L_i = L_{i0} * (1 + a_i * \Delta T) \qquad i = 1, 2, 3. \quad (1)$$

Here:

L_i (i=1,2,3) = part linear dimensions after undergoing temperature change ΔT

L_{i0} = original part linear dimensions

a_i = linear coefficients of thermal expansion associated with the three principal directions.

The volumetric CTE for a given material depends upon the basic macroscopic and microscopic structure and the type/volume fraction of any added filler. The volumetric CTE will to a lesser degree be affected by such factors as molding conditions, part geometry, and gate location, and here only to the extent that these factors effect a significant change in base resin structure. Such structural effects could include degree of crystallinity, but for many main-chain aromatic LCP's nearly all of the material is crystalline and little disordering occurs through the nematic transition. The volumetric CTE is related to the 3-D linear CTE's (7) by the following expression:

$$b = a_1 + a_2 + a_3, \text{ where } b = \text{volumetric CTE.} \quad (2)$$

The corresponding expression for volume as a function of change in temperature is (7):

$$V = V_0 * (1 + b * \Delta T). \quad (3)$$

Here V is the volume after change ΔT and V_0 is the initial volume before the temperature change occurred.

For a complex part geometry containing numerous flow interruptions, such as pins in a long connector, the 3-dimensional CTE's may be much closer together, i.e., the part is more isotropic. Where the part is completely isotropic the linear CTE, a, is identical for all three directions:

$$a = (a_1 + a_2 + a_3)/3 = b/3. \quad (4)$$

It is seen from equation (4) that the overall shrinkage for such a molded article could be dominated by that of the rarely measured but significant shrinkage contribution in the thickness direction.

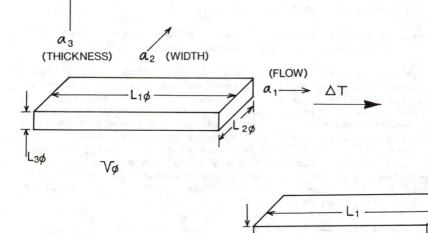

FIGURE 1: RELATIONSHIP BETWEEN MOLDED PART VOLUME CHANGE
AND CTE

a_1, a_2, a_3 = LINEAR COEFFICIENTS OF THERMAL EXPANSION ("/"-$^\circ$C)
V_ϕ = ORIGINAL SAMPLE VOLUME (IN.^{3})
V = FINAL SAMPLE VOLUME (IN.^{3})
ΔT = TEMPERATURE CHANGE ($^\circ$C)

Two extreme cases can be predicted for the shrinkage of an LCP molded part based upon 3-dimensional CTE data. In the first case the part is assumed to shrink from the melt temperature (Ts) to room temperature (Tf). Here it is postulated that no real part packing occurs during molding. For the second case, the part is assumed to cool from the mold temperature (Tm) to room temperature. For this case, maximum packing would occur. For cooldown from melt temperature, the equations are:

$$Li(Tf) = LiO(Ts) * (1 + ai * (Tf-Ts)) \qquad i = 1, 2, 3. \quad (5)$$

The corresponding expressions for part cooling from the mold temperature to ambient are:

$$Li(Tf) = LiO(Tm) * (1 + ai * (Tf-Tm)) \qquad i = 1, 2, 3. \quad (6)$$

RESULTS

Table I gives the summary of CTE measurements made on neat and filled XYDAR molded parts. These CTE measurements were performed in a temperature range from 0-150 $^{\circ}$C using a DuPont 942 thermomechanical analyzer (TMA). The CTE's shown in Table I are measured in the flow, transverse, and thickness directions for the formulation and part geometry specified. In a part of center-gated geometry, the radial direction is the flow direction and the circumferential direction is the width direction. All of the TMA testing was performed on as-molded samples (no annealing).

The descriptions for the XYDAR formulations shown in Table I are as follows. XYDAR resins increase in melting point (m.p.) as the series changes from 300 to 400 to SRT-300, the latter representing the highest m.p. material. A "G" designation indicates glass fiber filler, "M" mineral filler, and "MG" a combination of glass fiber and mineral fillers. The last two digits in the product designation define the total % filler (i.e., G-430 contains 30% glass fiber filler). The first digit ("3" or "4") indicates the XYDAR base resin. Table I also lists CTE data for a neat amorphous polymer Ultem polyetherimide, as well as neat Vectra A-950 and 30% glass filled Vectra A-130 LCP's. The XYDAR formulation denoted "experimental compound" has a high loading of filler.

Table I also shows two additional parameters, the "anisotropy index" (AI) and "normalized anisotropy index" (NAI). These indices give a quantitative measure of the molded part anisotropy and are defined as follows:

$$AI = ai(max.)-ai(min.), \text{ where } ai(max.) \text{ and } ai(min.) \text{ denote } (7)$$
maximum and minimum linear CTE's, respectively

$$NAI = AI/b = (ai(max.) - ai(min.))/b, \quad b = \text{volumetric CTE.} \quad (8)$$

It can be seen from the above expressions that molded part anisotropy will increase as either parameter AI or NAI increases.

TABLE I

EXPERIMENTAL COEFFICIENTS OF THERMAL EXPANSION FOR LCP COMPOUNDS

ITEM #	DESCRIPTION	LINEAR CTE (PPM/^{0}C) *			VOLUMETRIC CTE (PPM/^{0}C)
		FLOW	WIDTH	THICKNESS	
1	Ultem (Neat) Center Gated Plate	53.0	52.9	55.8	161.7
2	Vectra A-950 (Neat) Tensile Bar	5.0	84.8	114.2	204.0
3	Lower MP XYDAR Neat 300 Series Tensile Bar	43.4	67.3	167.5	278.2
4	Medium MP XYDAR Neat 400 Series Tensile Bar	6.8	91.2	161.5	259.5
5	High MP XYDAR Neat SRT-300 Tensile Bar	-2.0	59.0	237.0	294.0
6	High MP XYDAR Neat SRT-300 HDT Bar (1/4" Thk.)	52.0	67.0	155.0	274.0
7	Vectra A-130 ~ 30% Glass Fiber Tensile Bar	15.2	72.8	79.9	167.9
8	G-330 Center Gated Plate	9.1	21.9	192.0	223.0
9	G-430 Tensile Bar	10.3	61.5	159.5	231.3
10	G-430 Center Gated Plate	8.7	17.6	193.0	219.3
11	G-345 Tensile Bar	8.7	66.0	118.5	193.2
12	M-350 Tensile Bar	11.1	37.8	148.0	196.9
13	M-350 Center Gated Plate	15.5	15.3	159.5	190.3
14	MG-350 Tensile Bar	12.9	52.8	133.0	198.7
15	13253-38-2 XYDAR Experimental Cmpd. Tensile Bar	15.6	31.1	46.2	92.9

* NOTE: Linear CTE values cover range from 0 - 150 ^{0}C.

TABLE I: CONTINUED

ITEM #	DESCRIPTION	ANISOTROPY INDEX (PPM/°C)	NORMALIZED ANISOTROPY INDEX
1	Ultem (Neat) Center Gated Plate	2.9	0.02
2	Vectra A-950 (Neat) Tensile Bar	109.2	0.54
3	Lower MP XYDAR Neat 300 Series Tensile Bar	124.1	0.45
4	Medium MP XYDAR Neat 400 Series Tensile Bar	154.7	0.60
5	High MP XYDAR Neat SRT-300 Tensile Bar	239.0	0.81
6	High MP XYDAR Neat SRT-300 HDT Bar (1/4" Thk.)	103.0	0.38
7	Vectra A-130 ~ 30% Glass Fiber Tensile Bar	64.7	0.39
8	G-330 Center Gated Plate	182.9	0.82
9	G-430 Tensile Bar	149.2	0.65
10	G-430 Center Gated Plate	184.3	0.84
11	G-345 Tensile Bar	109.8	0.57
12	M-350 Tensile Bar	136.9	0.70
13	M-350 Center Gated Plate	144.0	0.76
14	MG-350 Tensile Bar	120.1	0.60
15	13253-38-2 XYDAR Experimental Cmpd. Tensile Bar	30.6	0.33

* NOTE: Linear CTE values cover range from 0 - 150 °C.

Table II gives a comparison of actual XYDAR molded part shrinkages vs shrinkages predicted upon cooling from melt and mold temperatures for 45% glass fiber (G-445) and 50% mineral/glass fiber (MG-350) compositions. These predicted values were estimated using equations (5) & (6) in the theory section.

Figures 2-4 and 5-7 give raw flow, width, and thickness direction TMA data for an M-350 center-gated plate and neat 300 series tensile bar, respectively.

Figure 8 gives melt specific volume as a function of temperature for neat 300 series resin. The DSC peak melting point for this material is about 345 $^\circ$C. These data were generated on a Sieglaff-McKelvey capillary rheometer using a blank (no hole) capillary.

DISCUSSION

Inspection of entries 1-6 in Table I indicates the large anisotropy present in moldings of neat LCP resins. The remarkable result here is the very large CTE for the thickness direction. It is noteworthy that as a series the neat LCP's possess a volumetric CTE greater than that of even neat Ultem polyetherimide below its glass transition temperature. The equivalence of the three linear CTE's for Ultem confirm the amorphous nature of this polymer. This comparison underscores the importance of examining 3-D linear and volumetric CTE's for LCP materials and designing parts as well as locating gates so as to take advantage of anisotropy. If the customary comparison between the neat LCP's and Ultem were made based upon flow and width direction CTE values, an erroneous conclusion regarding the volumetric shrinkage would be reached.

Entries 1-6 in Table I reveal an approximate constancy of volumetric CTE through the XYDAR 300, 400, and SRT-300 series resins. This result is not surprising, since the molecular structures of these materials are similar. However, the CTE values for a molded part reflect macroscopic structure (skin/core) as well as domain or molecular level morphology. The LCP molded part must therefore be regarded as a composite structure. These considerations as well as differences in basic molecular composition may explain the lower volumetric CTE for Vectra A-950 relative to the neat XYDAR resins, at least over the measured temperature range of 0-150 $^\circ$C. These factors may also be responsible for the lower anisotropy of the neat XYDAR 300 series resin.

Comparison of entries 5 and 6 for neat SRT-300 in Table I highlights the dependence of anisotropy on part thickness. In both cases the volumetric CTE's are nearly identical, but the thinner (1/8") tensile bar has a much higher degree of anisotropy. This result is probably due to a greater core thickness in the 1/4" HDT bar, this core possibly having a 90 degree orientation relative to the flow direction. Thus, the flow direction CTE for the 1/4" HDT bar is very similar to that of the width direction.

Inspection of entries 7-14 in Table I reveals that LCP formulations can retain anisotropy despite high loadings of fibrous

TABLE II

COMPARISON OF ACTUAL AND PREDICTED SHRINKAGES FOR XYDAR MOLDED PARTS

ITEM #	DESCRIPTION	FLOW DIRECTION SHRINKAGE ("/")		
		PRED. 1**	PRED. 2***	ACTUAL
1	G-445 .125" Thick Flex Bars	0.0033	0.0004	0
2	MG-350 .125" Thick Flex Bars	0.0033	0.0004	0.0008

ITEM #	DESCRIPTION	WIDTH DIRECTION SHRINKAGE ("/")		
		PRED. 1	PRED. 2	ACTUAL
1	G-445 .125" Thick Flex Bars	0.0260	0.0032	0.0140
2	MG-350 .125" Thick Flex Bars	0.0202	0.0027	0.0120

ITEM #	DESCRIPTION	THICKNESS DIRECTION SHRINKAGE ("/")		
		PRED. 1	PRED. 2	ACTUAL
1	G-445 .125" Thick Flex Bars	0.0500	0.0061	0.0380
2	MG-350 .125" Thick Flex Bars	0.0510	0.0068	0.0370

** NOTE: PRED. 1 based upon 388 $^\circ$C melt temperature for G-445 cooling to 21 $^\circ$C, based upon 360 $^\circ$C melt temperature for MG-350 cooling to 21 $^\circ$C.

*** NOTE: PRED. 2 based upon 66 $^\circ$C mold temperature cooling to 21 $^\circ$C.

NOTE: 0-250 $^\circ$C range CTE's used for G-445 and MG-350 shrinkage predictions:

COMPOUND	LINEAR CTE (PPM/$^\circ$C)		
	FLOW	WIDTH	THICKNESS
G-445	9.1	70.4	135.5
MG-350	9.7	59.5	150.0

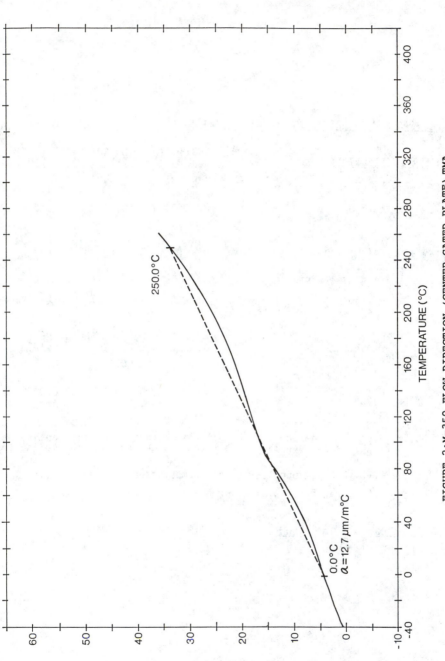

FIGURE 2:M-350 FLOW DIRECTION (CENTER-GATED PLATE) TMA

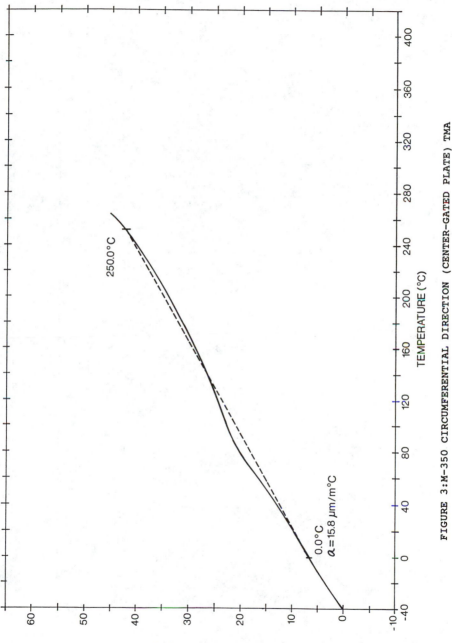

FIGURE 3:M-350 CIRCUMFERENTIAL DIRECTION (CENTER-GATED PLATE) TMA

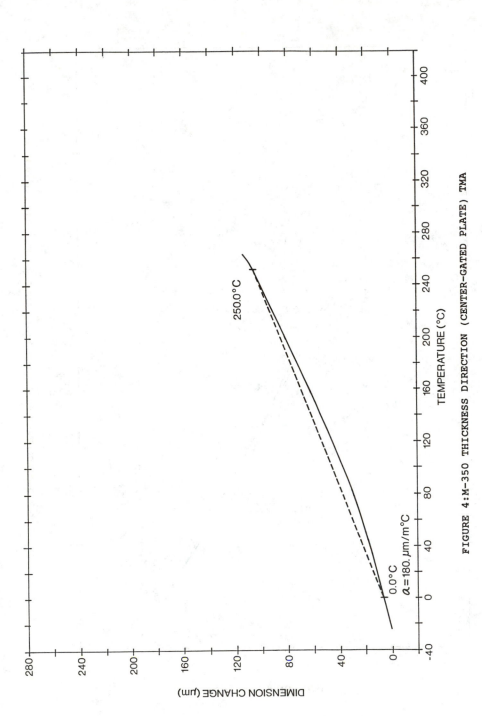

FIGURE 4: M-350 THICKNESS DIRECTION (CENTER-GATED PLATE) TMA

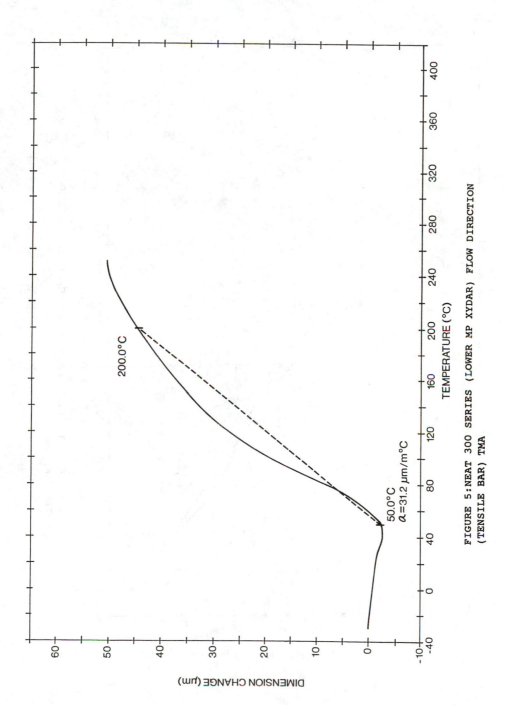

FIGURE 5: NEAT 300 SERIES (LOWER MP XYDAR) FLOW DIRECTION (TENSILE BAR) TMA

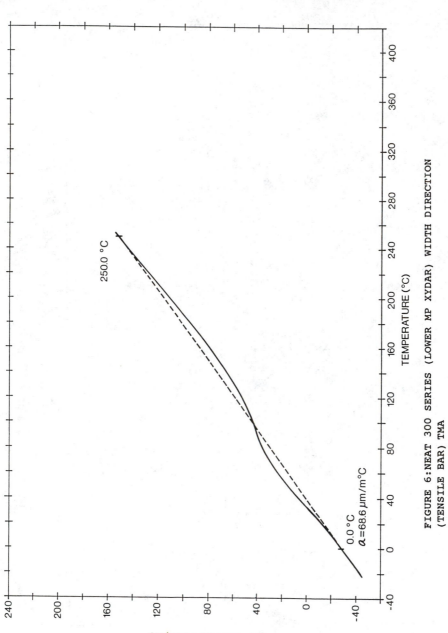

FIGURE 6:NEAT 300 SERIES (LOWER MP XYDAR) WIDTH DIRECTION
(TENSILE BAR) TMA

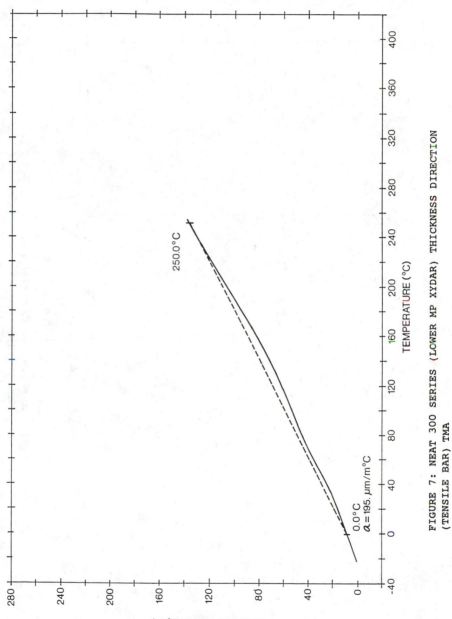

FIGURE 7: NEAT 300 SERIES (LOWER MP XYDAR) THICKNESS DIRECTION (TENSILE BAR) TMA

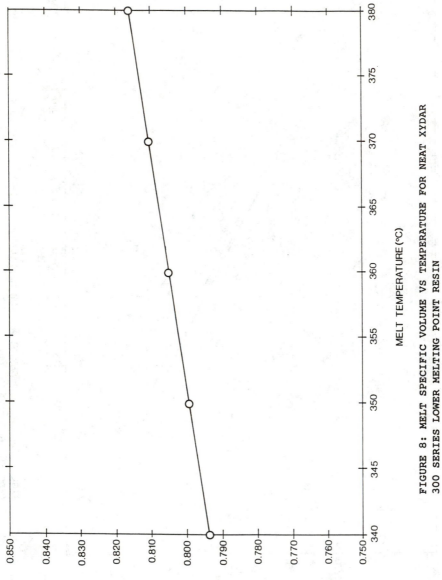

MELT TEMPERATURE (°C)

MELT SPECIFIC VOLUME (CC/GM)

FIGURE 8: MELT SPECIFIC VOLUME VS TEMPERATURE FOR NEAT XYDAR
300 SERIES LOWER MELTING POINT RESIN

or particulate fillers. Comparison of volumetric CTE's for any given compound for different part and gating geometries indicates that these CTE's are approximately constant. The exception again is Vectra A-130 which exhibits a lower volumetric CTE than the comparable 30% glass fiber loaded G-330 over the temperature range of 0-150 °C. The reasons for this difference are not presently clear, but as discussed above both molecular and macroscopic structures may play a role. The conservation of volumetric CTE for a given filler and type loading indicates that it should be possible to predict a linear CTE for a third dimension if two linear CTE's and the volumetric CTE are known.

Comparison of volumetric CTE's for products of identical filler loading (i.e., G-330 and G-430) indicates that base resin does not have a significant effect. This is in agreement with the observations made for the neat compositions, where the volumetric CTE's for the XYDAR 300, 400, and SRT-300 materials were found to be equivalent.

Inspection of the Table I AI and NAI values suggests that the center gated plate may be more anisotropic than a tensile bar. This is the result of lower flow (radial) and width (circumferential) direction CTE's and a larger thickness direction CTE for the plate geometry. This situation apparently results from the biaxial flow in a center gated part.

As expected, the anisotropy and volumetric CTE for the highly filled experimental compound (entry 15 in Table I) are much lower than the other XYDAR formulations. This material is particularly suited for applications where lower anisotropy and volumetric CTE match are needed and highlights the importance of adjusting filler loading in LCP's to meet shrinkage and CTE requirements for an end-use application.

Examination of Table II reveals that the actual molded part shrinkages for 45% glass fiber and 50% mineral/glass fiber 1/8" thick flex bars are very similar. It can be seen from this data that the directional shrinkages parallel the behavior of the linear CTE's, i.e., shrinkage in the thickness direction is greatest. It should be noted here that shrinkage is not generally measured in the thickness direction for a molded part because the packing process during injection molding can significantly affect this dimension. This effect probably becomes less as the absolute thickness of the part is increased. Despite this limitation, the Table II data underscores the need to measure thickness direction shrinkage in LCP molded parts.

Table II also indicates that the actual part shrinkages fall between the limits predicted by the linear CTE's for the melt temperature and mold temperature cooling to room temperature cases. In the flow direction, shrinkage appears to be best modeled as part cooling from the mold temperature. For the transverse and thickness directions, cooling from the melt temperature may more closely approximate the shrinkage. This is an important result because a part of more complex geometry may be more isotropic in nature and could, as a consequence of conservation of volumetric CTE, have considerably larger shrinkage than that predicted from flow and transverse direction shrinkages alone.

The shrinkage predictions made in Table II are based upon the
assumption that the linear CTE's remain constant through the
nematic transition. For most semi-crystalline polymers this would
not be a good assumption since the volumetric CTE usually changes
significantly through the melting transition. For LCP's, however,
the change appears to be small because of retention of a high
degree of order through the nematic transition. As a consequence,
both the enthalpy of "melting" and the associated specific volume
change are small. It is seen from Figure 8 that the melt specific
volume changes in a linear and continuous fashion through the
nematic transition.

The Figure 2-7 TMA plots indicate that the dimensional changes
of the XYDAR molded parts are fairly linear over a wide temperature
range. The applicable range for TMA measurements on these
materials appears to be up to their heat distortion temperature
(HDT). For the 300 series formulations shown in Figures 2-7 the
HDT is in the 200-250 $^{\circ}$C range.

As a further check on the validity of linear CTE data
extrapolation through the nematic transition, equation (3) in the
theory section was applied using the volumetric CTE for neat 300
series resin (Table I, entry 3). Here an initial room temperature
(21 $^{\circ}$C) specific volume of 0.7194 cc/gm was assumed along with a
final melt temperature of 360 $^{\circ}$C. Equation (3) gave an
extrapolated melt specific volume of 0.787 cc/gm vs an actual
measured value of 0.805 cc/gm from Figure 8. Thus, the volumetric
CTE extrapolation appears to be reliable, but further melt specific
volume data from other LCP formulations will be needed to verify
this.

It must be emphasized that the above conclusions are based
upon a very limited amount of experimental data. Shrinkage studies
are required on parts of many different sizes, geometries, and
gating arrangements to provide a more comprehensive picture.

Field experience with smaller molded parts such as electrical
connectors indicates that shrinkages in the flow and width
directions are very small and that the mold may be cut virtually
"to size" for these dimensions. This suggests that the dynamics of
the molding process, i.e., packing, may be very important for this
type of geometry. Since most electrical connectors are also
thin-walled parts, the complexity associated with the skin-core
composite structure may be substantially reduced as a result of the
part being mostly a skin structure. Further shrinkage measurements
in the thickness direction are needed to better quantify this
situation. However, the few presented examples highlight the
deficiencies in applying the standard flow and transverse direction
CTE/shrinkage analysis to LCP molded parts.

As the size of the part increases, the anisotropic nature of
LCP shrinkage becomes more critical because the absolute magnitude
of the shrinkage increases. Here gating geometry also becomes
important to minimize molded-in stresses and attendant warpage.

An additional important area of future study will be to
measure linear CTE's on annealed LCP parts as well as the effect of

mold temperature. This would remove some effects due to molded-in stresses and may or may not change the CTE behavior. In actual practice, however, CTE data obtained from the as-molded part is probably more important since same will be more representative of actual molding processing and end-use application.

CONCLUSION

The importance of considering the anisotropic nature of molded LCP parts for proper mold design has been highlighted. Knowledge of the 3-D linear and volumetric CTE's as well as 3-D shrinkages is required. Volumetric CTE appears to be preserved for a given LCP formulation, independent of part and gating geometry. A method for predicting molded part shrinkage from linear 3-D CTE data assuming cooldown from melt and mold temperatures to ambient has been suggested. Further 3-D shrinkage and CTE measurements must be made on parts of various sizes, shapes, and gating arrangements to confirm the validity of this approach.

LITERATURE CITED

1. Rubin, I. Injection Molding Theory and Practice; John Wiley & Sons, New York, 1972, pp. 270-301
2. Frayer, P. Polymer Composites 1987, December, Vol. 8, No. 6, pp. 379-395
3. Menges, G.; Schacht, T.; Becker, H.; Ott S. Intern. Polymer Processing 2 1987, 2, pp. 77-82
4. Weng, T.; Hiltner, A.; Baer, E. Journal of Materials Science 1987, 21, pp. 744-750
5. Suokas, E.; Sarlin, J.; Tormala P. Mol. Cryst. Liq. Cryst. 1987, Vol. 153, pp. 515-524
6. Duska, J. Plastics Engineering 1987, December, pp. 39-42
7. Sears, F.; Zemansky, M. University Physics; fourth edition; Addison Wesley 1970, pp. 220-222

RECEIVED April 3, 1990

APPLICATIONS OF LIQUID-CRYSTALLINE POLYMERS: BLENDS

Chapter 28

Blending of Polymer Liquid Crystals with Engineering Polymers

The Importance of Phase Diagrams

Witold Brostow[1], Theodore S. Dziemianowicz[2], Michael Hess[1,3], and Robert Kosfeld[3]

[1]Center for Materials Characterization and Department of Chemistry, University of North Texas, Denton, TX 76203–5371
[2]Himont U.S.A., Inc., 800 Greenbank Road, Wilmington, DE 19808
[3]FB6–Physikalische Chemie, University of Duisburg, D–4100 Duisburg 1, Federal Republic of Germany

We study phase diagrams of polymer liquid crystals (PLC) in function of LC concentration in copolymers and also phase diagrams of blends of PLCs with engineering polymers. Most PLC systems are multiphasic. Moreover, there are some non equilibrium phases with high longevity; such phases have to be included in the phase diagrams to make intelligent processing possible. We report phase diagrams or their parts for copolymers of poly(ethylene terephthalate) (PET) with p-hydroxybenzoic acid (PHB) in function of mole fraction x of PHB; PET/xPHB + poly(bisphenol-A-carbonate) (PC). Several techniques have to be used, since the diagrams are complex and some techniques are more sensitive to certain transitions while other techniques will not necessarily show a given a transition at all.

We know well that mechanical and other properties of polymer-based materials can be improved by introduction of reinforcements, for instance of chopped glass fibers into an epoxy. The resulting class of materials has been called macroscopic composites (1) or else heterogeneous composites (2). The use of such composites causes on occasions serious problems - due to insufficient

0097–6156/90/0435–0402$06.00/0
© 1990 American Chemical Society

adhesion between the fibers and the matrix. In service, under effects of external mechanical forces, pullout of individual fibers occurs, and in layer structures delamination of entire layers is possible.

At least two good ways out exist. One consists of the dispersion at the molecular level of rigid polymer molecules between flexible chains - the concept of molecular composites of Helminiak, Hwang e.a.(3, 4). The other involves the use of polymer liquid crystals (PLCs). As discussed by Witt (5), compared to widely used engineering thermoplastics, PLCs show clear superiority with regard to chemical resistance, low flammability, high modulus, low isobaric expansivity (or even zero, depending on the direction), and often unusual ease of processing.

Monomers for PLC synthesis are often available in small quantities only, with evident consequences for PLC prices. Hence our objective: blending of PLCs with ordinary engineering polymers in such a way that good thermophysical, rheological, mechanical and other properties of PLCs are preserved to a large extent, while costs lowered considerably. Pertinent work along these lines was done by several groups, including Weiss and collaborators (6-8), Kiss (9), DeMeuse and Jaffe (10-11) and also in our laboratories (2, 12-14).

PLCs typicaly form several phases, as do multicomponent systems containing them (2, 6-25). Multiplicity of phase transitions appears intrinsic to all liquid crystals - monomeric (MLCs) as well as PLCs. This work was aimed at better understanding of phase diagrams of PLC-containing blends. Complexity of the systems studied has brought about a collaboration of three laboratories.

Systems studied

There is approximately a dozen of classes of PLCs known today, differing in location (in the backbone, in the side-chain, etc.) and shape (rod, cross, disc, etc.) of the mesogenic groups. The first comprehensive systematical classification was proposed in (26) and somewhat amplified later (Brostow, W. *Polymer* 1990, in press), see also (27).

Because of the importance of longitudinal PLCs - class "alpha" according to (26) - in the development of high-modulus materials, a polyester easy to obtain and to modify was chosen as the rigid-rod type PLC component: the Jackson and Kuhfuss (15) copolyester based on poly(ethylene terephthalate) (PET) and p-hydroxy-benzoate (PHB). The polyester can be prepared and

analyzed quite easily with different compositions, that
is with different degrees of rigidity. It shows a
nematic mesophase above 30 mole % PHB. The notation
PET/xPHB will be used where x is the mole percent of
PHB.

The polymer had been kindly supplied by Eastman
Kodak and had a weight-average molar mass $<M_w> \rightleftharpoons 2.0\ 10^4$
g/mole, varying little with x, with almost but not quite
statistical distribution of the PHB-units in the
backbone. Considerable work on this material has been
done before (17, 28-39).

As the second component in the blend either PET from
Eastman Kodak or a bisphenol-A based poly(carbonate)
(PC) generously provided by the Bayer AG were used. The
weight-average molar mass was $<M_w> \rightleftharpoons 3.2\ 10^4$ g/mole for
each. The quotient $<M_w>/<M_n>$ for all polymers studied
was $<M_w>/<M_n> \rightleftharpoons 2$.

PET and PC were chosen for two reasons. First, both
have certain features common to all polyesters. Second
there is a difference: PET is less flexible and more
easily crystallizable while PC is more flexible and
typically nearly completely amorphous (40).

Sample preparation

It is advantageous to prepare a blend by copre-
cipitation from a mutual solution in a common solvent.
Very good dispersion is thus achieved even when e.g.
transesterification or degradation processes during
mechanical mixing and in the melt are possible. PET/xPHB
with x between 10% and 40% and PC are soluble in $CHCl_3$.
In these cases the samples were prepared by coprecipi-
tation followed by a subsequent annealing at 210°C/4MPa.
More information on these types of ternary systems and
their phase behavior is given in (13) and will be
considered in a later paper in conjunction with
theoretical predictions.

For blends containing PET/xPHB with x > 40% a mixed
solvent consisting of 20 volume % CF_3COOH and $CHCl_3$ was
used. In all cases the polymers were precipitated by
addition of acetone to the solution.

Experimental procedure

Following is a list of the kinds of operations
performed: drying the materials for at least 4 hours at
160°C in a dehumidyfying, recirculating oven; injection molding in an Engel machine (barrel temperatures 274-288°C, mold temperature 52°C, overall cycle
time $\rightleftharpoons$ 1 min.); mechanical testing (flexural modulus

according to ASTM D790 procedure, tensile yield stress and stress elongation according to ASTM D638, Izod impact test according to ASTM D256 Condition A); differential scanning calorimetry (DSC, du Pont Thermal Analyzer and Perkin-Elmer DSC-2); melt rheology (Rheometrics System 4 mechanical spectrometer at 260°C); electron microscopy of fracture surfaces (SEM, JEOL 35C scanning electron microscope); thermomechanical analysis (TMA, Perkin-Elmer TMS-2); automatic torsional pendulum (ATP, Myrenne ATM-3); wide angle X-ray scattering (WAXS, $\lambda = 154.18$ pm Cu-K, 25°C); 1H-NMR and 13C-NMR (Bruker WM-300 spectrometer with 200 mg of material dissolved in $CHCl_3$ + hexafluoroisopropanol in the ratio 40 : 1).

Mechanical Testing and Anisotropy of Mechanical Properties

Rigidity of the mesogens play the central role for most of the properties of these types of polymers (27). Jackson and Kuhfuss (15) have shown already in 1976 how easily acquire PLCs orientation during processing. Influence of the type of flow on orientation and texture of PLC copolyesters such as ours was studied by Ide and Ophir (47) and by Viola e.a. (46).

Figure 1 visualizes the influence of overall chain rigidity produced by variation of the PHB-content x in the copolymer on the elastic modulus parallel and transverse to the applied injection flow.

Since PET/60PHB shows particulary good mechanical properties, in the following we shall focus on this material and blends containing it.

The tendency to crystallize is quite poor in the pure PET/60PHB, as easily demonstrated by calorimetric measurements. Small-angle X-ray investigations on compression-molded samples show that only after annealing for several hours at 220°C reflexes due to crystallites may be identified. Nevertheless, a proper estimate of the degree of crystallinity was not possible; the peaks hardly emerged from the amorphous halo. Point-measurements at different areas of the sample revealed that there was an orientation of the crystallites present in small areas of about 1 mm^2. In general, these areas were randomly oriented, hence the sum over all areas present was virtually zero. These results are supported by SEM analysis of corresponding specimens fractured under liquid nitrogen, showing that compression-molded samples contain randomly oriented thin fibers. The crystallinity is mainly due to PHB units rather than to PET units (43).

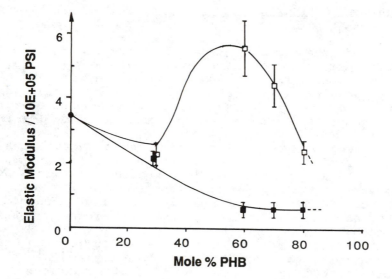

Figure 1. Elastic modulus of PET/xPHB copolymers in
function of x determined in the direction parallel
(□) and transverse to the flow (■).

In order to control anisotropy and hence properties, it is necessary to know which phases occur during processing, which result after processing has been completed, their stability, and their structure. Consequently, the first object of investigation has to be the phase diagram and the phase structure if an intelligent processing has to be conducted.

We note that the phase diagrams for PLC-containing blends are more complicated than those of ordinary engineering plastics. For PLCs additional phases have to be taken into account since the PLCs themselves may exhibit complex polymorphism.

Phase diagrams

It was expected that the PET/60PHB system will be relatively easy to interpret, particularly since a number of properties of PET/60PHB + PET blends had been studied before (2): calorimetry, melt rheology, several mechanical properties, and scanning electron microscopy of fracture surfaces. It turned out that even the phase diagram for PET/xPHB copolymers - no blending - in function of x is not exactly simple. While phase diagrams of PLC-containing systems have been little investigated so far, we recall an important result of Lin and Winter (25) that a PLC forms a high melting crystal from a nematic melt.

The phase diagram of pure PET/xPHB in function of x resulting from different methods and different authors is shown in figure 2; non-equilibrium phases and the corresponding transitions are included: glass transitions which occur in the system separately for PET and PLC, and also a cold crystallization curve. We believe that in general non-equilibrium phases have to be included in the PLC phase diagrams, since some such phases exhibit quite high longevity; disregarding them would have obvious consequences for processing procedures and for properties of the products. Moreover, from the behavior of glass transition temperatures conclusions can be drawn on miscibility of the systems.

The diagram in figure 2 contains at least one phase such that its existence has been noted only recently (Brostow, W.; Dziemianowicz, T.S., Hess, M., Saboe, Jr., S.H. Liquid Crystals, submitted), namely the quasi-liquid (q-l). This is the material which was in the amorphous state below its glass transition T_g, but now is in the temperature range between T_g and the melting transition. Thus, q-l can undergo crystallization - the so-called "cold crystallization" - in distinction to an ordinary liquid which cannot.

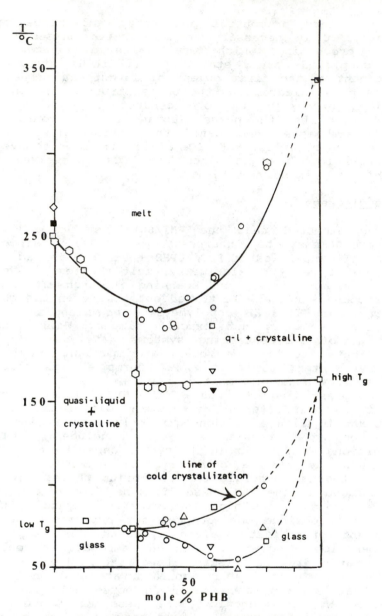

Figure 2. The phase diagram (essential part) for PET/xPHB copolymers in function of x. The sources of the results are denoted as follows:
o - our results, independently of the technique used (see above) and independently of location (U.S.A., Germany); □ - Meesiri e.a. (28); ■ - Jeziorny (29); ◇ - Chou e.a. (30); △ - Benson and Lewis (32); ▽ - Gedde e.a. (33); ▼ - Hedmark (34); ◯ - Jackson and Kuhfuss (15); and ◣ - Kricheldorf and Schwarz (35).

Further, the presence of another component below its glass transition and/or of crystallites prevents the quasi-liquid from flowing like a liquid does; the mobility is fairly low. Moreover, the notion of a "liquid" brings about a mental association of a material which upon heating can only undergo vaporization, or, if it is a polymer melt, no further transition at all. By contrast, the material containing a q-l phase has to undergo at least two more phase transitions: melting and isotropization at the clearing point. If more than one liquid-crystalline phase is formed, say smectic A and nematic, then there will be even more transitions.

We now go on to the PET/60PHB + PC phase diagram. A quite significant part of it is shown in figure 3. In particular, we would like to draw attention to a small approximately triangular region just below the glass transition of PC (154°C) in the range between 80% and 90 mole % PC. Starting from pure polycarbonate and going inside the diagram, we find a bifurcation of the glass transition line, confirmed by experiments by both U.S. and German groups. Hence in this concentration range we observe three glass transition lines, the third for PET at 54°C. The top line corresponds evidently to a phase dominated by PC. The rapid fall of the middle line between 10% and 20% PET/60PHB demonstrates considerable miscibility of constituents. PET is not a partner in the miscibility occurrence; if it were, given its low concentration in this range, the separate PET glass transition would not show. Hence the partners must be PHB and PC; there is enough of PC to show its own glass transition and simultaneously to mix in part with PHB.

The interaction between PHB and PC has consequences not only for glass transitions but also for crystallization as well as for mechanical properties. In figure 4 we show a projection of a three-dimensional WAXS surface in function of both scattering angle and temperature for x = 80%. It appears that PHB and PC aid each other to crystallize, and the effect increases along with temperature increase. In figure 5 we show the flexular modulus E_F in function of concentration for two different preparation conditions (annealing times of the samples). Starting as before from pure PC, in both cases we see a clear increase in modulus beyond 10% PET/60PHB, that is at the concentrations where we have observed miscibility in the glassy phase and co-crystallization of PHB and PC. The connection between the phase structures, the phase diagram and properties is clearly operative.

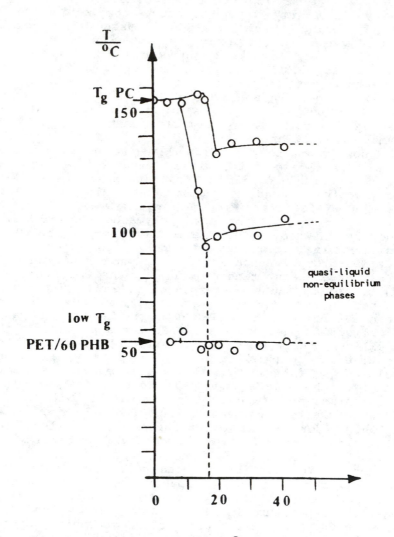

Figure 3. The phase diagram (essential part) for PET/0.6PHB + polycarbonate (PC) in function of PC concentration.

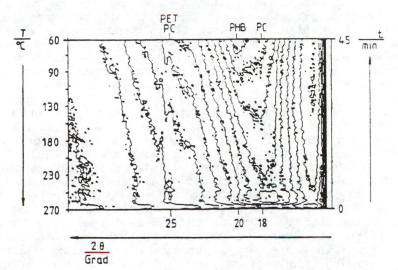

Figure 4. Two-dimensional projection from a WAXS surface in function of the scattering angle and temperature for PET/0.8 PHB showing independent crystallization of the different crystalline phases.

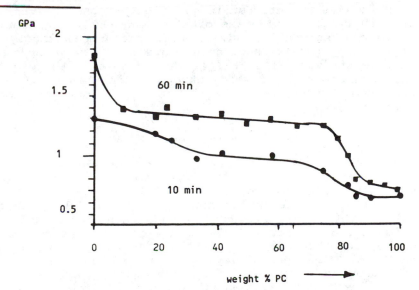

Figure 5. Flexural modulus for blends of PET/0.6PHB with polycarbonate (PC) in function of PC concentration.

To conclude this Section, let us return to the experimental procedures for the determination of phase diagrams. We find that transition temperatures determined by different techniques approximately coincide. However, some techniques are more sensitive than other for certain phases, and also for certain concentration regions. In relatively infrequent but dangerous cases, a technique might miss a transition entirely. In particular, dynamic mechanical testing (G' and G") seems to be overall more sensitive than DSC. In addition, it turned out that the quite simple analysis of the thermomechanical behaviour is a very useful tool: the penetration of a quartz rod with a flat tip under a small load gives an additional, independent information which very often appeared to be essential for proper interpretation of the results obtained from the other methods mentioned above, as this method seems most sensitive to all types of thermal transitions. In other words, the thermomechanical analysis very quickly gives an overview on almost all transitions occuring. Additional application of dynamic-mechanical analysis and calorimetry helps to identify the types of transitions and confirms the results. Therefore, a combination of techniques is recommended.

Solubility in organic liquids and ternary phase diagrams

We also have an interest in solution processing, that is in ternary phase diagrams of the type PLC + engineering polymer + solvent. A statistical mechanical theory of rigid-rod systems has been developed by Flory; he started the work in 1956 (44) when most of the present applications of PLCs were unknown and continued it more than two decades later (45). The Flory configurational partition function is

$$Z_M = Z_{comb} \cdot Z_{orient} \qquad (1)$$

There exists also a theory of Maier and Saupe (46, 47) which takes into account the stability of aligned molecules, for instance in the nematic state, due to the anisotropy of the dispersion forces. The Flory theory covers this as well, but in addition it takes into account two further effects: anisotropy of the molecular shape, and the influence of that anisotropy on molecular packing. The separability of factors in equation (1) has been well justified by subsequent results (48), (Flory, P.J. private communication to W.Brostow, August 20, 1985). The Flory theory has been applied, among others, to ternary systems of the type rigid rod polymer

+ flexible polymer + solvent (spherical molecules). If we apply this version of the theory to our blends with chloroform as the solvent, we obtain only qualitative and not quantitative agreement with the experiment. This is clearly due to the fact that our PLC is not a rigid rod polymer, but consists of rigid sequences intercalated with flexible ones. Work on extension of the Flory theory to such systems is ongoing. We expect to report later theoretical results in conjunction with the appropriate experimental phase diagrams.

Acknowledgments

PLC samples were kindly provided by Dr. W.J. Jackson, Jr. of Eastman Kodak Co., Kingsport, TN. W.B. began work on this problem at the Department of Materials Engineering of Drexel University in Philadelphia and appreciates discussions there with Prof. Marvin Garfinkle. Financial support was provided by the Deutsche Forschungsgemeinschaft, Bonn, for the research as well as the stay of M.H. at the University of North Texas and by the National Science Foundation, Washington, DC. M.H. is indebted to the Minister für Wissenschaft und Forschung, Nordrhein-Westfalen, for approving a sabbatical leave; to the HASYLAB at the DESY, Hamburg, for providing some time on their equipment; and to the coworkers of Prof. H.-G. Zachmann, University of Hamburg, for technical support.

Literature Cited

1. Krause, S.J.; Haddock, T.; Price, P.G.; Lehnert, P.G; O'Brien, J.F.; Helminiak, T.E.; Adams, W.W. J. Polymer Sci. 1986, 24, 1991
2. Brostow, W.; Dziemianowicz, T.S.; Romanski, J.; Werber, W. Polymer Eng. & Sci. 1988, 28, 785.
3. Husman, G.; Helminiak, T.; Adams, W.; Wiff, D.; Benner, C. Am. Chem. Soc. Symp. Ser.1980, 132, 203.
4. Hwang, W.-F.; Wiff, D.R.; Benner, C.L.; Helminiak, T.E. J. Macromol. Sci. Phys. 1983, B22, 231.
5. Witt, W. Kunststoffe-German Plastics 1988, 78, 795.
6. Huh, W.; Weiss, R.A.; Nicolais, L. Polymer Eng.& Sci. 1983, 23, 779.
7. Weiss, R.A.; Huh, W.; Nicolais, L. Polymer Eng.& Sci. 987, 27, 684.
8. Kohli, A.; Chung, N.; Weiss, R.A. Polymer Eng.& Sci. 1989, 29, 573.
9. Kiss, G. Polymer.Eng.& Sci. 1987, 27, 410.

10. DeMeuse, M.T.; Jaffe, M. <u>Molec. Cryst. Liq. Cryst.</u> 1988, <u>157</u>, 535.
11. DeMeuse, M.T.; Jaffe, M.   in the present volume.
12. Kosfeld, R.; Hess, R.M.; Friedrich, K. <u>Mater.Chem. & Phys.</u> 1987, <u>18</u>, 93.
13. Schubert, F.; Friedrich, K.; Hess, M.; Kosfeld,R. <u>Molec. Cryst. Liq. Cryst.</u> 1988, <u>155</u>, 477.
14. Friedrich, K.; Hess, M.; Kosfeld, R. <u>Makromol.Chem. Symp.</u> 1988, <u>16</u>, 251.
15. Jackson, Jr., W.J.; Kuhfuss, H.F. <u>J. Polymer Sci. Phys.</u> 1976, <u>14</u>, 2043.
16. Menczel, J.; Wunderlich, B. <u>J. Polymer Sci. Phys.</u> 1980, <u>18</u>, 1433.
17. Menczel, J.; Wunderlich, B. <u>Polymer</u> 1981, <u>22</u>, 778.
18. Meesiri, W.; Menczel, J.; Gaur, U.; Wunderlich, B. <u>J. Polymer Sci. Phys.</u> 1982, <u>20</u>, 719.
19. Noel, C. In <u>Polymeric Liquid Crystals</u>; Blumstein, A. (Ed.) Plenum: New York 1985; p.21.
20. Freidzon, Ya.; Boiko, N.I.; Shibaev, V.P.; Tsukruk, V.V.; Shilov, V.V; Lipatov, Yu.S. <u>Polymer Commun.</u> 1986, <u>27</u>, 19
21. Sawyer, L.C.; Jaffe, M. <u>J. Mater. Sci.</u> 1987 , <u>21</u>, 1897
22. Pelzl, G.; Latif, I.; Diele, S.; Novak, M.; Demus, D.; Sackmann, H. <u>Molec. Cryst. Liq. Cryst.</u> 1986, <u>139</u>, 333
23. Lin, Y.G.; Winter, H.H. <u>Macromolecules</u> 1988, <u>21</u>, 2439
24. George, E.R.; Porter, R.S.; <u>J.Polymer Sci. Phys.</u> 1988, <u>26</u>, 83.
25. Buchner, S.; Chen, D.; Gehrke, R.; Zachmann,H.-G, <u>Molec. Cryst. Liq. Cryst.</u> 1988, <u>155</u>, 357.
26. Brostow, W., <u>Kunststoffe - German Plastics</u> 1988, <u>78</u>, 411.
27. Brostow, W. In: <u>Polymer Liquid Crystals: From Structures to Applications</u>; Collyer, A.A.(Ed.) Elsevier Applied Science, London - New York, to appear in 1990; Chapter 1.
28. Meesiri, W.; Menczel, J.; Gaur, U.; Wunderlich, B. <u>J. Polymer Sci. Phys.</u> 1982, <u>20</u>, 719.
29. Jeziorny, A.; <u>Polimery</u> 1989, <u>34</u>, 210.
30. Chou, Ch.; Clough, S.B. <u>Polymer Eng.& Sci.</u> 1988, <u>28</u>, 65.
31. Viney, C.; Windle, A.H. <u>J. Mater. Sci.</u> 1982, <u>17</u>, 2661.
32. Benson, R.S.; Lewis, D.N. <u>Polymer Commun.</u> 1987, <u>28</u>, 289.
33. Gedde, U.W.; Buerger, D.; Boyd, R.H. <u>Macromolecules</u> 1987, <u>20</u>, 988.

34. Hedmark, P. Ph.D.Thesis, Royal Institute of Technology, Stockholm, 1988.
35. Kricheldorf, H.R.; Schwarz, G. Makromol.Chem. 1983, 184, 475.
36. Lenz, R.W.; Jin, J.I.; Feichtinger, K.A. Polymer 1983, 24, 327.
37. Nicely, V.A.; Dougherty, J.T.; Renfro, L.W. Macromolecules 1987, 20, 573.
38. Zachariades, A.E.; Economy, J.; Logan, J.A. J.Appl.Polymer Sci. 1982, 27, 2009.
39. Krigbaum, W.R.; Salaris, F. J.Polymer Sci.Phys.Ed. 1978, 16, 883.
40. Saechtling, H.J. Kunststofftaschenbuch Carl Hanser Verlag: München-Wien 1979, p.290.
41. Ide, Y.; Ophir, Z. Polymer Eng.& Sci. 1983, 23, 261.
42. Viola, G.G.; Baird, D.C.; Wilkes, G.L. Polymer Eng.& Sci. 1985, 25, 888.
43. Schubert, F.; PH.D. thesis, University of Duisburg, Duisburg, 1989.
44. Flory, P.J.   Proc. Royal Soc. A 1956, 234, 60.
45. Flory, P.J. e.a., Macromolecules 1978, 11, 1119, 1122, 1126, 1134, 1138, 1141.
46. Maier, W.; Saupe, A. Z. Naturforschung A 1959, 14, 882.
47. Maier, W.; Saupe, A. Z. Naturforschung A 1960, 15, 287.
48. Flory, P.J. In: Polymer Liquid Crystals; Cifferi, A., (Ed.) Academic Press: New York, 1982; Chap. 4.

RECEIVED April 3, 1990

Chapter 29

Novel Composites from Blends of Amorphous and Semicrystalline Engineering Thermoplastics with Liquid-Crystalline Polymers

Donald G. Baird and Tong Sun

Department of Chemical Engineering and Polymeric Materials and Interfaces Laboratory, Virginia Polytechnic Institute and State University, Blacksburg, VA 24061

This paper is concerned with the idea of generating films which are reinforced on an in-situ basis by means of the addition of thermotropic copolyesters. Various liquid crystalline copolyesters (LCP's) were used depending on the particular thermoplastic matrix which was selected. In this work three high performance matrices were used: polyether-etherketone (PEEK), polyphenylenesulfide (PPS), and Ultem, a polyetherimide. The processing window for the matrix/LCP pair was established using the temperature dependence of the dynamic mechanical properties of the materials. Under appropriate processing conditions fibrils of the LCP phase were generated and molecular orientation within the LCP phase was generated by drawing. An improvement of the torsional modulus of the films by a factor of 10 to 40 over that of the matrix resin was observed with as little as 10 wt % LCP. Torsional moduli as high as 40 to 75 GPa were observed. Dynamic tensile moduli were also measured but found to be in the range of 9 to 13 GPa which was lower than the torsional moduli range. It is not clear as to the origin of this difference, but it may be due to the fact the films are very thin (75 to 200 microns) and there may be an edge effect.

The field of composites is still dominated by the use of thermosetting resins based on epoxy and unsaturated polyester matrices loaded with inorganic reinforcements. High temperature amorphous and semicrystalline thermoplastic polymers such as polyetherimide (Ultem), polyetheretherketone (PEEK) and polyphenylenesulfide (PPS) whose properties when filled, match or exceed those of some thermosets, have recently entered the market posing a challenge to thermosets especially in high performance environments. Ways of further improving their engineering properties through the use of reinforcements have engaged the attention of researchers (1, 2). Furthermore, the very recent

0097–6156/90/0435–0416$06.75/0

availability of thermotropic liquid crystalline polymers (LCP) which have demonstrated value as high modulus/strength thermoplastics has also been reported (3). Blending of these LCPs with both amorphous and semicrystalline polymers such as polycarbonate (PC) and polyamide (PA) with a view of forming 'in-situ' thermoplastic composite prepreg has been studied in our laboratories (4, 5). The potential advantages that an all thermoplastic composite offers the processor are: 1) they may provide easier processability and wider range of properties and 2) they may lower the wear and tear on the processing equipment. This report will summarize the findings on the exploratory research conducted on processing films of blends of Ultem, PEEK, and PPS with several LCPs. An attempt is made to define under what combination of thermal and deformation history reinforcing fibrils of the LCP phase will be formed on an 'in-situ' basis in the matrix.

The objective of this work has been to generate films, tapes or ribbons which might serve as a prepreg from blends of either an Ultem or a PEEK or a high molecular weight PPS with various liquid crystalline polymers, to identify the parameters that control the formation of reinforcing microfibrils of LCP phase, and to study the mechanical properties of the composite films.

EXPERIMENTAL

Materials. The amorphous matrix polymer was polyetherimide (Ultem 1000) sold by the General Electric Company. The semicrystalline matrices were PEEK and a high molecular weight polyphenylenesulfide (PPS) provided by ICI and Phillips Petroleum Co. (Barthesville, OK.), respectively. The reinforcing phase was one of several LCPs: a liquid crystalline aromatic copolyester consisting of 73 % hydrobenzoic acid and 27 % 2-hydroxy 6-naphthoic acid moieties (Vectra A-900) sold by Celanese Corporation; another liquid crystalline aromatic copolyester consisting of terephthalic acid, phenyl hydroquinone and phenyl ethyl hydroquinone (Granlar) provided by Istituto Donegani (Montedison); and an experimental copolyester of parahydroxybenzoic acid and polyethylene terephthalate with 60 or 80 mole % of the hydroxybenzoic acid moiety (LCP60 or LCP80) supplied by Tennessee Eastman Company. A 50/50 blend of LCP60 and LCP80 was also used in this study.

Rheology. The rheological properties of the blends and their components were determined on a Rheometrics Mechanical Spectrometer (RMS 800). Three kinds of dynamic oscillatory measurements (i.e. temperature, time, and frequency sweeps) were carried out. All experiments were done by using a parallel plate attachment with a radius of 12.5 mm and a gap setting from 1.2 to 1.8 mm. There was no significant dependence of the experimental results on the gap setting.

For the temperature sweep experiments in the cooling mode (i.e. the cooling curves), a polymer sample to be studied was preheated at a temperature, usually 20°C to 30°C higher than its melting point or flow temperature for 5 minutes. Then the sample was cooled at a preset rate, usually 5 or 8 degrees per minute. The storage and loss moduli, G' and G", respectively, and the complex viscosity, $\eta*$, were measured at each temperature interval during cooling. Unless indicated otherwise, the angular frequency used was 10 rad/sec and the strain was 10 % for all tests. Although at lower temperatures this strain could exceed the linear viscoelastic limit, the measurements represented only the qualitative response of the materials. For the time sweep experiments the frequency and strain used were the same as in temperature sweeps. For the frequency sweep experiments the temperature range used for a given polymer will be specified in the text.

Processing. Pellets of the matrix polymers, Ultem, PEEK, and PPS were dried at 170°C, 150°C, and 110°C, respectively. Pellets of the liquid crystalline polymers Vectra, Granlar, LCP60, and LCP80 were dried at 150°C. All drying processes were carried out in a vacuum oven for at least 24 hours. The dried pellets of a given matrix polymer and a given liquid crystalline polymer were then tumbled together in predetermined weight ratios in a steel container before being blended in a screw extruder. A laboratory single screw extruder (Killion KL 100) connected to a 102 mm wide film die (die gap = 0.5 mm) was used to generate the films. Highly polished chrome-plated rolls were used to introduce some draw ratio into the films.

Structure and Morphology. Wide angle x-ray diffraction (WAXD) was carried out by using a Phillips 1720 table-top x-ray generator equipped with a fixed copper target x-ray tube and a Warhus camera. The diameter of the pin-hole collimator was 0.5 mm and the sample-to-film distance was 76 mm. The beam conditions were at 40 kv and 20 mA and the patterns were recorded under vacuum. The morphology of the drawn sheets was determined by means of scanning electron microscopy (SEM) (Cambridge model 200). All samples were fractured perpendicular or parallel to the draw direction in liquid nitrogen and the fracture surfaces were sputter coated to provide a gold layer of about 15 nm in thickness.

Mechanical Properties. Dynamic mechanical properties were determined both in torsion and tension. For torsional modulus measurements, a rectangular sample with dimensions of 45 by 12.5 mm was cut from the extruded sheet. Then the sample was mounted on the Rheometrics Mechanical Spectrometer (RMS 800) using the solid fixtures. The frequency of oscillation was 10 rad/sec and the strain was 0.1% for most samples. The auto tension mode was used to keep a small amount of tension on the sample during heating. In the temperature sweep experiments the temperature was raised at a rate of 5°C to 8°C per minute until the modulus of a given sample dropped remarkably. The elastic component of the torsional modulus, G', of the samples was measured as a function of temperature. For the dynamic tensile modulus measurements a Rheometrics Solid Analyzer (RSA II) was used. The frequency used was 10 Hz and the strain was 0.5 % for all tests.

RESULTS AND DISCUSSION

Rheology. In order to generate microfibrils of the LCP phase in a thermoplastic matrix it is necessary to establish a processing temperature window. For this purpose the upper and lower processing

temperature limits of each component polymer must be determined st
first. The overlapped temperature range of the processing window of
component polymers gives the width of the processing window of the
corresponding blend. For illustration the Ultem/Vectra blend system
is discussed below in detail. The processing windows of all other
blend systems studied in this paper were evaluated in a similar way.

The upper temperature limit of the processing window of a given
polymer can be determined rheologically. From a series of isothermal
time sweeps the highest temperature at which the rheological
behavior of a polymer melt still remains stable in a certain period
of time (at least longer than the residence time of the polymer in
the extruder) is defined as its upper rheologically stable
temperature, Tr. As seen from Figure 1, the storage modulus, G', of
Ultem remains almost unchanged as the melt is held at 390°C. (this
is the maximum attainable temperature in the RMS 800). The slight
change in viscosity at all temperatures other than 350°C in the
first few minutes is most likely due to the time required for the
temperature of the sample to reach equilibrium (because the
preheating temperature of the melt before all isothermal time sweeps
is 350°C).

In the case of Vectra, the complex viscosity, η^*, remains
almost unchanged with time when the temperature is below 350°C
(Figure 2). Above 350°C after a certain induction period the
viscosity increases with time. Meanwhile, the higher the
temperature the shorter the induction period. This increase of
viscosity may be due to further polycondensation of this
copolyester. Therefore, 350°C can be taken as its upper
rheologically stable temperature.

The lower temperature limit for the processing a polymer
depends on its rheological behavior as it is cooled down from the
melt. To determine the lower processing temperature for a polymer
melt the temperature at which G'=G" (or tan δ = 1) during cooling,
designated as Tx, was used as the criterion. This is taken as the
onset of a network in thermosetting systems. For the case of Ultem
which is amorphous, the normal activation energy controls the
recovery of the rheological properties as the material is cooled
from the extrusion temperature. Its Tx (Tx= 285°C) is actually not
affected by the initial temperature of the melt (Figure 3).
Similarly, the melt viscosity at a given temperature is also
independent of the initial temperature of the melt. For Vectra,
however, which is semicrystalline, the recovery of the rheological
properties as the melt cools down depends on the degree of
super-cooling which takes place. This is illustrated by the results
presented for both its T-tan δ plot (not shown here) as well as T
versus η^* plot (Figure 4). As the melt is cooled from 300°C, Tx, is
found to be 264°C which is 20°C lower than its melting point, Tm.
When the preheating temperature is raised to 310°C, the viscosity
drops about a decade and Tx is reduced to 255°C. The viscosity and
Tx then drop to a lesser extent as the preheating temperature goes
from 310°C to 360°C.

Based on the experimental data for Tr and Tx of Ultem and
Vectra, the width of the processing window of each neat polymer can
be determined. For Ultem and Vectra the temperature ranges are
285°C-390°C and 255°C-350°C, respectively. The overlapped
temperature range of the processing window of the Ultem/Vectra

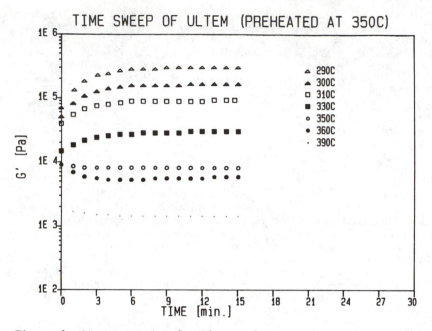

Figure 1. G′ versus time for Ultem melt at various temperatures.

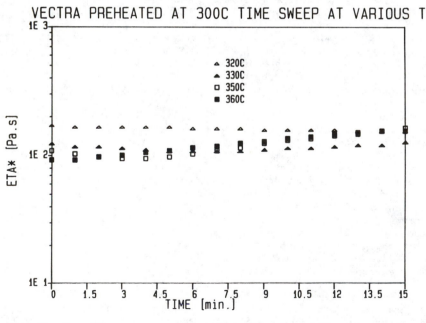

Figure 2. η* versus time for Vectra melt at various temperatures.

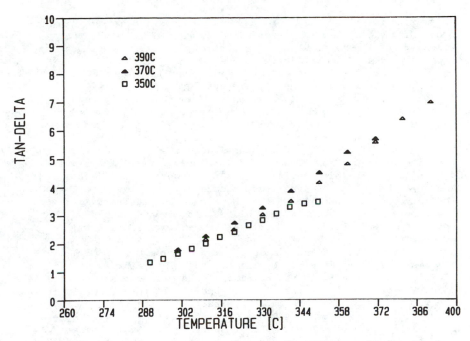

Figure 3. tan δ versus temperature for Ultem melt as it is cooled from various temperature.

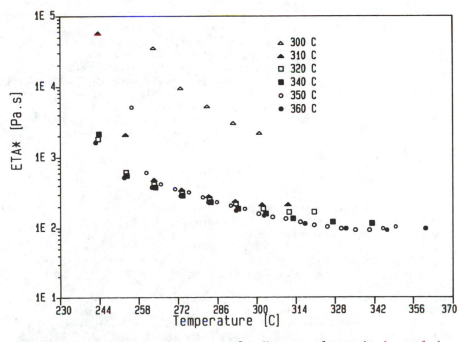

Figure 4. η* versus temperature for Vectra melt as it is cooled from various temperatures.

polymer pair (i.e. from 285°C to 350°C) serves as the width of the processing window for this system. Over this temperature the viscosity of Vectra is lower than that of Ultem and hence one can expect the formation of fibrils.

To understand the effect of blend composition on the rheological behavior of the Ultem/Vectra system the complex viscosity change during cooling was also monitored (Figure 5). For pure Ultem, Tx is around 285°C. This temperature is lowered to 280°C as 10 % of Vectra is added. It is also noticed that the viscosity of the Ultem/Vectra 90/10 blend is still dominated by Ultem. The Ultem/Vectra 70/30 blend shows a similar behavior. For the Ultem/Vectra 50/50 blend, $\eta*$ is somewhat lower than for other compositions. Tx also shifts to 275°C. However, the viscosity of the 50/50 blend is still at least a decade higher than for the pure Vectra. Therefore, based on the above results, it is more reasonable to use the Tx of the blend at a given composition (but not that of pure Ultem) as the lower temperature limit for processing.

Viscosity ratios are one of the key factors that determine if the disperse phase in a blend will elongate into fibrils or stabilize as droplets in the matrix. It is well known that the formation of fibrils of the disperse phase will only occur if it has a lower viscosity than the matrix phase (6-8). As seen from Figures 6 and 7, in the extrusion temperature range used, the viscosity of Ultem is 10 to 50 times more than that of Vectra over a relatively wide frequency range. Also the viscosities of both Ultem and PEEK (as matrix polymers) are higher than that of Granlar (as reinforcement polymer). One would expect that these facts would promote the in-situ generation of fibrils of the liquid crystalline polymer phase in the corresponding Ultem/Vectra, Ultem/Granlar, and PEEK/Granlar blend systems. Also noteworthy is the fact that if the Granlar is preheated at 365°C instead of at 350°C (Figure 7), both the viscosity at a given temperature and the Tx are lowered apparently. This thermal history dependant supercooling effect favors not only fibril formation, but also broadening the width of processing windows for Ultem/Granlar and PEEK/Granlar blend systems.

The viscosities of PPS and some liquid crystalline polymers are presented in Figure 8 as a function of shear rate (9). It should be mentioned here that the viscosities of LCP60 and LCP80 were difficult to measure at 300°C and hence are presented at 285°C and 310°, respectively. All the liquid crystalline polymers shear-thin in the range of shear rates studied whereas PPS seems to exhibit Newtonian behavior at rates below 10 sec^{-1} and shear thinning behavior above 10 sec^{-1}. Since all the liquid crystalline polymers exhibit lower viscosities than PPS, especially with increasing shear rate, it can be expected that this would favor the formation of fibrils of the reinforcement phase if the blends are processed at rates exceeding one sec^{-1}. The LCP60 seems to exhibit an order of magnitude lower viscosity than PPS in the range of rates measured and hence should form fibrils easily when processed as a blend with PPS.

The effect of deformation while cooling from above the melt temperature was also studied. The value of G' as a function of temperature at constant strain for the materials was followed. The results depicted in Figure 9 show that all the liquid crystalline

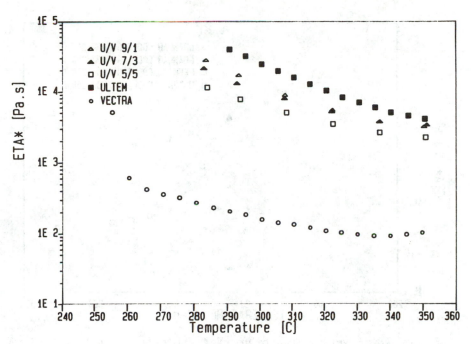

Figure 5. η^* versus temperature for the melts of Ultem, Vectra, and Ultem/Vectra blends (9/1, 7/3, and 5/5) cooling from 350°C .

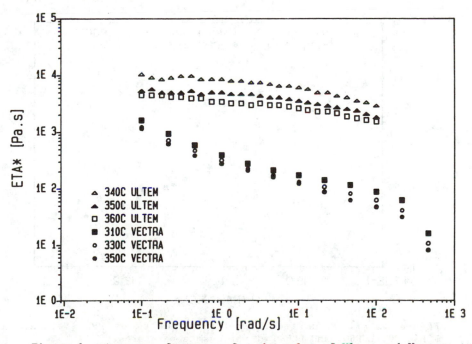

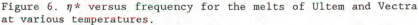

Figure 6. η^* versus frequency for the melts of Ultem and Vectra at various temperatures.

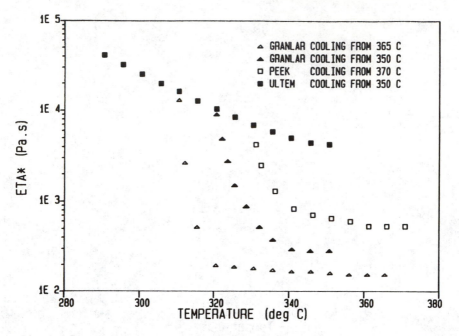

Figure 7. $\eta*$ versus temperature for the melts of Ultem, PEEK, and Granlar cooled from various temperatures.

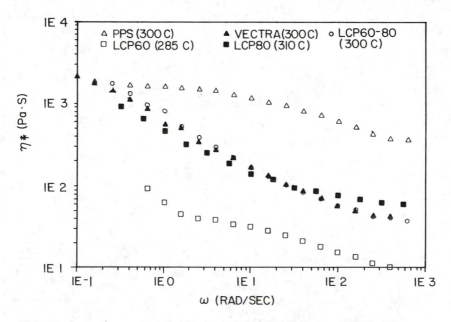

Figure 8. $\eta*$ versus frequency for the melts of PPS, Vectra, LCP60-80, LCP60, and LCP80 at various temperatures.

polymers were deformable well below their respective melt temperatures. However, PPS showed no abrupt change in the slope of G' indicating slow crystallization kinetics. The higher elasticity of the supercooled state of the liquid crystalline polymers, as indicated by the increase in G' with the lowering of temperature, promises some processing advantages especially in an operation such as film drawing where melt strength is of importamce. Such tests provide a clue to the minimum die temperature required to generate a film with sufficient elasticity for drawing.

Structure and Morphology. To reveal the effect of flow on fibril formation, the morphology of the extrudates obtained from different extrusion conditions was examined. When the Ultem/Vectra blends were extruded through a short capillary die (L/D= 12.5) using an Instron Capillary Rheometer (ICR) without melt drawing outside the die, distinct microfibrils in the extrudates were evident. However, if the Ultem/Vectra blend was extruded through a sheet die designed with a coathanger feed section connected to a single screw extruder, then no microfibrils were observed. The absence of Vectra microfibrils in the latter case is believed to be due to the lack of sufficient extensional deformation in the flow field in the sheet die and additional mixing in the land of the die which breaks up the fibrils. A possibility for creating an extensional flow field is to draw the extrudate outside the sheet die when it is still in a molten or deformable state. To examine this possibility and the effect of melt drawing on in-situ fibril generation, the morphological textures of fracture surfaces of Ultem/Vectra blend sheets for various composition ratios were studied as a function of melt draw ratio either in the first drawing zone (i.e. between the die and the first roll) or in the second drawing zone (i.e. between the first roll and the second roll).

The SEM micrographs of a series of Ultem/Vectra 70/30 blend films drawn to various degrees in the first zone are shown in Figure 10. For the sample with the lowest draw ratio, most of the Vectra domains are only slightly deformed. However, as the draw ratio is increased, the fibril formation of the Vectra phase is clearly displayed due to the higher extensional deformation. As the Vectra content is increased to 50%, some of the LCP domains coalesce and form a second continuous phase with a layer-like structure. A similar trend of the effect of draw on fibril generation could also be found for the extruded sheets drawn in the second zone (the speed of the first roller was adjusted so that the extrudate would pass through it without any drawing taking place in the first zone). Based on these results it is apparent that the formation of a fibrillar reinforcing phase is more easily obtained in an extensional flow field than in shear. Meanwhile, the higher the extensional deformation the thinner and the longer the microfibrils.

To obtain an insight into the effect of the draw ratio on the molecular orientation of the Vectra phase, wide angle x-ray diffraction, WAXD, was utilized and the results for the Ultem/Vectra 70/30 sheet drawn in the first zone are shown in Figure 11. For the sample with the lowest draw ratio, a sharp Debye ring corresponding to a d-spacing of 0.46 nm is observed. This observation is in agreement with previous data on Vectra reported in the literature (10). As the draw ratio increases, the azimuthal dependence of the

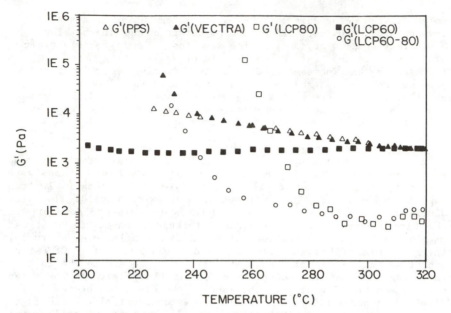

Figure 9. G' versus temperature for the melts of PPS, Vectra, LCP80, LCP60, and LCP60-80 cooled from 320°C.

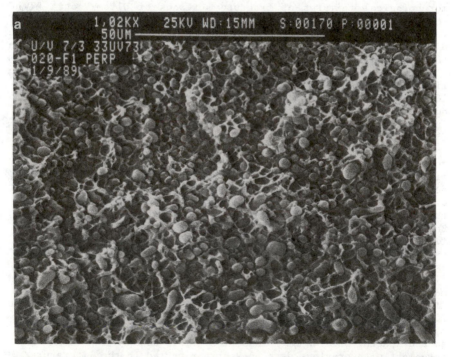

Figure 10. SEM micrographs of the fractured surfaces of Ultem/Vectra (70/30) blends drawn in the first zone with various draw ratios. (a) D.R. = 1.8. *Continued on next page.*

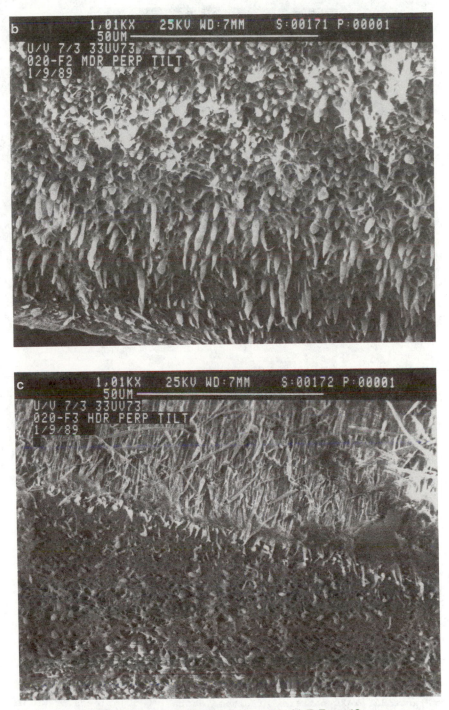

Figure 10. Continued. (b) D.R. = 6.3; (c) D.R. = 19.

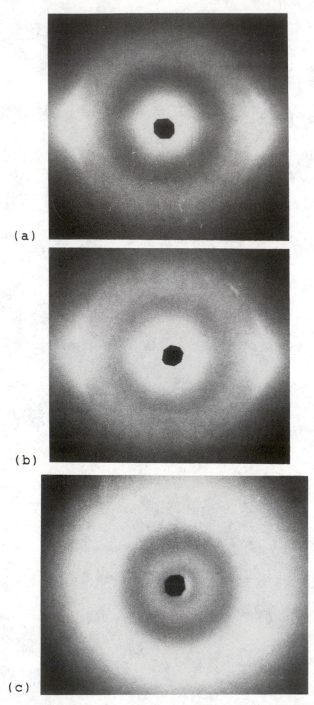

Figure 11. WAXD patterns of Ultem/Vectra (70/30) blends drawn in the first zone with various draw ratios, (a) D.R.=19, (b) D.R.=6.3, (c) D.R.=1.8.

scattering pattern indicates partial molecular alignment along the draw direction. In addition, the degree of orientation tends to increase with an increase in the draw ratio judging from the decrease in arc length. This molecular orientation after drawing should result in higher stiffness of the Vectra phase and consequently, the reinforcing effect should improve. This conclusion is supported by the results of the mechanical properties to be discussed below.

In comparison with the Ultem/Vectra blend, an interesting feature of the Ultem/Granlar blend is that the morphology of the latter shows a significant dependance on the blend composition ratio. As revealed from DSC and dynamic mechanical spectroscopy (11), Ultem and Granlar are partially miscible. Furthermore, both methods suggested that the limit of this miscibility is between 15 to 20 wt % of Granlar loading. Although the viscosity of the Ultem melt is much higher than that of Granlar at the extrusion temperature ($360°C-370°C$), at a 10 wt % of Granlar loading no phase separation, and consequently, no microfibril formation could be observed in the Ultem/Granlar 90/10 blend film, no matter how high a draw ratio was applied. As the Granlar content was increased to 20 wt % or higher, a distinct phase of Granlar was observed. However, the in-situ formation generation of fibrils in the drawing step was not very effective for the case in which drawing took place in the first or the second region. Most of the disperse liquid crystalline polymer phase in these cases was elongated but not to the extent of forming a thin fibril structure. The only case that showed good fibril formation for this Ultem/Granlar 70/30 system was by using the double-draw method (i.e., the film was drawn in both the first and the second regions, see Figure 12).

The morphology of composites from blends of semicrystalline engineering thermoplastics with various liquid crystalline polymers has also been studied. The purpose of this part of the research was to verify if the matrix polymer was switched to a semicrystalline one and had been successfully deformed into microfibrils, would it possess adequate interfacial adhesion with the matrix. When PEEK/Granlar 70/30 blends were extruded (barrel temperature = $370°C$, die temperature= $360°C$) and drawn in the first zone at various draw ratios ($DR_1 = 8$, 16, and 24), no microfibrils could be observed in the drawn film even at the highest draw ratio ($DR_1 = 24$). The disperse phase shows only some degree of fibril formation (Figure 13a). It seems most likely that the extrudate temperature was too high. One possible way for solving this problem is to lower the die temperature. But it was restricted by the high solidification temperature of the PEEK. Another approach is to lower the temperature of the extruded film during the drawing process. For the cases that the drawing takes place in the second zone ($DR_2 = 8$) or in both the first and the second zones ($DR_1 x DR_2 = 2 \times 8 = 16$), microfibril formation was displayed distinctly (Figures 13b and 13c). Judging from these SEM micrographs, it is interesting to notice that the interfacial adhesion seems to be better than that for the Ultem/Vectra system.

The fracture surface of PPS/LCP60 blend displays evidence of LCP60 fibrils of fine but non-uniform diamater embeded in the PPS matrix. This structure is similar to that observed for PC/LCP60 reported in earlier studies (3). The relative absence of voids and

Figure 12. SEM micrographs of the fractured surfaces of Ultem/Granlar (70/30) blends drawn in different zones: (a) first zone (b) second zone (c) both zones.

the higher degree of adherence between the phases, as compared with the other blends in this research, is surprising especially if one considers the wide differences in the melting points and rheological properties of the two components of the blend.

The morphology of the PPS/LCP60-80 blends is somewhat different from that of the PPS/LCP60 system in that the fibrils of the minor phase (LCP60-80) are of a relatively larger diameter perhaps indicating the possibility of a further reduction in domain size through processing. A reduction in fibril size has been correlated with improvement in properties in the PC/LCP60 systems in previous studies (3).

Figures 14 and 15 present evidence of undeformed droplets of the liquid crystal phase in the PPS matrix. The fracture surface of the PPS/LCP80 as depicted in Figure 14 shows the small diameter droplets of the LCP80 phase and suggests that drawing only resulted in, possibly, breaking up the large droplets rather than the elongation of the minor phase into fibrils. The large voids also indicate poor adhesion between the phases. Similarly, in Figure 15 is presented the morphology of the PPS/Vectra blend which shows no evidence of elongation of the minor phase at any degree of draw. The size and shape of the LCP80 phase ellipsoids indicate the possibility of altering the morphology to that of a fibrillar phase by changing the processing conditions. However, similar possibilities do not seem to exist with the fine sperical shaped droplets of Vectra. Moreover, the matrix structure in the PPS/Vectra blends seems similar to what a foam structure would look like, providing an indication perhaps of the escape of volatile gases generated during the processing of the blends.

Mechanical Properties. To reveal the reinforcing effect of liquid crystalline polymer microfibrils on the mechanical properties of the films both their dynamic torsional moduli and dynamic tensile moduli have been studied as a function of temperature using a Rheometrics Mechanical Spectrometer (RMS 800) and a Rheometrics Solids Analyzer (RSA II), respectively. For comparison purpose the modulus of neat matrix polymers and, in some cases, the modulus of carbon fiber and Kevlar fiber reinforced composites has also been measured.

As seen from Figure 16, the torsional modulus of the Ultem/Vectra 70/30 blend sheets increased as the draw ratio was increased. This trend is observed no matter what composition ratio (Ultem/Vectra 90/10, 80/20, and 50/50) is used and in which zone the drawing occurs. Hence, there is a significant indication that the LCP fibrils lead to a reinforcement of Ultem.

The concentration of Vectra in the blend also shows a pronounced effect on the drawability and the modulus of the Ultem/Vectra blend sheets. Under the same extrusion conditions, the drawability, as characterized by the maximum draw ratio, decreased with increasing content of the liquid crystalline polymer in the blend. However, when the draw ratio is similar the higher the content of the reinforcing phase, the higher the torsional modulus. The highest value of the torsional modulus was obtained for the materials with only 10 % Vectra because of the higher drawability of the system.

If the Ultem/Vectra blend sheets of equal composition were drawn in the second zone, both the maximum draw ratio and the

Figure 13. SEM micrographs of the fractured surfaces of PEEK/Granlar (70/30) blends drawn in different zones. (a) first zone (b) second zone (c) both zones $(DR_1 = DR_2, DR_2 = 8)$.

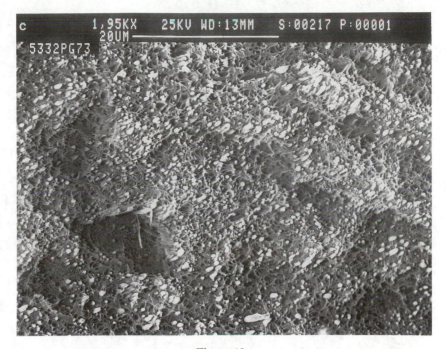

Figure 13c.

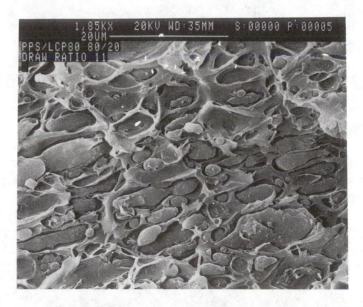

Figure 14. SEM micrographs of the fractured surface (parallel to draw direction) for PPS/LCP80 (80/20) blend.

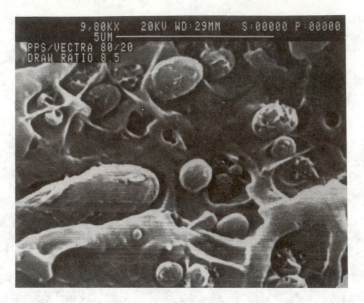

Figure 15. SEM micrographs of fractured surface (parallel to draw direction) for PPS/Vectra (80/20) blend.

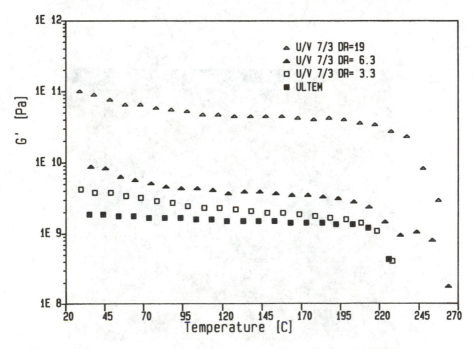

Figure 16. G' versus temperature for Ultem/Vectra (70/30) blend sheets drawn at various draw ratios in the first zone.

modulus were decreased in comparison with those of the sheets drawn in the first zone. Here, the lower drawability is, of course, not due to the difference in compositions, but due to the lower temperature in the second zone than that in the first zone. It is most likely that due to the remarkable increase of viscosity or even the solidification of the Vectra phase at the lower temperature, it becomes more difficult to elongate the Vectra droplets and to generate the long thin microfibrils and, consequently, a lower modulus results.

From the torsional modulus measurements, it is clear that by blending Vectra with Ultem the torsional modulus of the Ultem sheets can be greatly enhanced. G' of the neat Ultem sheet is around 1.5 GPa, but G' of the Ultem/Vectra blend sheets (with a Vectra content ranging from 10% to 30%) is in the range of 50 to 70 GPa, i.e. about 30 to 45 times higher than that of neat Ultem sheet. The existence of Vectra microfibrils improves G' of the blend sheet for temperatures up to the glass transition temperature of Ultem.

A similar trend of the effect of draw ratio on G' has also been found both for Ultem/Granlar and PEEK/Granlar blend systems (see Figures 17 and 18). One point worthy of note is that in contrast with Ultem, PEEK is a semicrystalline polymer. However, judging from the WAXD results for the PEEK/Granlar blend films, the PEEK matrix is still mainly amorphous due to the rapid cooling of the film. Therefore, further increase of the modulus for PEEK/Granlar blend films would be expected after promoting crystallization in the matrix phase by thermal annealing.

After drawing the torsional storage modulus of the blend of LCP60-80 with PPS is remarkably improved in comaprison with that of neat PPS film and its magnitude is comparable with that for the PPS/carbon fiber composite (Figure 19). On the other hand, for the blends of PPS with LCP80 there was no significant enhancement of the properties as the draw ratio was increased. This is due to the lack of fibrils in the latter case.

The dynamic tensile moduli, E', of the Ultem/Vectra 70/30 and the PEEK/Granlar 70/30 blend films have also been measured. The dynamic torsional moduli of the films are also plotted for comparison in Figure 20. The E' of these two films is 9.2 GPa for Ultem/Vectra 70/30 and 13 GPa for PEEK/Granlar 70/30, respectively. This high modulus also reflects the apparent reinforcing effect of the fibrils embeded in the matrix of the blend films. However, the value of the storage tensile modulus is significantly lower than the corresponding storage torsional modulus. This difference is not clearly understood at this point. It may be due to an edge effect which makes the thin films seem stiffer than they really are in torsion. Further efforts to understand this difference are being carried out.

CONCLUSION

Several liquid cryatalline polymers were melt blended with an amorphous (Ultem) and two semicrystalline (PEEK and PPS) engineering thermoplastics in a single screw extruder. Flat film was processed with different degrees of stretch imparted while the film was being cooled. In the case of Vectra, which was thought to be ideally suited to be blended with PPS based on thermal and rheological

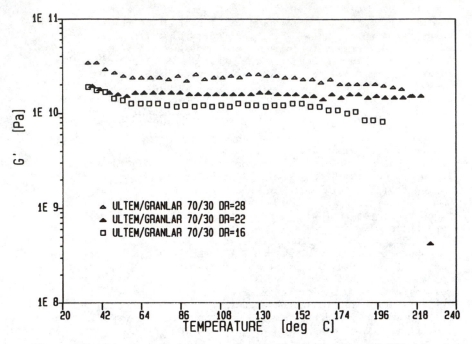

Figure 17. G' versus temperature for Ultem/Vectra (70/30) blend
sheets drawn at various draw ratios.

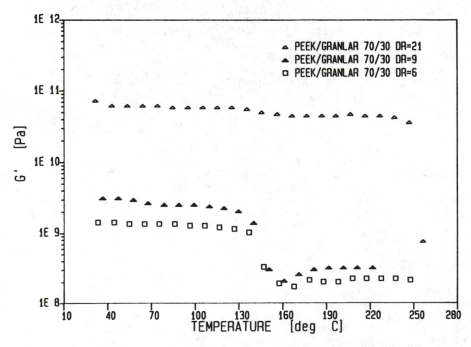

Figure 18. G' versus temperature for PEEK/Granlar (70/30) blend
sheets drawn at various draw ratios.

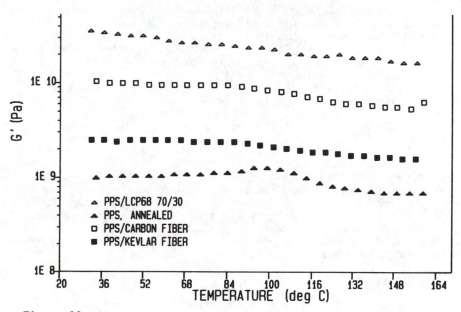

Figure 19. G' versus temperature for PPS/LCP60-80 (70/30) blend sheet, PPS/carbon fiber composite, PPS/Kevlar fiber composite and annealed PPS sheet.

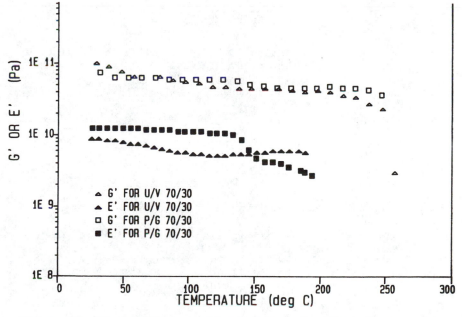

Figure 20. G' and E' versus temperature for Ultem/Vectra and PEEK/Granlar blend sheets.

considerations, but it was found to react with the PPS under the
conditions of extrusion and hence resulted in a poor match. It is
possible that the additives in Vectra caused the mismatch. In the
other blend systems studied, liquid crystalline polymer microfibrils
could be successfully generated as suitable compositions and drawing
conditions were chosen. In general, the higher the draw ratio, the
thinner the fibrils formed, the higher the degree of molecular
orientation in the reinforcing phase, and consequently, the higher
the modulus. For most blend systems an increase in the storage
torsional modulus of 10 to 40 times over that of the corresponding
neat matrix polymer could be achieved. A disagreement in the
magnitude of the dynamic torsional and tensile modulus was observed.
This unusual behavior is not well understood yet, but it may be due
to the discontinuity of the reinforcing phase and the highly
anisotropic nature of these materials. Despite this fact, the
tensile moduli of films reinforced with LCP are still a factor of 3
to 4 higher than those of the neat resin. The attempt of making
reinforced thermoplastic composite film from chosen polymer pairs
seems to be a realistic possibility. In the future, we will look at
laminating and thermoforming these materials.

Acknowledgments

    Support of this research was provided by Istituto Guido
Donegani, the Center On Innovative Technology (Grant No. MAT-88-019)
and the Army Research Office (Grant No. DAALO3-88-K-0104). Their
support is sincerely appreciated. Authors would thank also Dr. H. H.
Huang for SEM measurements.

LITERATURE CITED

1.   Lopez, L.C., Ph.D. Thesis, Virginia Polytechnic Institute and
     State University, Blacksburg, VA, 1987.
2.   Cogswell, F.N., Natl. SAMPE Symp., 1983, 28, 528.
3.   Jackson, W. J., Jr.; and Kuhfuss, H.F., J. Poly. Sci., 1976,
     14, 2043.
4.   Ramanathan, R.; Blizard, K. G.; and Baird, D. G. 45th SPE-ANTEC
     1987, 1399.
5.   Blizard, K. G.; and Baird, D. G. Polym. Eng. Sci. 1987, 27,
     653.
6.   Utracki, L. A. 45th SPE-ANTEC 1987, 1339.
7.   Ramanathan, R.; and Baird, D. G. Polym. Engr. Sci., 1988, 28,
     17.
8.   Baird, D. G., and Ramanathan, R. ACS Polymer Preprints, 1989,
     30, 546.
9.   Baird, D. G.; Sun T.; Done, D. S.; and Ramanathan, R. ACS
     Polymer Preprints, 1989, 30, 546.
10. Biswas, A.; and Backwell, J. Macromolecules, 1989, 21, 3158.
11. Sun, T.; Huang, H. H.; Baird, D. G.; and Wilkes, G. L.
     (unpublished work).

RECEIVED March 26, 1990

Chapter 30

Investigations into the Structure of Liquid-Crystalline Polymer–Liquid-Crystalline Polymer Blends

Mark T. DeMeuse[1] and Michael Jaffe[2]

[1]Enimont America, Inc., 2000 Princeton Park Corporate Center, Monmouth Junction, NJ 08852
[2]Hoechst Celanese Corporation, 86 Morris Avenue, Summit, NJ 07901

Blends in which both of the components are capable of forming liquid crystalline phases in the melt have been investigated. Rheological data as well as solid state data is presented which suggests that not all of the blends behave the same. This observation is in contrast to ideas of small molecule liquid crystals. The results are interpreted in terms of present theories regarding blends.

Much work has appeared recently in the polymer literature concerning blending of polymers as a mechanism for tailoring physical properties to obtain a certain performance level. Many of these efforts have focused on determining miscibility criteria for the blends. Variables such as molecular weight, temperature and composition of the two component materials have been shown to be important in understanding the phase behavior present in these systems (1).

The majority of the work done thus far has dealt with mixing of two random coil polymers. The Flory–Huggins formalism is usually used to describe the expected free energy of mixing. As such, miscibility is normally dominated by the enthalpic part of the free energy.

The one major exception to this observation has involved studies of molecular composites (2,3). In those systems, the two polymers being mixed have vastly different molecular conformations. As predicted by Flory (2,3), phase separation is easily induced in such systems. It should be recognized that this separation is based solely on entropic effects.

0097–6156/90/0435–0439$06.00/0

Two extreme cases have, then, been established in
the literature. Random coil/random coil mixing is
dominated by enthalpic effects while random coil/rigid
rod mixing is dominated by entropic effects. The present
investigation will be part of an ongoing study to bridge
the gap between these two extremes by examining mixtures
of two liquid crystal polymers. The conformation of
these polymers is best described as being semi-flexible.

By analogy with small molecule liquid crystals,
where the type of liquid crystal formed is used as a test
for miscibility, it is expected that all polymer
molecules that form the same type of liquid crystalline
phase will be miscible (4). This is in contrast to more
traditional polymers where miscibility is the exception
rather than the rule. The present work will suggest
which of these concepts is applicable to liquid crystal
polymer blend systems.

A related study has recently been reported by
Ciferri, et al (5) using the system cellulose acetate and
(hydroxypropyl) cellulose dissolved in
N,N-Dimethylacetamide (DMAC). The two polymers exhibit
similar conformations and both exist in anisotropic
phases above a certain critical concentration. It was
found that when the ternary systems were examined, two
anisotropic phases exist above the critical volume
fraction. Further, each of the stable anisotropic phases
contain pure polymer. The authors conclude that
miscibility between polymers forming the same type of
mesophase isn't necessarily observed, in contrast to low
molecular weight liquid crystals.

The purpose of the present work is to extend the
above studies to blend systems consisting of two
components, each of which is capable of forming a liquid
crystalline phase in the melt. Such blends become of
increasing importance with the recently reported finding
(6) that it is possible to observe synergisms in
mechanical properties in these systems. Our previous
studies in this area have suggested that the theories of
traditional polymer blend systems (7,8) are applicable to
these blends. This paper is a further study of the
applicability of these concepts.

Experimental Procedure

The samples used in the present study are (1) a copolymer
consisting of p-hydroxybenzoic acid (HBA) and
6-hydroxy-2-naphthoic acid (HNA) in the mole fraction
ratios 73:27 and (2) a polymer which contains HBA, HNA,
terephthalic acid (TA) and hydroquinone (HQ) in the
ratios 57:41:1:1. It will be shown later that the
incorporation of the small amount of TA and HQ into the
latter structure makes the material less shear-thinning
than the copolymer containing only HBA and HNA, when
examined rheologically. It is generally believed that a
random sequence distribution of monomers occurs in both

of these polymers. This conclusion has been reached based on X-ray diffraction work done by both Stamatoff (9) and Blackwell, et al (10,11) which has shown that the meridonal scattering pattern of these copolymers is consistent with a random copolymer concept.

The blends were produced via a melt blending procedure. The equipment used for this procedure was a Haake—Buchler Rheocord. The sample size was 70.0 grams which was determined by the density of the sample. Since the samples used in this work are capable of undergoing hydrolytic degradation (12), drying before any melting process is required. The samples in the present study were dried overnight at 120 C prior to melting.

Standard mixing conditions were to melt blend the samples at 300 C for five minutes at a rotor speed of 40 RPM. The effect of mixing history on the structure of the LCP/LCP blends has not been investigated in the present work. Such a study has been reported by Mehta and Baird (13).The samples were, then, given two minutes at 5 RPM followed by programming the rotor speed to 100 RPM in a two minute period. The speed is, then, instantaneously lowered back down to 5 RPM in two minutes. This is followed by two minutes at 5 RPM to determine whether the ramp test has affected the sample in any way.

The X-ray diffraction work which is reported was performed on strands of both the neat materials and a 50/50 composition blend. Meridonal scans were performed using a focused beam on a high resolution Huber diffractometer utilizing CuK$_\alpha$ radiation on a high intensity rotating anode. The beam, focused at the sample position, was about 1 X 1 mm in size. A θ-2θ coupling was performed using the total external reflection of a glass slide. A step scan was performed in transmission mode from 5° to 60°(2θ) with 0.1 step size for a period of 60 seconds and for expanded scans from 8° to 16°(2θ), the stepping time was 120 seconds per step.

DSC experiments were performed by heating powdered samples of both the blends and component materials in a nitrogen atmosphere at 99 C/min to 300 C, letting the samples ambiently cool and rescanning using a heating rate of 20 C/min. Both crystallization traces upon cooling and melting traces upon reheating were recorded.

Dynamic mechanical experiments were performed on compression molded bars which had been molded from powders. Compression molding of the samples produced specimens adequate for solid state evaluation but not sufficient for physical property evaluation. Samples were molded using a standard Wabash press and a molding temperature of 300 C.

Evaluation of the molded samples was performed using a Polymer Laboratories DMTA operating in the bending mode. The samples were evaluated in the dual cantilever

mode under conditions of constant strain and frequency.
Typical heating rates were either 2 or 5 C/min.

Rheology

As demonstrated previously (7), the ramping of rotor
speeds can be used to obtain similar rheological
information as is obtained via more traditional capillary
rheometer methods. Specifically, it was shown that the
following relation is applicable to torque and rotor
speed data

$$M = K \cdot C(n) \cdot S^n \tag{1}$$

In this relation, M is the torque, K is the familiar
consistency index from viscosity versus shear rate
relationships, S is the rotor speed, and n is the
power-law index, also familiar from viscosity versus
shear rate plots. It can be seen from Equation 1 that a
log-log plot of torque versus rotor speed should yield a
straight line with intercept given by log $(K \cdot C(n))$ and
slope given by n, the power-law index.
    The power-law values which are obtained for the
present blends are displayed in Figure 1. The values
which are reported there are averages of values obtained
on two separate runs on the same composition blends.
Also, on the same run, values for the up part of the
ramping procedure as well as the down part have been
averaged. In essence, then, the values reported
correspond to averages of four data points. In general,
the power-law index values differed by no more than 0.03
on the duplicate experiments.
    Several interesting points can be observed from an
analysis of the data. First, in data not presented here,
it can be shown that the value of Log $(K \cdot C(n))$ passes
through a minimum at the 50/50 composition blend.
Recalling that the value of $K \cdot C(n)$ really corresponds to
the torque generated by the samples at a rotor speed of 1
RPM, this result suggests that the torque generated at
very low rotor speeds passes through a minimum at the
50/50 composition blend. In more familiar rheology
terms, the viscosity should pass through a minimum at low
shear rates at the 50/50 composition blend.
    Another interesting point which is noted from the
plot of the power-law index versus weight percent
displayed in Figure 1 is that the power-law index values
reach a maximum at the 50/50 composition blend. The
power-law value of about 0.7 obtained for this blend is
higher than for either of the component materials. This
result indicates that the blend has a flow curve more
closely approximating Newtonian in shape than either of
the component materials.
    It is possible to get further insight into how
blending affects the rheological response of the two

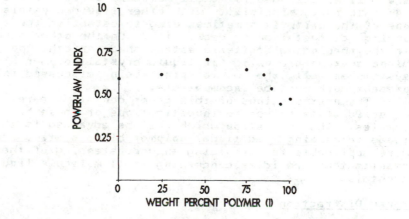

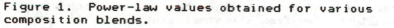

Figure 1. Power-law values obtained for various composition blends.

materials by considering the actual measured torque
values which are generated at a rotor speed of 5 RPM.
What is observed is that the torque values again pass
through a minimum at the 50/50 blend composition, very
similar to the behavior observed for the calculated
torque values at 1 RPM.

Han and Kim (14) have observed a similar phenomenon
in a completely different system consisting of
polypropylene blended with polystyrene. The conclusion
of that work was that the polystyrene and polypropylene
have no interaction at the interface and do not form an
interphase. Han and coworkers (14) have studied other
systems which display minima in the viscosity versus
composition graphs and have concluded that there is
little chemical interaction between the two phases.

The suggestion that two liquid crystal polymers are
incompatible with each other is contrary to ideas which
are well established for small molecule liquid crystals
(4). In fact, miscibility with other liquid crystals is
one of the criteria sometimes used to establish the type
of liquid crystal being dealt with. On the other hand,
if the rheological criteria established for other polymer
blend systems are valid for liquid crystal polymer blend
systems as well, the two materials being discussed in the
present work must be incompatible.

The next sections of this paper contain the results
of solid state structure investigations of these blend
samples. The suggestion which will be advanced is that
ideas concerning traditional polymer blend systems are
more applicable for providing an understanding of these
results than are ideas concerning small molecule liquid
crystals.

X-ray Diffraction

X-ray patterns were obtained on strands which were
manually pulled from the samples after the mixing
operation was completed. As shown by Gutierrez, et al
(9), significant changes only occur in the first and
second meridonal peaks for copolymers which contain only
HBA and HNA. Since component (2) contains only one mole
percent TA and HQ, it is expected that similar
observations hold true for components (1) and (2) in the
present blends. Due to this fact, the present
examination will focus on the changes in the first two
peaks which occur upon blending.

Figure 2 contains the meridonal scans of component
(1), (2), and the 50/50 blend. Figure 3 contains an
expanded plot of the 6-7 Å region of the diffraction
patterns. It is seen there that the position of the
second peak in the component materials is too close to
make any definite statement about what's happening in the
blend. The position of the first peak is separated by
0.44 Å in the component materials. To see if this is a

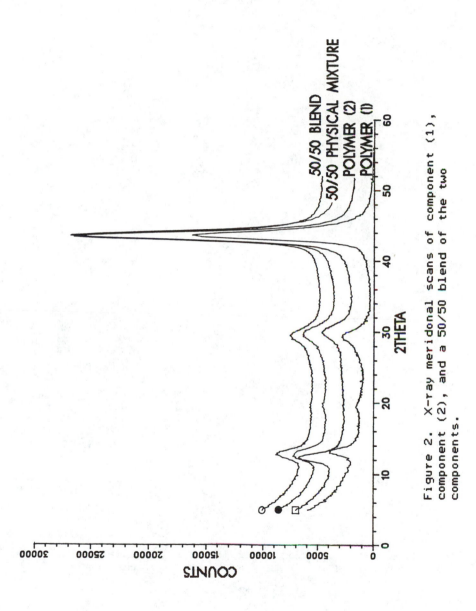

Figure 2. X-ray meridonal scans of component (1), component (2), and a 50/50 blend of the two components.

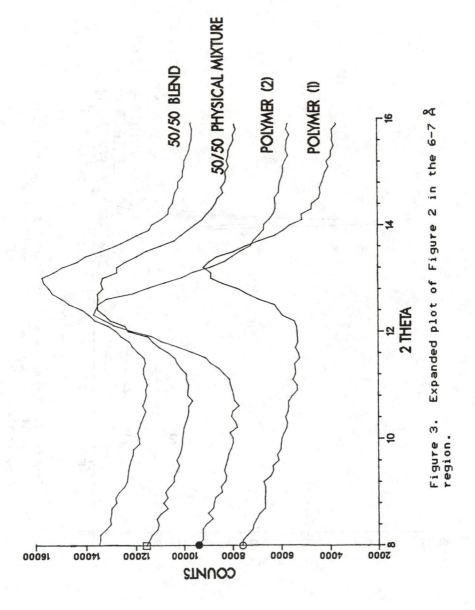

Figure 3. Expanded plot of Figure 2 in the 6-7 Å region.

large enough difference to be observed, a physical
mixture of the two component materials was made by
physically placing fibers next to each other in the
diffractometer. In the physical mixture, it is difficult
to distinguish the two overlapping peaks based on peak
position alone. However, the width of the peak showed an
increase from 0.76 to 0.95 . This suggests that the
physical mixture indeed consists of two overlapping
peaks. Since the peak width in the 50/50 blend is 0.76
rather than 0.95 , as predicted from the physical mixture
results, the 50/50 blend is suggested to have a single
peak in this region of the diffraction pattern.

One suggestion for what may occur when blends of two
liquid crystal polymers are produced is that an ester
exchange reaction will occur and a single copolymer with
intermediate composition will result. Such an issue has
been addressed in the work of Mehta and Baird (13).The
present results suggest that such is not the case for
this system. If a single random copolymer of
intermediate composition was being produced in the
present system, a peak centered at 12.56 Å is expected.
Instead, a peak centered at 12.85 Å is observed. The
difference between the expected result and the observed
result is outside of experimental error.

It can be speculated that the ester exchange
reaction has already occurred but complete randomization
of the monomers has not yet occurred. In that case, a
blocky structure would be expected. However, as will be
demonstrated later, the observed DSC traces are not
consistent with the occurrence of a blocky structure.
Thus, the conclusion is reached that the present data
suggests that transesterification is not a dominant
mechanism in determining the behavior of the present
blends.

DSC Results

Figures 4 and 5 contain the DSC traces which are obtained
upon second heating for each of the component materials.
The small endotherm observed is typical of these
materials. This is attributed to the small entropy
change which occurs at the crystal–nematic transition
temperature (15). It should also be noted that the two
polymers display transition temperatures which differ by
about forty degrees. Thus, in the blend samples, if two
endotherms are present, there should be no problem in
discerning them.

A typical cooling curve for a blend sample is shown
in Figure 6. There is typically an undercooling of about
forty degrees which is observed for these samples. In
all cases, only a single crystallization exotherm is
noted, suggesting that cocrystallization is occurring.

The subsequent reheat of samples that had been
ambiently cooled from the melt yielded only a single
endotherm. The position of this endotherm as a function

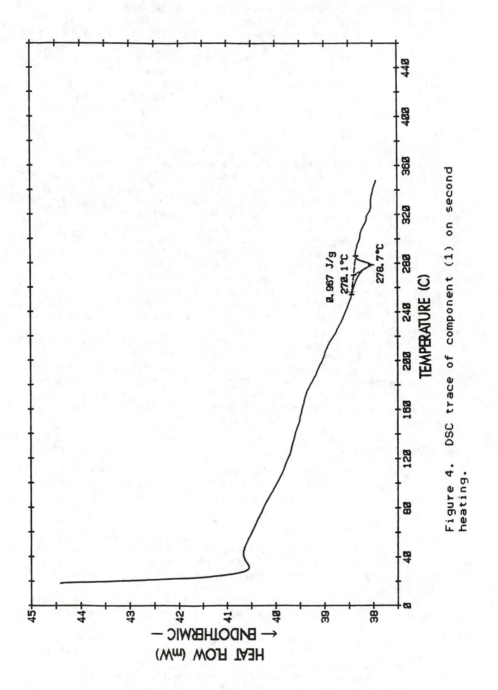

Figure 4. DSC trace of component (1) on second heating.

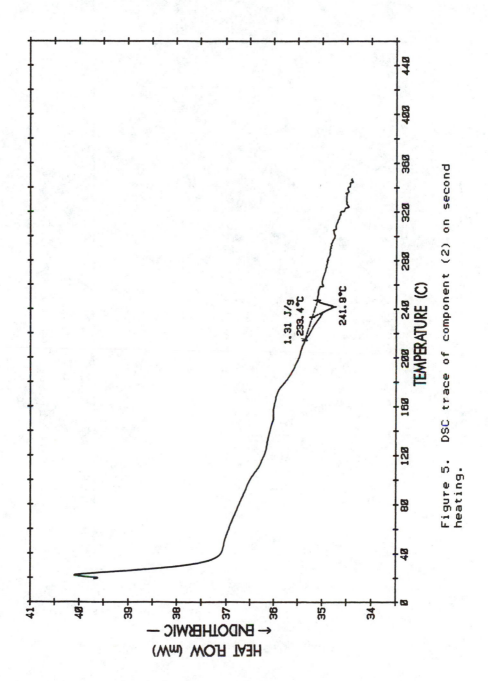

Figure 5.   DSC trace of component (2) on second heating.

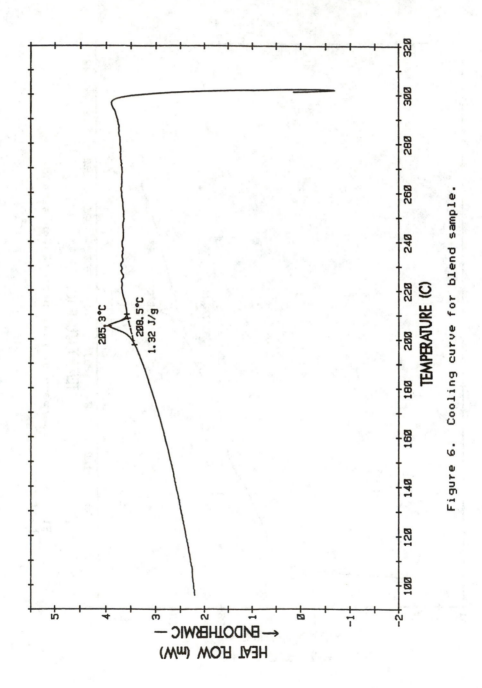

Figure 6.   Cooling curve for blend sample.

of blend composition is tabulated in Table I and displayed in Figure 7. Also displayed in Figure 7 is the line which corresponds to what is expected from a weighted average of the amount of component material values. It is observed from this plot that as the amount of copolymer is increased in the blend, the position of the DSC endotherm tends toward what is expected from a weighted average calculation. However, for all compositions, only a single endotherm is observed.

Table I.   Position of DSC endotherm as a function of blend composition.

Weight % Polymer (1)	DSC Endotherm (C)
0	241
25	248
50	256
75	268
85	272
90	275
95	276
100	278

The present DSC results also argue against the possibility of chemical reaction being the dominant mechanism in the blends. If the two component materials had completely reacted to form a single random copolymer, a DSC endotherm for the 50/50 blend at about 270 C is expected. Instead, for this blend, an endotherm at 256 C is observed. The other possibility is that the two materials have reacted but the reaction has not progressed to the level of a completely random copolymer. This also does not seem to be the case because such a block copolymer would be expected to display two DSC endotherms, something which is not observed.

Very little work has appeared in the literature which deals with blends in which the component materials can cocrystallize. It is generally believed (16,17) that a requirement for cocrystallization is that there must be a close matching of the polymer chain conformations and of crystalline dimensions. Also, some level of miscibility should exist between the two polymers and the crstallization kinetics cannot be very different. Certainly, in the case of liquid crystalline polymers, in general, these requirements would be expected to be met. Some of our recent work (8) has suggested, however, that not all liquid crystal polymers do cocrystallize. The present work suggests that in certain cases it may be possible to achieve this effect.

Dynamic Mechanical Results

The dynamic mechanical response of liquid crystalline
polymers has received a great deal of attention in the
recent literature. Yoon and Jaffe ([18]) examined the
response in tension of annealed, highly oriented strands
of various composition liquid crystalline polymers. More
recently, Ward, et al ([19,20]) studied the dynamic
mechanical response in both tension and torsion of these
same polymers. The following brief summary takes into
account the results of both of those works.
        The liquid crystal polymers of interest to the
present investigation display three transitions in the
temperature range -100 to 200 C. There is a low
temperature transition which occurs at about -70 C and is
very weak in intensity. This has been attributed to
motion of the p-phenylene units. A transition which is
highly dependent on composition occurs in the region 20 -
80 C. This transition has been attributed to
reorientation of the naphthalene groups. Finally, there
is a transition with a large activation energy in the
range 100 - 150 C. This transition is analogous to the
glass transition of traditional polymers and has been
attributed to delocalized orientation.
        Most of the samples for these studies were
compression molded bars. In the case of the blend which
contains 25 percent of component (2), a large enough
sample for injection molding evaluation was prepared. A
special note will be made of these samples when they are
discussed. The main difference between the injection
molded and compression molded samples is the absolute
magnitude of the modulus values which are obtained.
        The main focus of the present work will be on the
α-transition, which has characteristics similar to a
glass transition. The criteria which will be used for
miscibility is the appearance of a single α-transition
which is intermediate in temperature to the α-transition
of the component materials. The effect of blending on
the secondary transitions is outside the scope of the
present investigation.
        Figure 8 depicts the dynamic mechanical response of
the two component materials as well as a blend which
contains 75 weight percent of the copolymer component.
It can be seen from that figure that the α-transitions
of the two component materials are separated by about 20
degrees. If the blend is an immiscible mixture, it is
expected that there would be two transitions observed
which are 20 degrees apart, or, at least, it is expected
that a single significantly broadened transition would be
observed.
        It can be seen from Figure 8 that such is not the
case in the 75/25 blend. Instead, what is observed is a
single transition which is as sharp as in the two
component materials. This observation holds true for all
of the blend compositions studied. The position of the

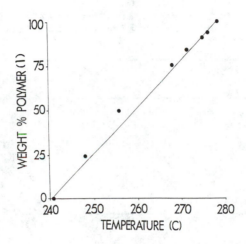

Figure 7.   Position of melting endotherm as a function of blend composition.

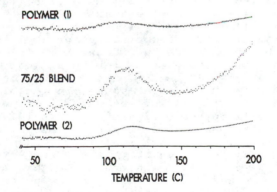

Figure 8.   Dynamic mechanical data for the two component materials as well as a blend which contains 75 weight percent of component (1).

α-transition in the blends varies systematically between
the component material values, as shown in Table II and
Figure 9. This observation as well as the fact that the
transition remains sharp in the blends indicates complete
miscibility for all blend compositions.

Table II.   Position of α-transition as a
function of blend composition.

Weight % Polymer (1)	α-transition (C)
0	119
25	115
50	111
75	107
85	105
90	104
95	103
100	102

Comparison to Other Systems

In our previous studies (7,8), we have explored the
applicability of traditional polymer blend concepts for
understanding the behavior of blends of two liquid
crystal polymers. What has been observed is that the
miscibility of these systems is dependent on both the
polymers being blended and the ratios of the blend. For
example, it has been shown by DSC and DMTA (8) that
blends of the same copolymer as in the present work with
a terpolymer of HNA, TA, and HQ are miscible only when
the terpolymer is the major component in the blend.
        The contrast of this work compared to findings for
mixtures of small molecule liquid crystals (4) suggests
that entropy plays an important part in determining
miscibility in liquid crystal systems. The entropic
effect could conceivably enter in two different ways.
The first is simply through molecular weight, as in the
usual Flory-Huggins theory. The second way for entropy
to enter into such systems is through conformational
differences between the two polymers. Only two extreme
cases have been adequately explored in the literature.
The first is the usual random coil/random coil mixing
where enthalpic effects dominate the thermodynamics. The
second case is "molecular composites", i.e., mixtures of
rigid rods and random coils. In these systems, entropic
effects which are based on conformational differences
control the mixing characteristics. Between these two
extremes lie the systems presently under investigation.

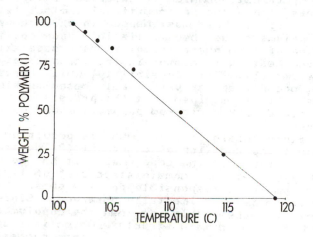

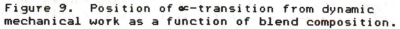

Figure 9. Position of ∝-transition from dynamic mechanical work as a function of blend composition.

The interplay between entropy and enthalpy which exists in these blends is not well understood.

Conclusions

This work is an extension of our previous studies aimed at investigating miscibility characteristics of liquid crystal polymer systems. The present study has shown that it is possible to significantly modify the rheological properties of one liquid crystal polymer by blending with another liquid crystal polymer.

Solid-state characterization of the structure of these blends indicates miscibility. This result, taken with the rheological observations, suggests two interpretive possibilities. The first possibility is that the blends are immiscible in the melt and that the rheology modification which is observed is very similar to what has been noted in traditional immiscible polymer blend melts (14). What must happen in this scenario is that the polymers must become miscible upon cooling. An observation of such phase behavior where separation occurs upon heating is common in traditional polymer blend systems (1). This possibility, while certainly feasible, doesn't appear very likely based on kinetic arguments. It is imagined that the process of forming miscible blends for rigid rod polymers would be a very slow process.

The other possibility for what is occurring in the melt is that the addition of component (2) destroys the "domain structure" of the copolymer. It has been proposed (21) that the "domain structure" of liquid crystal polymers is responsible for the shear-thinning behavior which is experimentally observed. Since the blends are less shear-thinning than the copolymer, it is suggested that the domain structure which is responsible for this effect has been broken up. This suggestion would indicate a certain miscibility in the melt which remains in the solid state.

It is clear that to differentiate between the two proposed mechanisms, in-depth studies of the structure present in the liquid crystalline melt must be performed. DSC studies in which the melt structure has been quenched might be helpful. The only difficulty is that the transitions observed in the DSC are often very difficult to detect.

A more promising method for observing the structure of the liquid crystalline phase is optical microscopy. The copolymer of interest to this study has a distinct "domain structure" which can be easily observed by optical methods. If blending has altered this "domain structure" in any way, it should be observable and perhaps, even measureable.

The present study is part of a larger effort to understand the miscibility characteristics of semiflexible and rigid polymer systems. What is emerging

from these studies is that traditional polymer blend
concepts seem more applicable to the understanding of
these systems than do ideas of small molecule liquid
crystals. The only difference is that the entropic
contribution to the free energy may play a larger part in
determining miscibility in these systems than random coil
polymers.

Literature Cited

1. See for example, Paul, D.R.; Newman, S. In Polymer
    Blends; Academic Press: New York, 1979.
2. Flory, P.J. Macromolecules 1978, 11, 1138.
3. Flory, P.J. Macromolecules 1978, 11, 1141.
4. Gray, G.W.; Windsor, P.A. In Liquid Crystals and
    Plastic Crystals; Halsted Press: N.Y., 1974; Vol. 1,
    p 20.
5. Marsano, E.; Bianchi, E.; Ciferri, A.
    Macromolecules 1984, 17, 2886.
6. Kiss, G. U.S. Patent 4 567 227, 1986.
7. DeMeuse, M.T.; Jaffe, M. Mol. Cryst. Liq. Cryst.
    Inc. Non-Lin Opt. 1988, 157, 535.
8. DeMeuse, M.T.; Jaffe, M. Accepted for Publication
    in Polymers for Advanced Technologies.
9. Gutierrez, G.A.; Chivers, R.A.; Blackwell, J.;
    Stamatoff, J.B.; Yoon, H. Polymer 1983, 24, 937.
10. Blackwell, J.; Gutierrez, G.A.; Chivers, R.A.
    Macromolecules 1984, 17, 1219.
11. Blackwell, J.; Lieser, G.; Gutierrez, G.A.
    Macromolecules 1983, 16, 1418.
12. Wissbrun, K.F.; Zahorchak, A.C. J. Polym. Sci.
    A-1 1971, 9, 2093.
13. Mehta, A.; Baird, D.G. ANTEC Preprints 1989, 35,
    1716.
14. Han, C.D. In Rheology of Polymer Processing;
    Academic Press: New York, 1976.
15. Calundann, G.W.; Jaffe, M. Proceedings of the
    Welch Conference on Chemical Research XXVI, 1982,
    p 247.
16. Allegra, G.W.; Bass, I.W. Adv. Polym. Sci. 1969,
    1, 549.
17. Wunderlich, B. In Macromolecular Physics; Academic
    Press: New York, 1973.
18. Yoon, H.N.; Jaffe, M. National ACS Meeting,
    Seattle, 1983.
19. Ward, I.M.; Davies, G. In Strength and Stiffness
    of Polymers: Part II; Porter, R.S.; Zachariades,
    A.E., Eds.; To be Published.
20. Troughton, M.J.; Davies, G.R.; Ward, I.M. To be
    published.
21. Wissbrun, K.F. Private communication.

RECEIVED March 1, 1990

Chapter 31

Kinetics of Phase Segregation in Thermotropic Liquid-Crystalline Copolyester and Polyether Imide Blends

Johnny Q. Zheng and Thein Kyu

Institute of Polymer Engineering, University of Akron, Akron, OH 44325

Phase segregation in the mixtures of polyethylene terephthalate and polyhydroxybenzoic acid copolymer (PET-PHB) and polyether imide (PEI) has been examined by means of differential scanning calorimetry, optical microscopy and laser light scattering. The solvent cast blends from mixed solvents of phenol and tetrachloroethane (60/40) show a single glass transition (Tg) and is optically clear. Thermally induced phase separation takes place upon heating above the Tg of PEI. A phase diagram was established on the basis of cloud point determination which is reminiscent of a lower critical solution temperature (LCST). However, we are unable to confirm the reversibility of the phase diagram. Several temperature jump (T-jump) experiments were carried out from ambient to two-phase temperature regions. Phase separation occurs via spinodal decomposition and is dominated by non-linear behavior. The time-evolution of scattering curves were analyzed in the context of non-linear and dynamical scaling theories.

In a previous paper (1), phase segregation by spinodal decomposition in mixtures of polyethylene terephthalate and polyhydroxybenzoic acid copolymer (PET-PHB) and polycarbonate (PC) has been investigated. It was shown that thermally induced phase segregation takes place above the Tg of PC and exhibits a lower critical solution temperature (LCST). However, the phase separated domains do not grow until the temperature exceeds 255°C. Some disclinations developed within the liquid crystal rich regions. Even in the pure PET-PHB component, four dark brushes with negative sense of disclinations form around 240°C, indicating the presence of nematic liquid crystals. Paci and coworkers (2) claimed that a smectic-nematic transition exists near 270°C in this liquid crystalline copolyester.

0097–6156/90/0435–0458$06.00/0

In this paper, we continue our study on the kinetic aspects of thermally induced phase segregation in PET-PHB/polyether imide (PEI) blends. We select PEI because it is easier to prepare a thicker miscible film of PET-PHB blended with PEI as compared to that with PC. Solvent induced crystallization of PC (3,4) often presents added difficulty in the analysis of phase segregation. This problem can be avoided in the present case because PEI is amorphous under the present experimental conditions. Several temperature (T)-jump experiments were carried out from ambient to 260, 270 and 280°C. The kinetic results are compared with those obtained for polymer blends (5-9), molecular composites (10), and thermotropic liquid crystals containing polymer alloys (11,12).

EXPERIMENTAL

Copolyester of p-hydroxybenzoic acid with ethylene terephthalate (PHB-PET, 60/40) was supplied by Tennessee Eastman Kodak Co., whereas polyether imide (PEI) was provided by General Electric. Co. (Ultem 1000). These polymers were dissolved together in a mixed solvent of phenol and tetrachloroethane in the ratio of 60/40 by weight at 80°C for about a week. The polymer concentration of the solution was 2 wt%. Various PHB-PET/PEI films were cast on glass slides at ambient temperature, then dried in a vacuum oven at 60°C for two weeks. Thicker films were prepared in Petri dishes for differential scanning calorimetric (DSC) studies.

A cloud point phase diagram was established at 1°C/min, using small-angle light scattering (SALS). The SALS set-up consists of a He-Ne laser light source with a wavelength of 632.8 nm. The scattered intensity was monitored by means of a two-dimensional Vidicon camera (Model 1254, EG & G Co.) interlinked with an Optical Multichannel Analyzer (OMA III). SALS pictures were photographed with a Polaroid instant camera (Land film holder 545). Optical micrographs were obtained on a Leitz optical microscope (Laborlux 12 Pol). Several temperature jump (T-jump) experiments were carried out from room temperature to 260, 270 and 280°C.

Differential scanning calorimetric runs were conducted on a Du Pont Thermal Analyzer (Model 9900) at an arbitrary heating rate of 10°C/min. Indium standard was used for temperature calibration.

RESULTS AND DISCUSSION

Miscibility Phase Diagram. The solvent casting films of PHB-PET/PEI mixtures were transparent and scattered no light, suggestive of miscible character. Figure 1 shows DSC thermograms for various blend compositions. The neat PHB-PET reveals a small Tg in the vicinity of 65°C and a small endotherm around 195°C. There is no clearcut view on this melting transition. Viney and Windle (13) postulated that this peak represents melting out of regions of local order which, at low temperature, pin down the microstructure. The authors noted that the polymer has gained molecular mobility above this transition. The disclinations with four-fold brushes were observed under polarized microscope which rotates in the same direction of the polarizers, suggesting positive sense of 1. In a similar temperature region (240⁰C), Shiwaku et al. (12) observed the Schlieren textures with the strength of -1/2 and 1. On the other hand, Paci et al. (2) claimed a smectic-nematic transition around 270⁰C. Hence, there is no agree upon opinion on the nature of the mesophase region, but this melting transition may be associated with the crystal to mesophase transition (13).

In their blends with PEI, the melting transition of PHB-PET is no longer discerned. Instead, a single Tg appears in the intermediate blends and moves systematically with blend compositions. The movement of Tg with composition, as shown in Figure 2, appears to follow the prediction of Fox (14), indicating that PHB-PET and PEI may be a miscible pair or they merely form a kinetically entrapped single phase.

When the blend films were annealed at 250⁰C, the films turned translucent within a minute, suggesting the occurrence of thermally induced phase segregation. Tiny, but interconnected domain structures developed under optical microscope. A large, but very weak scattering halo was discerned in light scattering studies. These features are familiar characteristics of spinodal decomposition. As can be seen in Figure 3, the average size of the phase separated domains does not get larger with longer annealing time up to 10 h, implying that phase growth is prohibited at that annealing temperature.

However, when the temperature is raised instantaneously from ambient to 260⁰C, thermally induced phase separation occurs within a few second and the domain size increases subsequently. Figure 4 shows a typical evolution of domain structure of the 50/50 blends as a function of annealing time at 270⁰C. Within the liquid crystalline polymer rich regions, birefringent entities can be discerned. The Vv (vertical polarization with vertical analyzer) scattering shows a scattering halo corresponding to the periodic composition fluctuations of phase separated domains (Figure 5). The scattering halo becomes smaller with elapsed time as a result of phase growth. According to Viney and Windle (13), a small endotherm was observed around 250⁰C which is difficult to be reproduced within 5⁰C. We occasionally observe such a DSC peak in some scans of pure PHB-PET or its blends with PEI, but it is absent in the second runs. However, Paci and coworkers (2) observed a transition around 270⁰C and attributed to a smectic-nematic transition.

The cloud point measurement was undertaken for 40/60 PET-PHB/PEI at a heating rate of 1⁰C/min by measuring the scattered intensity at a scattering wavenumber q = 2. Here q is defined as $(4\pi/\lambda) \sin(\theta/2)$ with λ and θ being the wavelength of light and the scattering angle measured in the medium. The intensity begins to change its slope slightly above the Tg of PEI due to phase segregation. However, the increase of intensity is insignificant below 250⁰C. This is consistent with the optical microscopic investigation that the phase separated domains do not grow for some initial period, e.g. for about 2 h. At present, it is unclear to us why the domains are unable to grow below 250⁰C. It may be speculated that the moment the mixture undergoes phase separation, the LCP molecules tend to self-associate due to the large aspect

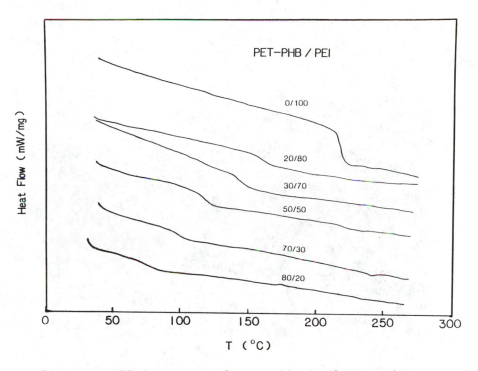

Figure 1. DSC thermograms of varous blends of PET-PHB/PEI, displaying a single glass transition for intermediate compositions.

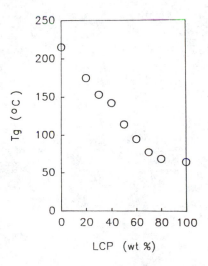

Figure 2. The variation of Tg as a function of blend composition.

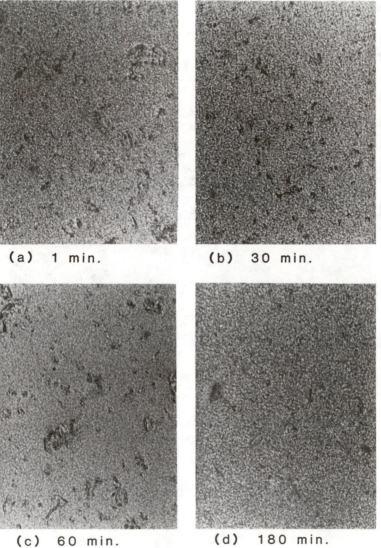

Figure 3. Optical micrographs of 50/50 blends for various isothermal annealing time at 250°C.

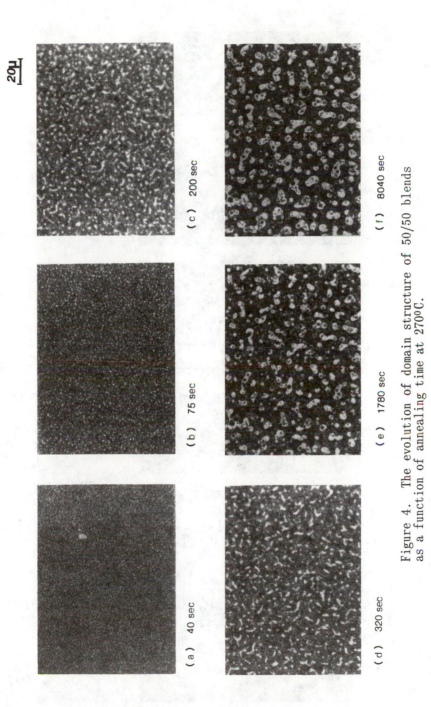

Figure 4. The evolution of domain structure of 50/50 blends as a function of annealing time at 270°C.

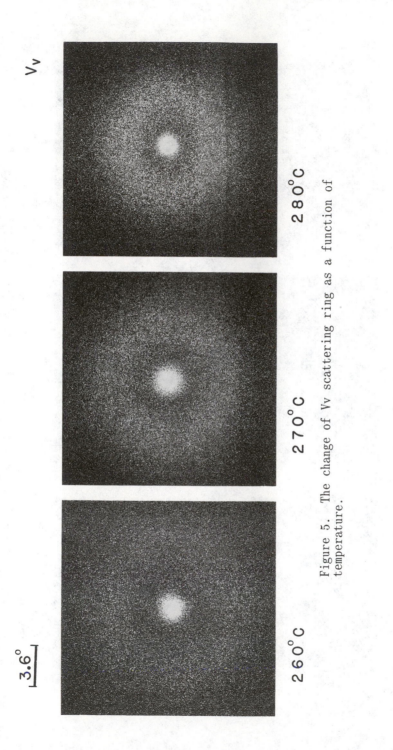

Figure 5. The change of Vv scattering ring as a function of temperature.

ratio of rigid molecules, thus aligning themselves to form anisotropic regions. This ordering process may suppress the molecular mobility of LCP regions, thereby restricting the phase growth process. When the temperature increases further beyond $260^\circ C$, the system gains additional mobility which would permit molecular transport so that the domains can grow further.

Similar experiments were conducted for other compositions and a cloud point phase diagram was established as shown in Figure 6. As expected the cloud point phase diagram resembles a lower critical solution temperature (LCST) in character. The formation of liquid crystal phase makes it difficult to confirm the reversibility of the phase diagram; thus it should be regarded as a virtual one.

Dynamics of Spinodal Decomposition. Several temperature (T)-jump experiments were undertaken for 50/50 PET-PHB/PEI from ambient to 250, 260, 270, and $280^\circ C$. At $250^\circ C$, a very weak scattering ring is observed, but the intensity does not appreciably increase for a considerable period of approximately 10 h. The peak position remains virtually constant, indicating that the phase separated regions are not getting larger with time which is exactly what was observed in the optical microscopic investigation. Figure 7 shows the time-evolution of scattering curves for 260 to $280^\circ C$, in which the scattering peak appears at relatively low wavenumbers, implying the large average size of concentration fluctuations. There is no initial period where the peak position is stationary (15). The lack of linear regime may be understandable because phase separated domains are already formed at relatively low temperatures, say $240^\circ C$ or less. Hence, we are probably detecting the late stages of the growth process exclusively at such high T-jumps. In this regime, it is difficult to distinguish whether the growth process occurs via spinodal decomposition or nucleation-growth. Judging from the optical micrographs and the scattering halo, although by no means conclusive, the SD mechanism may be appropriate. This non-linear growth character may be best explained in terms of power law relations of maximum scattering wavenumbers (qm) and the corresponding maximum scattered intensity (Im) versus phase separation time (t) as follows,

$$q_m(t) \sim t^{-\varphi} \tag{1}$$

and

$$I_m(t) \sim t^{\psi} \tag{2}$$

where the subscript m stands for the maximum values. The kinetic exponents φ and ψ are predicted to have various values depending on the time scale. On the basis of the non-linear statistical consideration, Langer, Bar-on, and Miller (LBM) (16), obtained a value of $\varphi = 0.21$. Binder and Stauffer (17) postulated a relationship $\psi = 3\varphi$ with the values of $\varphi = 1/3$ and $\psi = 1$ by considering the coalescence of cluster domains. On the other hand,

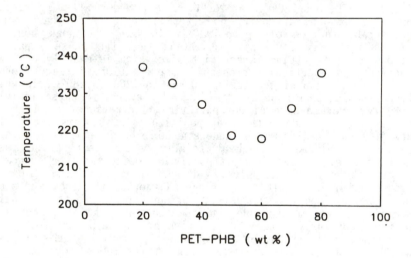

Figure 6.   Cloud point temperature versus composition phase diagram obtained at a heating rate of 1ºC/min.

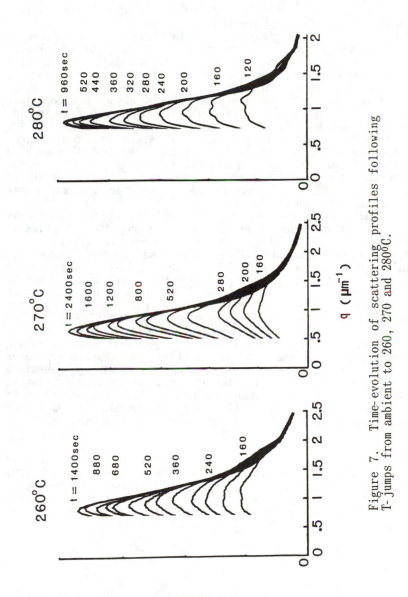

Figure 7.   Time-evolution of scattering profiles following
T-jumps from ambient to 260, 270 and 280°C.

Siggia (18) predicted the same equation from the percolation approach by taking into consideration the hydrodynamic flow, but with $\varphi = 1/3$ for the initial growth regime and $\varphi = 1$ for the intermediate flow stage.

Figures 8 and 9 exhibit the log-log plots of the maximum wavenumber and the corresponding intensity versus time, respectively. At the beginning of growth, the slopes are close to -1/3 and 1, which are consistent with the prediction of cluster dynamics, i.e., the coalescence of cluster domains. At a later time, the slopes become smaller which may be affected by the formation of anisotropic liquid crystalline region within the phase separated domains. The optical micrographs show no appreciable change in size or periodic distance in the same time scale. Although the LC mesophase transition and the phase growth are believed to be independent processes, it seems one process can interfere the other. That is to say, the kinetics of mesophase transition in pure LC may be different from that in the confined phase separated regions. Conversely, the growth of phase separated domains may be affected by the mesophase ordering process as the molecular mobility can be slow down.

It should be pointed out that the initial growth process is similar to that of Hasegawa et al. (19), who reported the kinetic exponent of $\varphi = 1/3$ for 50/50 mixtures of PET-PHB with polyethylene terephthalate (PET), following T-jumps above the crystal melting temperature of PET crystals. However, the latter result of the present studies is appreciably different from the above authors' observation, who found that the kinetic exponent varies with a slope of $\varphi = 1$ as the process reaches the percolation regime and eventually slows down with an exponent of -1/6 due to the pinning effect. They also demonstrated that the morphology changes from interconnected structure to spherical domains during intermediate to final stages of SD. However, our experiments never reach the percolation region. Chuah and co-workers (10) obtained the same -1/3 and 1 values for the molecular composites consisting of poly-p-phenylene benzobisthiazole (PBZT) rigid rod and flexible nylon 66 matrix. Since the diffusivity and the correlation length in the single phase solution are not known, the universal curve was not established.

Dynamical Scaling Tests. In the cluster regime, the structure is expected to become universal with time. The temporal scaling (20) of the scattering intensity $I(q,t)$ may be described in terms of a structure function $S(z)$ which is scaled with a single length parameter ζ in order to examine self-similarity during the coarsening process,

$$I(q,t) \sim V\langle\eta^2\rangle\zeta^3 S(z) \qquad\qquad (3)$$

where V is the scattering volume, $\langle\eta^2\rangle$ is the mean square fluctuation of refractive indices and $z = q\zeta = q/q_m$. Alternatively,

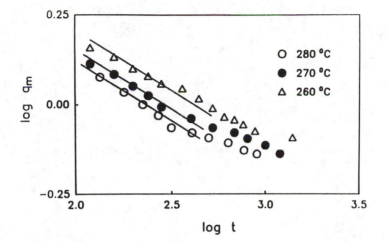

Figure 8. Log-log plots of the maximum wavenumber versus phase separation time for 50/50 blends.

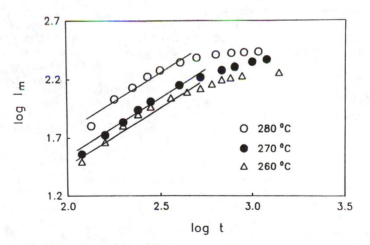

Figure 9. Log-log plots of the intensity maxima versus phase separation time for 50/50 blends.

$$S(z) \sim I(q,t)q_m^3 \tag{4}$$

As can be noticed in Figure 10, the superposition of the scaled scattering profiles is reasonably good at the late stages of SD, suggesting that the structure function is universal with time. There is no noticeable difference between the scaled structure functions for different T-jumps. This is in good accord with other studies including PBT/nylon 66 molecular composites ([10]). Furukawa ([21]) proposed a scaling law for the shape of structure function $S(q,t)$ in terms of the cluster size $R(t)$

$$S(q,t) \sim R(t)^d \, \tilde{S}(x) \tag{5}$$

where $\tilde{S}(x)$ is a universal scaling function which has been predicted to have the following form,

$$\tilde{S}(x) \sim x^2/(\gamma/2 + x^{2+\gamma}) \tag{6}$$

where $x = qR(t)$ and d represents the dimensional of the system. γ is further related to d as $\gamma = d+1$ for off-critical (cluster regime) and $\gamma = 2d$ for critical mixtures (percolation regime). For three dimensional growth, the structure function or scattering intensity is predicted to vary with an exponent of 2 at q<qm and -4 and -6 for off-critical mixtures, respectively. Figure 11 shows the log-log plots of intensity versus wavenumber for the 50/50 PET-PHB/PEI. The slopes of 2 and -4 were obtained in the regions of q<qm and q>qm, respectively, which are the values predicted for off-critical mixtures ([21]). Similar values have been reported in the demixing of Al-Zn and Al-Zn-Mg metal alloys ([22]), polystyrene (PS)/polymethyl phenyl siloxane (PMPS) oligomer mixtures ([23]), hydroxypropyl cellulose (HPC)/water,([24]) epoxy/rubber ([25]) and polycarbonate (PC)/polymethyl methacrylate (PMMA) blends ([26]). In the case of critical mixture of PS/polyvinyl methyl ether (PVME), Hashimoto and co-workers ([27]) obtained the slopes of 2 and -6 as predicted by the theory. Recently, Lim and Kyu ([28]) obtained the same value of 2 and -6 for the 40/60 blends of PC with moderately low molecular weight PMMA. However, Furukawa ([29]) comprehended that the slope of -6 has no physical basis and may only be valid at intermediate stage of SD. Hashimoto and coworkers ([27]) demonstrated that the Porod value of -4 recovers at very late stages of SD. Very recently, Oono and Puri ([30]) predicted on the basis of the cell dynamics that the exponent could be much steeper than -4 at intermediate wavenumbers in the percolation regime. We are very encouraged to learn that the asymptotic behavior of the scaled structure functions for various binary mixtures agree very well with dynamical theories.

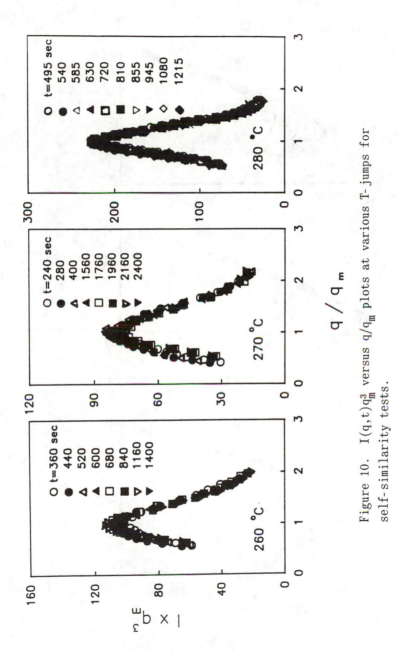

Figure 10. $I(q,t)q_m^3$ versus q/q_m plots at various T-jumps for self-similarity tests.

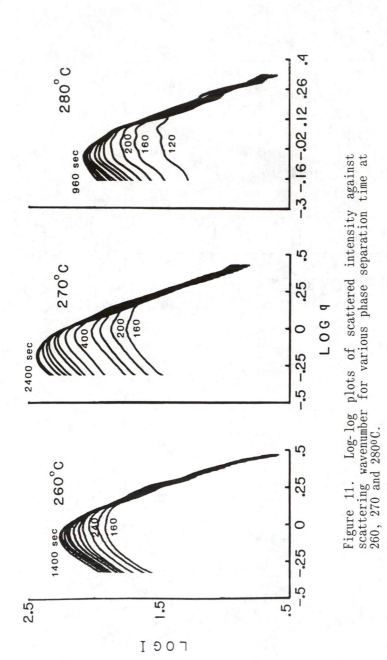

Figure 11. Log-log plots of scattered intensity against scattering wavenumber for various phase separation time at 260, 270 and 280°C.

CONCLUSIONS

We have demonstrated miscible blends of PET-PHB/PEI can be formed by rapid solvent casting from the mixed solvent of phenol and tetrachloroethane. The miscibility was confirmed by the systematic movement of Tg in the DSC studies. However, the blend is unstable and undergoes thermally induced phase separation with a miscibility window reminiscent of LCST. The dynamics of spinodal decomposition is non-linear in character and obeys the power law with kinetic exponents of -1/3 and 1 in accordance with the cluster dynamics of Binder and Stauffer as well as of Furukawa. In the temporal scaling analysis, the structure function exhibits universality with time, suggesting temporal self-similarity of the system.

ACKNOWLEDGMENTS

Support of the present work by the Edison Polymer Innovation Corporation (EPIC) is gratefully acknowledged.

LITERATURE CITED

1. Kyu, T. and Zhaung, P. Polym. Commun. 1988, 29, 99.
2. Paci, M., Barone, C. and Magagnini, P. J. Polym. Sci., Polym. Phys. Ed. 1987, 25, 1595.
3. Varnell, D.F., Runt, J.P. and Coleman, M. Macromolecules, 1981, 14, 1350.
4. Saldanha, J.M. and T. Kyu, Macromolecules, 1987, 20, 2840.
5. MacMaster, L.P. Adv. Chem. Ser. # 142, 1975, 43.
6. Hashimoto, T., Kumaki, J. and Kawai, H., Macromolecules, 1983, 16, 641.
7. Okada, M. and Han, C.C. J. Chem. Phys., 1986, 85, 5317.
8. Hill, R.G., Tomlins, P.E. and Higgins, J.S. Macromolecules, 1985, 18, 1985.
9. Kyu, T. and Saldanha, J.M. Macromolecules, 1988, 21, 1021.
10. Chuah, H.H., Kyu, T. and Helminaik, T.E. Polymer, 1989, in press.
11. Nakai, A., Shiwaku, T., Hasegawa, H. and Hashimoto, T. Macromolecules, 1986, 19, 3010.
12. Shiwaku, T.; Nakai, A.; Hasegawa, H.; Hashimoto, T. Polym. Commun. 1987, 28, 174.
13. Viney, C. and Windle, A.H. J. Mater. Sci. 1982, 17, 266.
14. Fox, T.G. Bull. Am. Phys. Soc. 1956, 2(2), 123.
15. Cahn, J.W. J. Chem. Phys. 1965, 42, 93.
16. Langer, J.S., Bar-on, M. and Miller, H.S. Phys. Rev. A 1975, 11, 1417.
17. Binder, K. and Stauffer, D. Phys. Rev. Lett. 1973, 33, 1006.
18. Siggia, E.D. Phys. Rev. A 1979, 20, 595.
19. Hasegawa, H., Shiwaku, T., Nakai, A. and Hashimoto, T. In Dynamics of Ordering Processes in Condensed Matter; Komura, S. and Furukawa, H. Eds., Plenum, New York, 1989.
20. Furukawa, H. Phys. Rev. Lett. 1979, 43, 136.
21. Furukawa, H. Physica A 1984, 123, 497.
22. Komura, S., Osamura, K., Fujii, H. and Takeda, T. Phys. Rev. B 1984, 30, 2944; ibid 1985, 31, 1278.

23. Nojima, S., Ohyama, Y., Yamaguchi, M. and Nose, T. Polym. J.,
     1984, 14, 907.
24. Kyu, T. and Mukherjee, P. Liq. Cryst. 1988, 3, 631.
25. Lee, H.S. and Kyu, T. Macromolecules, 1990, 23, 459.
26. Kyu, T. and Saldanha, J.M. J. Polym. Sci. Polym. Phys. Ed.,
     1990, 28, 97.
27. Hashimoto, T., Itakura, M. and Shimizu, N. J. Chem. Phys.
     1988, 85, 6773.
28. Lim, D.S. and Kyu, T., J. Chem. Phys., in press.
29. Furukawa, H. J. Appl. Cryst. 1988, 21, 805.
30. Oono, Y. and Puri, S. Phys. Rev. Lett. 1987, 58, 836.

RECEIVED April 10, 1990

Chapter 32

Polymer-Dispersed Liquid Crystals

John L. West

Liquid Crystal Institute, Kent State University, Kent, OH 44242

Polymer-dispersed liquid crystals (PDLCs) are electro-
optic materials that modulate light through electrical
control of the refractive index similar to other liquid
devices. Like dynamic scattering and smectic displays,
PDLCs switch between scattering and clear states. PDLC-
type devices consist of droplets of low-molecular weight
liquid crystals dispersed in a solid polymer binder. They
do not require polarizers and have a number of other
unique advantages: ease of fabrication, suitability for
large area devices, environmental stability and fast
switching speeds. PDLCs may be tailored for a wide
variety of applications, ranging from architectural glass to
projection TV and shutters for infrared video cameras.

PDLCs are light-scattering materials belonging to a class of liquid
crystal devices that operate on the principle of electrically modulating
the refractive index of a liquid crystal to match or mismatch the
refractive index of an optically isotropic, transparent solid. The first
demonstration of this type of device consisted of micron-sized glass
particles dispersed in a liquid crystal film (Figure 1A) (1). Another
approach imbibed liquid crystals in microporous polymer films (Figure
1B) (2,3). Two recently developed approaches distribute the liquid
crystal as droplets in a solid polymer binder. The liquid crystal may be
encapsulated by standard microencapsulation or emulsification
techniques which suspend it in a solid polymer film. These materials,
termed "nematic curvilinear aligned phase" (NCAP), are shown
schematically in Figure 1C (4,5). The newest technique involves phase
separation of low-molecular weight liquid crystal from a prepolymer or
polymer solution to form droplets of uniform size and controlled density.
Materials formed by this final method are termed "polymer-dispersed
liquid crystals" (PDLC) (Figure 1D), the subject of this review (6,7).
Only the NCAP films, being developed by the Taliq and Raychem
Corporations, and the PDLC films, being developed by Kent State

0097–6156/90/0435–00475$06.25/0
© 1990 American Chemical Society

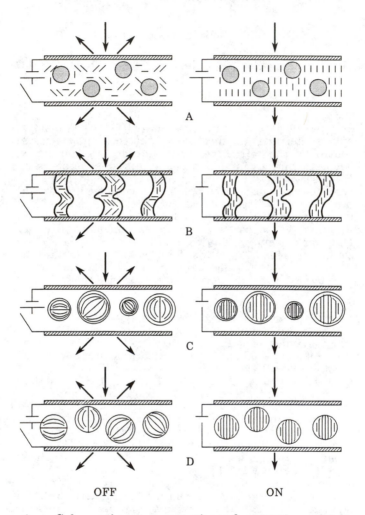

OFF   ON

Figure 1.   Schematic representation of operation of heterogeneous liquid crystal light shutters: A) glass beads in a liquid
crystal film, B) liquid crystal imbibed in a microporous film, C)
encapsulated liquid crystal, D) polymer dispersed liquid crystals.

University in cooperation with industry, have shown promise for commercial application. These films can be sandwiched between conducting plastic films to form continuous sheets. They are highly scattering in the OFF state and window-glass clear in the ON state. They do not use surface alignment layers, polarizers or cell seals. They are durable and aesthetically pleasing. A new class of electro-optic materials, they are superior in many ways to conventional liquid crystal shutters and offer exciting new applications.

Principle of Operation

PDLCs consist of micron-size droplets of a low-molecular weight nematic liquid crystal dispersed in a polymer binder. Figure 2 shows a scanning electron microphotograph of a cross section of a PDLC film. The size of the nematic droplets is on the order of the wavelength of light. Because of the droplet size and the refractive index mismatch between the liquid crystal in the droplet and the polymer binder, the films are highly scattering with a white, opaque appearance. A PDLC film sandwiched between substrates having a transparent conducting electrode, such as indium tin oxide, form a shutter. Upon application of a voltage across the electrodes of the shutter, it switches from an opaque, light scattering state to a clear, transparent state. The applied electric field aligns the droplets so that their refractive index matches that of the polymer, substantially reducing the light scattered by the droplets. The droplets return to their original alignment and the film returns to the scattering state upon removal of the field.

The configuration of the liquid crystal in the droplets depends on the elastic constants of the liquid crystal, droplet size and shape, and surface anchoring of the liquid crystal at the droplet wall. The bipolar configuration is the most common; it occurs in droplets where the molecules are anchored tangentially to the droplet wall (Figure 3A). The director is assigned as the average orientation of the liquid crystal in the droplet. The bipolar droplet is birefringent with an extraordinary refractive index, n_e, for light polarized parallel to the director and an ordinary refractive index, n_0, for light polarized perpendicular to the director. The liquid crystal is usually selected to have a positive dielectric anisotropy and therefore aligns parallel to an applied electric field. The polymer binder is selected to have a refractive index, n_p, essentially equal to n_0. In the absence of an applied field the directors are randomly oriented and because of the mismatch of the refractive index of the droplets and the polymer the films scatter light. Application of an electric field aligns the directors normal to the film surface and since $n_0 = n_p$, the films are transparent for light incident normal to the film.

The radial configuration occurs when the liquid crystal molecules are anchored with their long axes perpendicular to the droplet wall (Figure 3B). The radial droplet is not birefringent. Application of an external field switches the radial droplet to an axial configuration. As with the bipolar case the films switch from scattering to transparent upon application of an electric field if $n_p = n_0$.

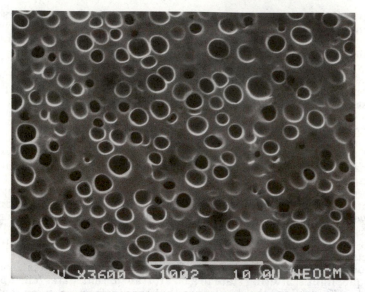

Figure 2.   Scanning electron microscope photograph of a cross section of a PDLC film.

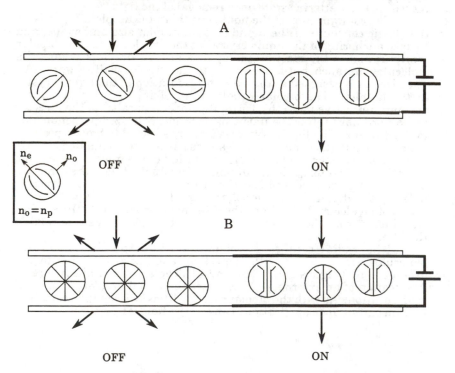

Figure 3.   Schematic of droplet configuration and PDLC device operation: A) bipolar configuration, B) radial configuration.

PDLC Formation

PDLC materials are formed by phase separation of low-molecular weight liquid crystals from a homogeneous solution with a prepolymer or polymer. The liquid crystal forms droplets whose size, shape and density depend on the techniques used. The polymer binder gels around the droplets, locking in their morphology. It is possible to prepare PDLC materials with uniform droplet sizes ranging from 40 μm to less than 0.01 μm in diameter (8,9). For display applications droplet diameters in the range of 0.3 to 3 μm are usually desired, whereas for shuttering in the infrared droplet diameters up to 25 μm are required. Three general techniques have been developed for forming PDLCs.

Polymerization-Induced Phase Separation (PIPS). Polymerization-induced phase separation (PIPS) generally utilizes polymers formed by a step growth reaction. A low-molecular weight liquid crystal is dissolved in a prepolymer solution. Polymerization occurs either thermally or photochemically, changing the chemical potential of the solution and reducing the solubility of the liquid crystal (10,11). The PIPS process is illustrated in Figure 4. The solution passes through the miscibility gap and the liquid crystal phase separates into droplets. Phase separation occurs by either spinodal decomposition or droplet nucleation and growth (West, J. L.; Tamura-Lis, W. "Phase Separation of Low-Molecular Weight Liquid Crystals Dissolved in a Polymer Melt," 12th International Liquid Crystal Conference, Frieburg, Germany, 1988). The droplets continue to grow until polymer gelation locks in the droplet morphology.

Epoxies, polyurethanes and a variety of photopolymers have been used as binders in the PIPS process (12-14). The epoxies have been the most studied of the thermoset polymers because of the large number of commercially available epoxy resins and cure agents. The resins and cure agents can be blended to form copolymers with specified physical properties such as refractive index. Photopolymerization utilizes either free-radical chain reaction or a step-growth reaction. A free-radical chain reaction results in the high-molecular weight polymer phase separating from the low-molecular weight polymer precursor/liquid crystal solution. A polymer ball morphology results and the liquid crystal is the continuous phase. A step-growth reaction produces the desired morphology of liquid crystal droplets dispersed in a continuous polymer binder (15).

Two major factors affect droplet size and density in the PIPS process: types and relative concentration of materials used and cure temperature. The cure temperature influences the rate of polymerization, viscosity of the polymer, diffusion rate of the liquid crystal and solubility of the liquid crystal in the polymer. Each factor is affected differently by the cure temperature with the result that droplet size varies in a complex manner with cure temperature (Figure 5) and must therefore be empirically determined for each formulation.

Thermally-Induced Phase Separation (TIPS). Thermally-induced phase separation (TIPS) results from cooling a liquid crystal/thermoplastic melt. The liquid crystal and thermoplastic are chosen to

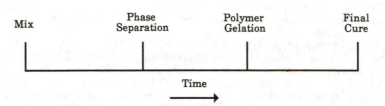

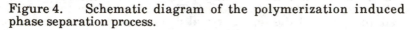

Figure 4.   Schematic diagram of the polymerization induced phase separation process.

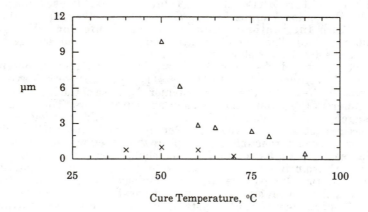

Figure 5.   Graph of average diameter as a function of cure temperature for:
            Δ = 1:1:1 mixture of epon 828, capcure 3800 and E7:
            X = 20% MK107, 11% epon 828, 28% capcure 3800 and 41% E7. (E7 is an cutectic mixture of cyanobiphenyls and cyanoterphenyls.)

form a homogeneous solution above the melt temperature of the polymer. Figure 6 illustrates the phase diagram of a liquid crystal, thermoplastic mixture. At point A, the mixture forms a homogeneous solution of the liquid crystal dissolved in the thermoplastic melt. As the solution cools it passes through the miscibility gap at point B, resulting in liquid crystal droplet formation and growth. Gelation of the polymer binder locks in the droplet morphology. Droplet size and density are controlled by the types and relative concentrations of liquid crystal and thermoplastic and by the rate of cooling. Figure 7 is a plot of the average droplet diameter vs cooling rate for a system composed of E7 and a thermopolastic epoxy formed by curing Epon 828 with t-butylamine. In general rapid cooling results in smaller droplets and more liquid crystal remaining dissolved in the binder.

Solvent-Induced Phase Separation (SIPS). Solvent-induced phase separation (SIPS) utilizes a liquid crystal and a thermoplastic dissolved in a common solvent. Evaporation of the solvent results in phase separation of the liquid crystal, droplet formation and growth, and polymer gelation. Figure 8 is a ternary phase diagram showing the SIPS process schematically. A system represented by point X consists of a polymer and a liquid crystal dissolved in a common solvent. Evaporation of the solvent moves the system along line XA. As the system crosses the miscibility gap, droplets of liquid crystals form and grow until gelation of the polymer locks in the droplet morphology. Point A represents the final composition of the PDLC film. Droplet size and density depend on the types and relative concentration of liquid crystal and thermoplastic, the type of solvent and the rate of solvent removal. Table I lists droplet formation time and the droplet size as a function of air flow rate over a thin film of a solution of E7 and polymethylmethacrylate dissolved in chloroform and coated on a glass substrate. The faster the rate of solvent removal, the smaller the droplets.

Table I. Droplet Size and Formation Time

Air Flow Rate (ml/min)	Time to Droplet Formation (min)	Droplet Size (µm)
100	34	≪1
20	150	3
3	720	12

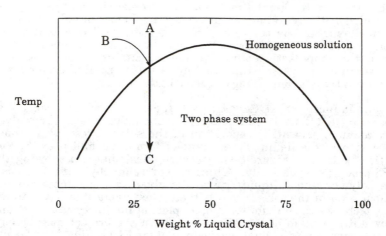

Figure 6.   Schematic phase diagram of a liquid crystal thermoplastic mixture.

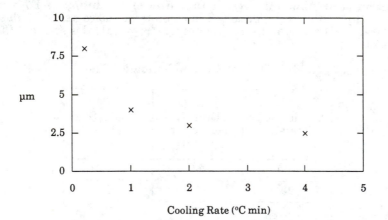

Figure 7.   Droplet diameter vs cooling rate of a thermoplastic
melt consisting of E7 dissolved in Epon 828 cured with t-butylamine.

PDLC Shutter Construction

A PDLC shutter "sandwiches" a film of the PDLC material between conducting substrates and utilizes no insulation or alignment layers. The substrates usually consist of either glass or plastic coated with a transparent conducting electrode such as a thin, vacuum deposited layer of indium tin oxide. The application of the transparent conducting electrode usually requires heating the substrate above 180°C, limiting the types of useable plastics.

For a shutter formed by the PIPS process, a solution of a prepolymer and liquid crystal is sandwiched between two substrates with glass or plastic spacers used to control thickness. The assembly is then heated or exposed to UV radiation to effect polymerization and PDLC formation. The polymers used in the PIPS process are usually highly crosslinked epoxies or polyurethanes, and the resulting shutter is stable over a wide temperature range.

Shutters can also be constructed using a combination of SIPS and TIPS processes. A solution of liquid crystal and thermoplastic dissolved in a common solvent is coated on glass or plastic substrates. Precise control of film thickness is possible utilizing standard film-forming techniques such as roll-to-roll coating, doctor blading, spray coating, etc. The solvent is rapidly removed with no control of the droplet morphology. The film is heated to redissolve the liquid crystal and soften the polymer; then a second substrate is laminated to the PDLC film. The resulting sandwich is then cooled at a controlled rate to produce the desired droplet morphology. The combination SIPS/TIPS process is especially well suited for the production of PDLC films in a roll-to-roll process.

Factors Affecting Electro-optic Performance

The electro-optic properties of PDLC films are controlled by the types of materials used, the droplet morphology and the method of film construction. Desireable properties include high clarity and transmission of the film in the ON and OFF states, low driving voltage, low power consumption, fast switching times and high film resistance. Since these properties are related, it is usually not possible to change them independently.

Clarity and Transmission. Precise definitions are required to quantify the clarity and transmission of PDLC films (16,17). "Clarity" is a measure of the sharpness of an image viewed through a film, and "transmission" is a measure of the efficiency of light passage thorough the film. Because PDLC films scatter rather than absorb light, it is possible to have high clarity with low transmission and vice versa. Transmission through the film is defined as the intensity of light transmitted by a film divided by the incident light intensity. A UV-visible spectro-photometer, equipped with an integrating sphere, measures both the directly transmitted and total transmitted light. Clarity is defined as the intensity of the light transmitted unscattered divided by the total light transmitted. It can be measured with a haze meter or with an integrating sphere.

The clarity of a PDLC film in the ON state depends on the match of n_o and n_p. The closer the match, the clearer the film in the ON state. Clarity greater than 95% is possible for light incident normal to the shutter surface. This is usually achieved by precisely adjusting n_p. The n_p of the epoxy resins are easily adjusted by mixing epoxy resins of high and low refractive indices. Figure 9 shows how the refractive index of an epoxy can be varied by adjusting the proportions of the resins used. Epon 828, the reaction product of epichlorohydrin and bisphenol A (Miller Stephenson Company) and MK 107, the diglycidyl ether of cyclohexanedimethanol (Wilmington Chemicals), are miscible in all proportions and can be cured with equivalents of Capcure 3-800, a trifunctional mercaptan terminated liquid polymer (Miller Stephenson Company). Thus, by adjusting the proportions of Epon 828 and MK 107, the refractive index of the resulting epoxy binder can be adjusted over the range 1.48-1.56. Many thermoplastics are suitable for use as binders for PDLCs. The thermoplastic selected should have a refractive index matched with n_o of the liquid crystals used.

Liquid crystal dissolved in the binder of a PDLC varies its refractive index (14). Also, the effective n_o of the droplet is not precisely equal to n_o of the bulk liquid crystal because the alignment is not parallel throughout the droplet. It is not practical to directly measure n_p and n_o of the PDLC film; however, it is possible to determine whether n_p is >, =, or < n_o from the electro-optic response of the film. Figure 10 shows the transmission through a PDLC film in the ON state as a function of incident angle for $n_p > n_o$, $n_p = n_o$ and $n_p < n_o$ for vertical (V) and horizontal (H) polarized light. If $n_p > n_o$ the film will not be of maximum clarity in the fully ON state for normally incident light. As the PDLC film rotates from the normal, the transmission for vertically polarized light increases because the effective refractive index of the droplet increases and more closely matches n_p. The transmission reaches a maximum at some angle and then decreases. For $n_p = n_o$ the maximum clarity occurs for normally incident light. For $n_p < n_o$ the film will also be at maximum clarity for normally incident light; however, the clarity will be reduced and the decrease with angle will be greater. The shape of the optical response to an applied voltage and the transmission vs applied voltage curve can also be used to determine the index match (13).

The OFF state clarity and transmission are determined by the size and density of the droplet and the birefringence of the liquid crystal. Maximum scattering and therefore minimum transmission and clarity are achieved when the droplet size and spacing is on the order of the wavelength of light. Highly birefringent liquid crystals offer the largest mismatch of refractive indices in the OFF state. Thicker films are also more scattering; however, they also reduce the clarity in the ON state. Little increase in scattering is achieved by increasing the thickness to more than 25 microns.

The refractive index match of the liquid crystal and the polymer is also temperature dependent. Because n_o tends to increase with temperature while n_p tends to decrease, it is usually not possible to have an exact match over the entire operating temperature range of the shutter (14). The polymer and liquid crystal are therefore selected to

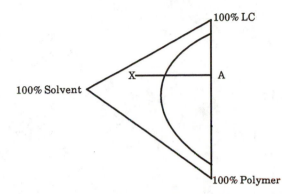

Figure 8. Ternary phase diagram of a liquid crystal, polymer, solvent mixture.

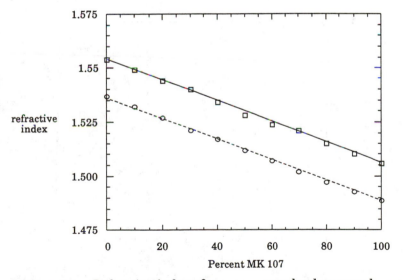

Figure 9. Refractive index of ———— cured polymer and ------ uncured resin for a mixture of MK 107 and Epon 828 resins cured with one equivalent of capcure 3800.

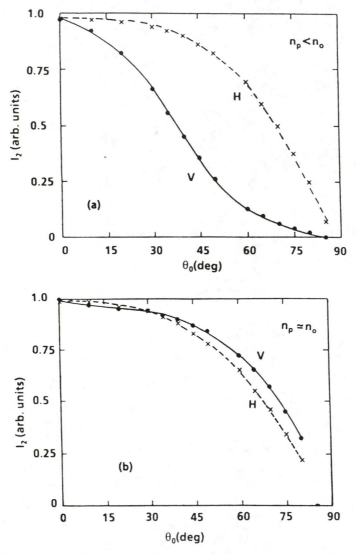

Figure 10. Angular dependence of the transmitted intensity of V and H polarizations through PDLC films in the field ON state for four samples: (a) Epon 812 with $n_p/n_o = 0.993$' (b) polyvinylformal with $n_p/n_o = 1.006$; (c) Epon 828 with $n_p/n_o = 1.020$, and (d) polycarbonate with $n_p/n_o = 1.046$. (Reproduced with permission from reference 13. Copyright 1987 American Institute of Physics.)

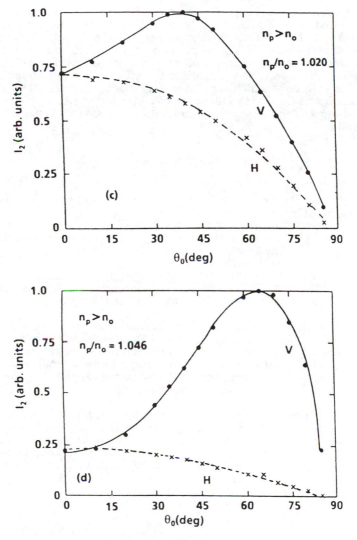

Figure 10. Continued.

have n_o and n_p matched at a temperature in the middle of the operating range.

In summary, PDLC shutters with a high contrast will therefore have a film thickness of about 25 microns, will have evenly spaced droplets of about 1 micron in diameter, will have $n_p = n_o$ and will use a highly birefringent liquid crystal.

Driving Voltage. For a perfectly spherical droplet, no elastic distortion is required to align a bipolar droplet with an electric field. In practice, the droplets in PDLC materials are never perfectly spherical and the random orientation of bipolar droplets in a PDLC film is caused by a distribution in the shapes and orientations of slightly elongated droplets. It is possible to calculate the effect of droplet shape on the driving voltage of a PDLC shutter with bipolar droplets (18,19) (Wu, B.-G.; Erdmann, J. H.; Doane, J. W. Liq. Cryst., to appear). The switching voltage of a bipolar droplet, $\overline{V}_B$, is given by the following formula:

$$V_B = \frac{d}{3a}\left(\frac{\rho_p + 2}{\rho_{\ell c}}\right)\left[\frac{K(\ell^2 - 1)}{\Delta\varepsilon_o}\right]^{\frac{1}{2}}$$

where
d = thickness of the PDLC film
ℓ = a/b, the ratio of the length of the semi-major axis, a, to the length of the semi-minor axis, b.
ρ_p = resistivity of the polymer
$\rho_{\ell c}$ = resistivity of the liquid crystal
K = $K_{11} = K_{22} = K_{33}$ = elastic constant of liquid crystal
$\Delta\varepsilon$ = dielectric anisotropy of the liquid crystal

This formula demonstrates that thinner films containing nearly spherical droplets of a liquid crystal with a large dielectric anisotropy have low driving voltage. Low driving voltages are achieved with a polymer binder with a low resistivity and a liquid crystal with a high resistivity. Driving voltage depends also on the droplet size, the larger the droplet the lower the driving voltage.

Response Time. Droplet shape is an important factor in determining the response time of a PDLC shutter (19) as demonstrated in Figure 11. The turn off time, τ_{off} is plotted as a function of temperature for both nearly spherical and elongated droplets (18). Droplet elongation results in a decrease in τ_{off} from >100 msec to <10 msec. The τ_{off} for a PDLC light light shutter is given by the formula:

$$\tau_{off} = \frac{\gamma_1 a^2}{K(\ell^2 - 1)}$$

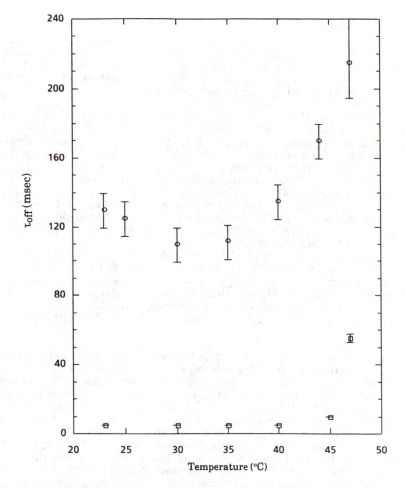

Figure 11. Relaxation time vs. temperature for an epoxy PDLC shutter for droplets which are nearly spherical and for droplets elongated by shearing the shutter substrate during the droplet formation process. (Reproduced with permission from Ref. 18. Copyright 1989 International Society of Optical Engineering.)

where:

$\gamma_1 = $ viscous torque $= 4 \times 10^{-2} \, kg \, m^{-1} s^{-1}$.
previous values for the other parameters.

An ellipsoidal droplet of $\ell = 1.1$ gives $\tau_{off} \approx 10$ msec, showing the effect droplet shape has on relaxation time.

PDLC Innovations

PDLC and NCAP films are just now reaching commercialization. Some applications, such as large area signs and switchable windows, could not be served by previous electro-optic materials. Recent innovations in PDLC films are expanding the possible applications for PDLC materials and improving their overall performance.

Dye Containing PDLCs. Colored PDLC shutters with improved contrast can be produced with the incorporation of dyes (20,21) (Erdmann, J. H.; Žumer, S.; Doane, J. W. Phys. Rev. Lett., submitted). Dyes of low order parameter and dichroic ratio, unsuitable for use in conventional guest-host systems, can be used to produce PDLC films with high contrast. The contrast results from the increased path length of light passing through the shutter in the OFF state (Vaz, N. A., private communication). The absorbance of an isotropic dye in the ON and OFF states of a PDLC film can be used to determine the average path length in the ON and OFF states and to estimate the scattering efficiency of the films (20).
     Incorporation of dyes with high order parameter and dichroic ratio can improve the color and contrast of a PDLC film. These dyes must be incorporated in the liquid crystal droplets of the PDLC to be effective. Utilizing highly crosslinked polymer binders, it is possible to achieve high segregation of the dichroic dye in the liquid crystal droplets (West, J. L.; Ondris, R.; Erdmann, M. A. SPIE, to appear, 1257).

Active Matrix Displays. As with twisted nematic liquid crystal displays, PDLCs require the use of an active matrix substrate to produce high-information content displays because the low slope of the transmission vs voltage curve of the PDLC materials prevents high level multiplexing. Recently, a PDLC film was coated on an amorphous silicon active matrix substrate to produce a black and white projection TV image (22). Although the film produced gray scales with little blurring, the resistivity of the PDLC was too low, $\sim 10^9 \, \Omega cm$, for the pixels to be turned fully ON, reducing the contrast of the display. Improved PDLC formulations will increase the resistivity of both the polymer and the liquid crystal to maintain a low driving voltage while increasing the resistivity of the PDLC.

Infrared PDLC Shutters. PDLC films operating in the infrared could be used as shutters for infrared video systems for thermal analysis and night vision applications. The infrared PDLC films operate on the same principles as those designed for visible applications; however, they must be tailored for optimum performance in the infrared. The films must have reasonable transmission in the ON state, be highly scattering in the OFF state and have fast switching speeds. Two wavelength regions are of particular interest, 2.5-5 µm and 8-14 µm. Polymers such as polymethylmethacryalte, polystyrene and polyvinyl-pyrrolidone (PVP) have acceptable transmission in the the wavelength regions of interest. Cyanobiphenyl liquid crystal eutectic mixtures, such as E7, also have acceptable infrared transmission (23,24).

Maximum scattering occurs when the droplet size is on the order of the wavelength of light. The liquid crystal droplet size is altered by warming the films to form a homogeneous solution, followed by cooling at a controlled rate. Table II lists droplet size as a function of cooling

Table II

Cooling Rate (°C/min)	Droplet Diameter (microns)
> >10	5-10
4	10-15
2	15-25
1	20-30
0.5	25-40

rate for a PDLC composed of E7/PVP in a 4:1 ratio. The ON and OFF state transmissions are shown as a function of wavelength for a 10 µm thick E7/PVP PDLC film sandwiched between NaCl substrates and processed to have 5-10 µm droplets (Figure 12A) and subsequently reprocessed to have 15-25 µm droplets (Figure 12B). It is clear from these figures that scattering efficiency depends on droplet size and wavelength. In the OFF state, the scattering is greater for the 5-10 µm film in the 2.5-5 µm region, whereas OFF state scattering is greater in the 8-14 µm region for films with 15-25 µm droplets.

With some improvements it seems likely that PDLC film can be developed for infrared video shuttering applications. Substrates with antireflection coatings can maximize transmisssion through the shutters. Thicker films will produce multiple scattering. Including an aperture in the system will effectively eliminate light scattered by the PDLC from the optical path in the system.

Haze-Free PDLC Films. Conventional PDLC films made with isotropic polymers are transparent only for a narrow cone of incident light; the refractive index of the liquid crystal droplet is angle dependent and can therefore be matched with an isotropic polymer for light incident over only a narrow range of angles. PDLC films consisting of low-molecular weight nematic droplets dispersed in a side-chain polymer liquid crystal matrix are transparent and non-scattering for all directions of

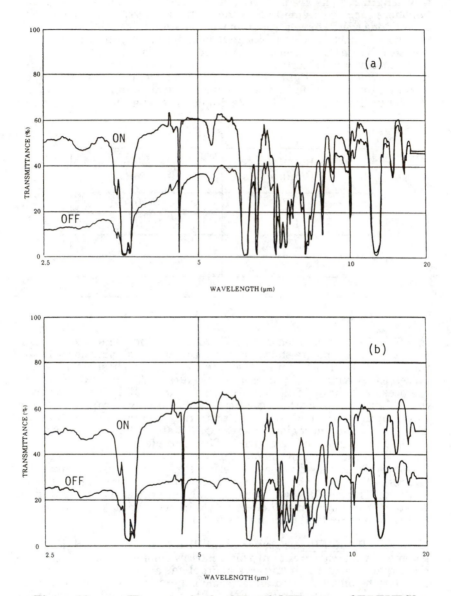

Figure 12.   IR spectra in the ON and OFF states of E7/PVPfilm
with A) 5-10 μm droplets and B) 15-25 μm droplets.

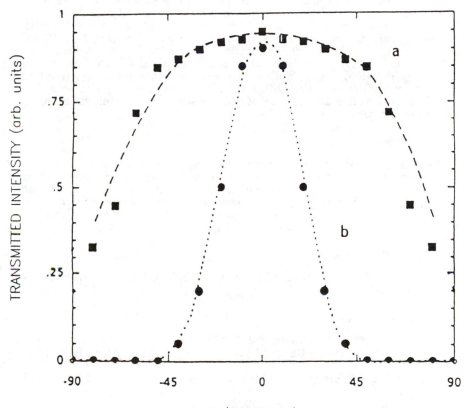

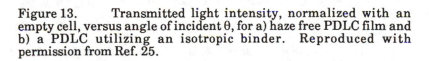

Figure 13. Transmitted light intensity, normalized with an empty cell, versus angle of incident θ, for a) haze free PDLC film and b) a PDLC utilizing an isotropic binder. Reproduced with permission from Ref. 25.

incident light in the ON state (25). The side-chain polymer liquid crystal binder is birefringent with principal refractive indices, n_0' and n_e', which are approximately matched with n_0 and n_e of the low-molecular weight liquid crystal in the droplets. In the ON state both the polymer side chains and the low-molecular weight liquid crystal are aligned parallel with the applied field and the refractive indices of the polymer and the liquid crystal are matched for all incident angles and the film is haze free. Upon removal of the field the low-molecular weight nematic droplets revert to a random orientation, the indices of the droplets and matrix do not match and the films scatter light.

Liquid crystal epoxies formed by curing epoxy resins with mesogenic amines have been used as binders for haze-free PDLC shutters (26). The angular transmission of a PDLC shutter constructed from an epoxy, formed by the reaction product of ethylene glycol diglycidyl ether and a p-amine alkoxy cyanobiphenyl curing agent and E7. a conventional PDLC film is given in Figure 13. The haze free PDLC films offer the potential of wide-viewing angle displays and haze-free windows.

Conclusion

Polymer-dispersed liquid crystals are the next generation of display materials. Initial research has outlined the basic principles of their operation The PDLC films combine the properties of plastics and liquid crystals producing display devices impossible with conventional materials. They will be used in applications ranging from architectural glass to projection TV and optical computing.

Acknowledgments

Research was supported in part by DARPA/ONR grant N00014-86-K-0766.

Literature Cited

1. Hilsum, C. U.K. Patent, 1 442 360, 1976.
2. Craighead, H. G.; Cheng, J.; Hackwood, S. Appl. Phys. Lett. 1982, 40, 22.
3. Beni, G.; Craighead, G.; Hackwood, S. U.S. Patent, 4 411 495, 1983.
4. Drzaic, P. S. J. Appl. Phys. 1986, 60, 2142.
5. Fergason, J. L. U. S. Patent, 4 616 903, 1986.
6. Doane, J. W.; Vaz, N. A. P.; Wu, B.-G.; Žumer, S. Appl. Phys. Lett. 1986, 48, 269.
7. Doane, J. W.; Chidichimo, G.; Vaz, N. A. P. U.S. Patent, 4 688 900, 1987.
8. Golemme, A.; Žumer, S.; Doane, J. W.; Nuebert, M. E. Phys. Rev. A 1988, 37, 559.
9. Golemme, A.; Žumer, S.; Allender, D. W.; Doane, J. W. Phys. Rev. Lett. 1988, 61, 2937.
10. West, J. L. Mol. Cryst. Liq. Cryst. 1988, 157, 427.

11. Vaz, N. A.; Smith, G. W.; Montgomery, Jr., G. P. Mol. Cryst. Liq. Cryst. 1987, 146, 17.
12. Vaz, N. A.; Smith, G. W.; Montgomery, Jr., G. P Mol. Cryst. Liq. Cryst. 1987, 146, 1.
13. Wu, B. -G.; West, J. L.; Doane, J. W. J. Appl. Phys. 1987, 62, 3925.
14. Vaz, N. A; Montgomery, Jr., G. P. J. Appl. Phys. 1987, 62, 3161.
15. Yamagishi, F. G.; Miller, L. J.; van Ast, C. I. Proc. SPIE 1989, 1080, 24.
16. Montgomery, Jr., G. P.; Vaz, N. A. Applied Optics 1987, 26, 738.
17. Lackner, A. M.; Margerum, J. D.; Ramos, E.; Wu, S. -T.; Lim, K. C. Proc. SPIE 1988, 958, 73.
18. Erdmann, J. H.; Doane, J. W.; Žumer, S.; Chidichimo, G. Proc. SPIE 1989, 1080, 32.
19. Drzaic, P. S. Liq. Cryst. 1988, 3, 1543.
20. West, J. L.; Tamura-Lis, W.; Ondris, R. SPIE 1989, 1080, 48.
21. Drzaic, P. S.; Wiley, R. C.; McCoy, J. SPIE 1989, 1080, 41.
22. Yaniv, Z.; Doane, J. W.; West, J. L.; Tamura-Lis, W. Proceedings, Society for Information Display, October 1989.
23. Wu, S. -T.; Efron, U.; Hess, L. D. Appl. Phys. Lett. 1984, 44, 1033.
24. West, J. L.; Doane, J. W.; Domingo, Z.; Ukleja, P. Polymer Preprints 1989, 30, 530.
25. Doane, J. W.; West, J. L.; Tamura-Lis, W.; Whitehead, Jr., Joe B. Pacific Polymer Preprints 1989, 1, 245.
26. West, J. L.; Chien, L. -C.; Tamura-Lis, W. Pacific Polymer Preprints 1989, 1, 247.

RECEIVED April 24, 1990

Author Index

496

Affiliation Index

Subject Index

Production: Donna Lucas
Indexing: Deborah H. Steiner
Acquisition: Cheryl Shanks

Books printed and bound by Maple Press, York, PA
Dust jackets printed by Sheridan Press, Hanover, PA

Paper meets minimum requirements of American National Standard
for Information Sciences—Permanence of Paper for Printed Library
Materials, ANSI Z39.48–1984   ∞